Nonlinear Optical Effects in Organic Polymers

NATO ASI Series

Advanced Science Institutes Series

A Series presenting the results of activities sponsored by the NATO Science Committee, which aims at the dissemination of advanced scientific and technological knowledge, with a view to strengthening links between scientific communities.

The Series is published by an international board of publishers in conjunction with the NATO Scientific Affairs Division

A Life Sciences	Plenum Publishing Corporation
B Physics	London and New York
C Mathematical and Physical Sciences	Kluwer Academic Publishers Dordrecht, Boston and London
D Behavioural and Social Sciences	
E Applied Sciences	
F Computer and Systems Sciences	Springer-Verlag
G Ecological Sciences	Berlin, Heidelberg, New York, London,
H Cell Biology	Paris and Tokyo

Series E: Applied Sciences - Vol. 162

Nonlinear Optical Effects in Organic Polymers

edited by

J. Messier

F. Kajzar

Commissariat à l'Energie Atomique,
Gif-sur-Yvette, France

P. Prasad

Department of Chemistry, State University of New York,
Buffalo, U.S.A.

and

D. Ulrich

Air Force Office of Scientific Research,
Washington, D.C., U.S.A.

Kluwer Academic Publishers

Dordrecht / Boston / London

Published in cooperation with NATO Scientific Affairs Division

Proceedings of the NATO Advanced Research Workshop on
Nonlinear Optical Effects in Organic Polymers
Sophia–Antipolis, Nice, France
19–20 May 1988

Library of Congress Cataloging in Publication Data

Nonlinear optical effects in organic polymers / J. Messier ... [et
al.].
        p.   cm. -- (NATO ASI series. Series E, Applied sciences ; vol.
162)
    Proceedings of a workshop sponsored by the NATO Scientific Affairs
Division and other organizations.
    Includes index.
    ISBN 0-7923-0132-3
    1. Polymers and polymerization--Optical effects--Congresses.
2. Nonlinear optics--Congresses.   I. Messier, J. (Jean), 1930-
II. North Atlantic Treaty Organization. Scientific Affairs
Division.   III. Series: NATO ASI series. Series E, Applied sciences
; no 162.
QD381.9.O66N65   1989
621.36--dc19   88-37517

ISBN 0-7923-0132-3

Published by Kluwer Academic Publishers,
P.O. Box 17, 3300 AA Dordrecht, The Netherlands.

Kluwer Academic Publishers incorporates the publishing programmes of
D. Reidel, Martinus Nijhoff, Dr W. Junk and MTP Press.

Sold and distributed in the U.S.A. and Canada
by Kluwer Academic Publishers,
101 Philip Drive, Norwell, MA 02061, U.S.A.

In all other countries, sold and distributed
by Kluwer Academic Publishers Group,
P.O. Box 322, 3300 AH Dordrecht, The Netherlands.

printed on acid free paper

TABLE OF CONTENTS

Preface   ix

List of Participants   xi

THEORY OF NONLINEAR POLARIZABILITY IN ORGANIC LOW DIMEN-
SIONAL SYSTEMS

Impact of Dimensionality in the Optical Nonlinearities
Ch. Flytzanis   1

Theoretical Design of Organic Molecules and Polymers for Optoelectronics
J. Delhalle, M. Dory, J.G. Fripiat and J.M. Andre   13

Structural Relaxation and Nonlinear Zero-Point Fluctuations as the Origin
of the Anisotropic Third-Order Nonlinear Optical Susceptibility in Trans-$(CH)_x$
M. Sinclair, D. Moses, K. Akagi and A.J. Heeger   29

NONLINEAR SUSCEPTIBILITY OF ORGANIC SYSTEMS

Third-Order Nonlinear Susceptibility in Semiconducting Polymers
J. Messier   47

Recent Nonlinear Optical Studies of MNA and Conjugated Linear Chains
C. Grossman, J.R. Heflin, K.Y. Wong, O. Zamani-Khamiri and A.F. Garito   61

Quadratic Nonlinear Behaviour of Various Langmuir-Blodgett Molecules
I. Ledoux, D. Josse, J. Zyss, T. McLean, R.A. Hahn, P.F. Gordon, S. Allen,
D. Lupo, W. Prass, U. Scheunemann, A. Laschewsky and H. Ringsdorf   79

Prediction of Third Order Nonlinear Optical Activity in Ordered Polymers
I.J. Goldfarb and J. Medrano   93

Effect of π-Electron Delocalization on the Second-Order Polarizability of
Disubstituted Hydrocarbons
R.A. Huijts and G.L.J. Hesselink   101

SYNTHESIS AND NONLINEAR OPTICAL PROPERTIES CHARACTERIZATION
OF NEW 1-D CONJUGATED POLYMERS

Studies of Structure-Property Relations
G.R. Meredith and S.H. Stevenson   105

Synthesis of new Nonlinear Optical Ladder Polymers
L.R. Dalton   123

Nonlinear Optical Properties of poly(p-Phenylene Vinylene) thin Films
C. Bubeck, A. Kaltbeitzel, R.W. Lenz, D. Neher, J.D. Stenger-Smith and
G. Wegner   143

The Synthesis and Properties of some Novel Diacetylene Monomers and
Polymers
G.H.W. Milburn   149

Design and Optical Properties of a low Energy gap Conjugated Polymer:
Polydithieno (3,4-b:3',4'-d)Thiophene
C. Taliani, R. Zamboni and G. Ruani   159

Synthetic Approaches to Stable and Efficient Polymeric Frequency Doubling
Materials. Second-Order Nonlinear Optical Properties of Poled, Chromophore-
Functionalized Glassy Polymers
C. Ye, N. Minami, T.J. Marks, J. Yang and G.K. Wong   173

Langmuir-Blodgett Films of Rigid rod Polymers with Controlled Lateral
Orientation
C. Bubeck, D. Neher, A. Kaltbeitzel, G. Duda, T. Arndt, T. Sauer and
G. Wegner   185

Polymers and Molecular Assemblies for Second-Order Nonlinear Optics
D.J. Williams, T.L. Penner, J.J. Schildkraut, N. Tillman, A. Ulman and
C.S. Willand   195

Spectral Properties and Second-Harmonic Generation of Hemicyanine dye
in Langmuir-Blodgett Films
P. Winant, A. Scheelen and A. Persoons   219

EXPERIMENTAL METHODS

Cubic Susceptibility of Organic Molecules in Solution
F. Kajzar   225

Measurement of the Non-Linear Refractive Index of some Metallocenes by
the Optical Power Limiter Technique
C.S. Winter, S.N. Oliver and J.D. Rush   247

Polymerization and X-ray Structure of new Symmetrical and Unsymmetrical
Diacetylenes
M. Bertault and L. Toupet   253

APPLICATION OF ORGANIC POLYMERS IN NLO DEVICES

Third-Order Nonlinear Guided-Wave Devices
George I. Stegeman, Ray Zanoni, K. Rochford and Colin T. Seaton   257

Organic Integrated Optical Devices
R. Lytel, G.F. Lipscomb, M. Stiller, J.I. Thackara and A.J. Ticknor   277

Orientationally Ordered Nonlinear Optical Polymer Films
J.E. Sohn, K.D. Singer, M.G. Kuzyk, W.R. Holland, H.E. Katz, C.W. Dirk,
M.L. Schilling and R.B. Comizzoli   291

Overview-Nonlinear Optical Organics and Devices
Donald R. Ulrich   299

Non Linear Optical Polymers for Active Optical Devices
A. Buckley and J.B. Stamatoff   327

Linear Electrooptic Coefficient of a Ferroelectric Polymer
Y. Levy, V. Dentan, M. Dumont, P. Robin and E. Chastaing   337

Application of Third-Order Non-Linearities of Dyed PVA to Real-Time Holography
R.A. Lessard, J.J.A. Couture and P. Galarneau   343

DEVICE APPLICATIONS AND MATERIALS REQUIREMENTS

Ultrafast Third-Order Non-linear Optical Processes in Polymeric Films
Paras N. Prasad   351

Picosecond Phase Conjugation in Yellow Polydiacetylene Solutions
J.M. Nunzi, J.L. Ferrier and R. Chevalier   365

Picosecond Studies of Optical Stark Effect in Polydiacetylenes
F. Charra and J.M. Nunzi   369

FOUR WORKING GROUPS REPORTS

Ultrafast Phenomena in Conjugated Polymers
P.N. Prasad   375

Device Applications and Materials Requirements
R. Lytel and G.I. Stegeman   379

Polymer Synthesis for Nonlinear Optics and Characterization
L.R. Dalton   383

Characterization Techniques
G.R. Meredith   385

Index   389

PREFACE

Photonics, the counterpart of electronics, involves the usage of photons instead of electrons to process information and perform various switching operations. Photonics is projected to be the technology of the future because of the gain in speed, processing and interconnectivity of network. Nonlinear optical processes will play the key role in photonics where they can be used for frequency conversion, optical switching and modulation. Organic molecules and polymers have emerged as a new class of highly promising nonlinear optical materials which has captured the attention of scientists world wide. The organic systems offer the advantage of large nonresonant nonlinearities derived from the π electrons contribution, femtosecond response time and the flexibility to modify their molecular structures. In addition, organic polymers can easily be fabricated in various device structures compatible with the fiber-optics communication system. The area of nonlinear optics of organic molecules and polymers offers exciting opportunities for both fundamental research and technologic development. It is truly an interdisciplinary area.

This proceeding is the outcome of the first NATO Advanced Research Workshop in this highly important area. The objective of the workshop was to provide a forum for scientists of varying background from both universities and industries to come together and interface their expertize. The scope of the workshop was multidisciplinary with active participations from chemists, physicists, engineers and materials scientists from many countries.

The talks presented by leading scientist, main lectures and contributions were divided into six parts :

- Theory of nonlinear polarizability in organic low dimensional systems
- Nonlinear susceptibility of organic systems
- Synthesis and nonlinear optical properties characterization of new 1-D conjugated polymers
- Experimental methods
- Application of organic polymers in NLO devices
- Ultrafast optical phenomena in polymers.

The presentations and discussions allowed an on time insight in this rapidly developing area and an introduction to the round table discussions organized around four themes :

- Experimental methods
- Polymer synthesis for NLO and characterization
- Ultrafast phenomena in conjugated polymers
- Device applications and materials requirements.

A poster session was organized during the workshop offering a very good opportunity for extended discussions. The workshop was concluded by a panel discussion giving a summary of different themes and problems treated during the meeting and also some guide lines for future developments.

The volume contains lectures, contributions, some selected posters as well as working groups reports. First part is devoted to the theory of nonlinear susceptibility in organic systems with a special emphasis on dimensionality effects as well as hyperpolarizability calculations in organic systems. Different experimental techniques used in nonlinear susceptibility determination are reviewed and discussed in Part II. The synthesis of novel class of polymers and other organic materials with large $\chi^{(3)}$ values is described in Part III. Different experimental techniques used in hyperpolarizability determination are reviewed and discussed in Part IV. Conjugated polymers with excitonic origin optical nonlinearities are characterized by fast response time. These aspects are discussed in part V. Due to the unique physico-chemical properties organic polymers can be used either as a matrix for orientation of higly efficient $\chi^{(3)}$ molecules or as active nonlinear $\chi^{(3)}$ materials, both in thin film configuration. Different aspects of application in nonlinear optical devices, especially in integrated optics where the peculiar properties of polymers can be used are presented and discussed in Part VI.

We are highly indebted to the NATO Scientific Affairs Division, the main sponsor of the workshop as well as to the following cosponsors : Commissariat à l'Energie Atomique, Centre National d'Etudes des Télécommunications, Coherent Scientific, Eastman Kodak, Foster-Miller, Hoechst Celanese, Quantel and Rhône-Poulenc. The success of the workshop is due to the high level of well prepared lectures and contributions as well as an active collaboration of all participants. We would like also to express our warm thanks to Mr Jacques Normand for an excellent management of the workshop and Mrs Nicole Gambier for a very efficient and smiling secretarial work.

Saclay, September 1988

The organizing committee,

Jean Messier   Paras Prasad

François Kajzar   Donald Ulrich

LIST OF PARTICIPANTS

Prof. BERTAULT M.
GROUPE DE PHYSIQUE CRISTALLINE
BAT. B
CAMPUS DE BEAULIEU
35042 RENNES CEDEX
FRANCE

Prof. BLAU W.
TRINITY COLLEGE
University of Dublin
 DUBLIN 2
IRLANDE

Prof. BUBECK Ch.
MAX-PLANCK-INSTITUT FUR POLYMERFORSCHUNG
POSTFACH 3148
D-6500 MAINZ
RFA

Dr. BUCKLEY A.
HOECHST
CELANESE RESEARCH DIVISION
86 MORRIS AVENUE
07922 SUMMIT, NEW JERSEY
USA

Mr. BYRNE H.
TRINITY COLLEGE
PHYSICS DEPARTMENT
DEPT. OF PURE AND APPLIED PHYSICS
 DUBLIN 2
IRLANDE

Mr. CHARRA F.
CEA - CEN SACLAY
DEIN
LPEM
91191 GIF-SUR-YVETTE CEDEX
Workshop's

Dr. CHOLLET P.
CEA - CEN SACLAY
DEIN
LPEM
91191 GIF-SUR-YVETTE CEDEX
Workshop's

Prof. DALTON L.R.
UNIVERSITY OF SOUTHERN CALIFORNIA
University Park
Department of Chemistry
620 Seaver Science Center
 LOS ANGELES, Cal. 90089-0482
USA

Prof. DEGIORGIO V.
UNIVERSITA DI PAVIA
Dipartimento di Elettronica
Sezione Fisica Applicata
Via Abbiategrasso 209
27100 PAVIA
ITALIE

Prof. DELHALLE J.
FACULTE UNIVERSIT. NOTRE DAME DE LA PAIX
Departement de Chimie
61 rue de Bruxelles
B-5000 NAMUR
BELGIQUE

Dr. DEROUINEAU P.
QUARTZ ET SILICE
BP 95
77792 NEMOURS CEDEX
FRANCE

Dr. DULTZ W.
INSTITUT FUR PHYSIK II-FESTKORPERPHYSIK
Universitatsstrabe 31
Postfach 397
8400 REGENSBURG
RFA

xii

Dr. DUMONT M.
INSTITUT D'OPTIQUE THEORIQUE ET APPLIQUEE
BP 43
91406 ORSAY CEDEX
FRANCE

Dr. FICHOU D.
CNRS
2 rue Henri Dunant
94320 THIAIS
FRANCE

Prof. FLYTZANIS C.
ECOLE POLYTECHNIQUE
Laboratoire d'Optique Quantique
91128 PALAISEAU CEDEX
FRANCE

Prof. GARITO A.
UNIVERSITY OF PENNSYLVANIA
Department of Physics
 PHILADELPHIA PA 19104-6396
USA

Dr. GOLDFARB I.
AFWAL/MLEP
Wright-Petteson Air Force Base
45433 DAYTON, OHIO
USA

Prof. HEEGER A.
UNIVERSITY OF CALIFORNIA
Department of Physics
93106 SANTA BARBARA, California
USA

Dr. HUIJTS R.A.
AKZO CORPORATE RESEARCH
DEPT. APPLIED PHYSICS
AKZO RESEARCH LAB. ARNHEM
P.O. Box 9300
6800 SB ARNHEM
PAYS-BAS

Dr. KAJZAR F.
CEA - CEN SACLAY
D.LETI/DEIN
LPEM
91191 GIF-SUR-YVETTE CEDEX   FRANCE
Workshop's

Dr. KAPTAN Y.
HACETTEPE UNIVERSITY
Faculty of Engineering
Departmert of Physics Engineering
06532 BEYTEPE - ANKARA
TURQUIE

Dr. LE MOIGNE J.
CNRS
ICS
6, RUE BOUSSINGAULT
67083 STRASBOURG CEDEX
FRANCE

Dr. LEDOUX I.
CNET
196 avenue Henri Ravera
92220 BAGNEUX
FRANCE

Prof. LESSARD R.A.
UNIVERSITE DE LAVAL
LROL DEPARTEMENT DE PHYSIQUE
SAINTE FOY
G1K7P4 QUEBEC
CANADA

Prof. LEVY Y.
INSTITUT D'OPTIQUE THEORIQUE ET APPLIQUEE
BP 43 - BAT. 503
91406 ORSAY CEDEX
FRANCE

Dr. LUPO D.
HOSCHST ET.
Applied Physics
6230 FRANKFURT 80 Bdg 864
RFA

Dr. LYTEL R.
LOCKHEED MISSILES & SPACE COMPANY INC.
Research and Development Division
3251 Hanover St
CA94304 PALO ALTO
USA

Dr. MEREDITH G.
E.I. DUPONT DE NEMOURS AND CO INC.
Central Research and Development Depart.
Experimental Station
Delaware
19898 WILMINGTON
USA

Dr. MESSIER J.
CEA - CEN SACLAY
D.LETI/DEIN
LPEM
91191 GIF-SUR-YVETTE CEDEX   FRANCE
Workshop's

Dr. MEYRUEIX R.
RHONE-POULENC RECHERCHES
85 avenue des Freres Perret
BP 62
69192 SAINT FONS
FRANCE

Dr. MEYRUEIX P.
SCHLUMBERGER INDUSTRIES
CENTRE DE RECHERCHES SMR
BP 620-05
92542 MONTROUGE CEDEX
FRANCE

Prof. MILBURN G.H.W.
UNIVERSITY OF EDINBURGH
Dept. of Applied Chemical Sciences
Colinton Road
EH10 5DT EDINBURGH
GB

Mr. NEHER D.
MAX-PLANCK-INSTITUT FUR POLYMERFORSCHUNG
Postfach 3148
D-6500 MAINZ
RFA

Prof. NICOUD J.F.
CRM STRASBOURG
6, rue Boussingault
67083 STRASBOURG CEDEX
FRANCE

Dr. NUNZI J.M.
CEA - CEN SACLAY
DEIN
LPEM
91191 GIF-SUR-YVETTE CEDEX
Workshop's

Dr. PERRY J.W.
JET PROPULSION LABORATORY
CALIFORNIA INSTITUTE OF TECHNOLOGY
MAIL STOP 67.201
4800 OAK GROVE DR.
CA91109 PASADENA
USA

Prof. PERSOONS A.
UNIVERSITY OF LEUVEN
Department of Chemistry
Lab. of chemical & biological Dynamics
Celestijmenlaan 200 D
B-3030 HEVERLEE
BELGIQUE

Mr. PRATESI G.
IROE
VIA PANC ATICHI 64
50127 FIRENZE
ITALIE

Mrs ROBALO M.P.
CENTRO QUIMICA ESTRUTURAL IST
AV. ROVISCO PAIS
1096 LISBOA CEDEX
PORTUGAL

Prof. STEGEMAN G.
UNIVERSITY OF ARIZONA
Optical Sciences Center
Arizona Research Laboratories
Arizona
85721 TUSCON
USA

Dr. ULRICH D.
AIR FORCE OFFICE OF SCIENTIFIC AIR FORCE B
Directorate of Chemical and
Atmospheric Science
Room B21
 WASHINGTON DC 20332-6448 USA
Workshop's

Dr. WILLIAMS D.
EASTMAN KODAK COMPANY
Research Laboratories
NY14650 ROCHESTER
USA

Prof. WONG G.K.
COLLEGE OF ARTS AND SCIENCES
DEPT. OF PHYSICS AND ASTRONOMY
NORTHWESTERN UNIVERSITY
60208 EVANSTON, ILLINOIS
USA

Prof. PRASAD P.N.
STATE UNIVERSITY OF NEW-YORK
Department of Chemistry
Faculty of Natural Sciences & Mathematics
Acheson Hall
Buffalo
14214 BUFFALO - USA
Workshop's

Dr. RIKKEN G.
PHILIPS RESEARCH LABORATORIES
Building WB 2
PO Box 80.000
5600 JA
PAYS-BAS

Dr. SOHN J.E.
AT&T LABORATORIES
PO Box 900
NJ08540 PRINCETON
USA

Prof. TALIANI C.
CNR
Instituto di spettroscopia Molecolare
Via de Castagnoli 1
40126 BOLOGNA
ITALIE

Dr. VON PLANTA C.
F. HOFFMANN - LA ROCHE INC.
Grenzacherstr. 124
Building 65
 BALE
SUISSE

Dr. WINTER C.S.
BRITISH TELECOM RESEARCH LAB.
RT2333/BTRL
B55/122
MARTLESHAM HEATH
 IPSWICH
GB

IMPACT OF DIMENSIONALITY IN THE OPTICAL NONLINEARITIES

Ch. FLYTZANIS

Laboratoire d'Optique Quantique du C.N.R.S.
Ecole Polytechnique, 91128 Palaiseau cédex, FRANCE

1. INTRODUCTION

New developments in nonlinear optics and implementation [1] of several nonlinear effects in devices is conditionned by a substantial improvements of several characteristics of the nonlinear optical materials. These are of very diverse nature but the ones most relevant for the nonlinear process are :

    a) the susceptibility $\chi^{(n)}$ (magnitude, phase or sign)

    b) its recovery time τ

    c) the real index of refraction n

    d) the absorption coefficient α

    e) the relevant frequency domain $\Delta\omega$.

Certainly in principle, each one of these quantities can be calculated for any given material but with formidable computational effort which in effect obscurs their trends and interrelations that one intuitively expects among them and does not facilitate the comparison of different nonlinear materials. A closer inspection however of the physical origin of these quantities hints at the working of some simple but powerful functional interdependence which can be extracted by simple models that contain the gross features of the charge distribution, the ones that really persist after summing and averaging over all quantum states.

Below we briefly review some implications of such simple models regarding the nonlinear susceptibilities $\chi^{(n)}$ and summarize some interrelations between the linear and nonlinear coefficients. For infinitely extended systems by exploiting their periodicity we show that the nonlinear susceptibilities $\chi^{(2n+1)}$, like the linear one $\chi^{(1)}$, in their transparency region can be cast in the form of scaling laws, essentially power laws of an effective parameter, closely connected with charge delocalization, with exponents that critically depend on the dimensionality of the charge distribution in the material ; for the even order susceptibilities $\chi^{(2n)}$ a similar analysis does not lead to any relevant relations and one must resort to a different approach. For finite size systems, also termed zero dimensionality systems, like short polymer chains, one can derive certain interrelations but for the most interesting cases of semiconductor and metal microcrystallites it is advantageous to exploit certain broad resonances that result from quantum and dielectric confinements and the analysis must take into account then the dynamics of the photoexcited charges.

2. NONLINEAR OPTICAL SUSCEPTIBLILITIES

The starting point is the fact [2] that the nonlinear optical properties of a material, like its linear ones, in the visible and near visible are essentially determined by the ability of the valence electrons to respond to an intense electric field of optical frequency. Accordingly the relevant quantity is the <u>optical nonlinearity per valence electron</u> and it is of importance to single out its most salient features.

For extended periodic systems the electron states are represented in terms of Bloch band states and the expressions of the optical susceptibilities $\chi^{(n)}$ using the Genkin-Mednis approach [2,3], the most appropriate for our purpose, are ;

1

J. Messier et al. (eds.), Nonlinear Optical Effects in Organic Polymers, 1–12.
© *1989 by Kluwer Academic Publishers.*

$$\chi_{xx}^{(1)} = \frac{4e^2}{hV} \int_{B.Z} \Omega_{vc} S_{cv}\, dk \tag{1}$$

$$\chi_{xxxx}^{(3)} = \frac{8e^4}{h^3 V} \int_{B.Z} \left[\frac{1}{\omega_{cv}} \left(\frac{\partial S_{cv}}{\partial k} \right) \left(\frac{\partial S_{vc}}{\partial k} \right) - \Omega_{vc} S_{cv} S_{vc} S_{cv} \right] dk \tag{2}$$

where $\hbar\omega_{cv} = \varepsilon_c - \varepsilon_v$, Ω_{vc} is the transition dipole moment matrix element between the highest valence (v) and lowest conduction (c) bands, $S_{vc} = \Omega_{vc}/\omega_{vc}$ and ε_v and ε_c are the band energies for the v-and c-band, respectively ; all quantities are function of the wavevector k which is let to vary over the entire Brillouin zone (B.Z.) and V is the normalization volume. In (2.1) ard (2.2) we have introduced the two band approximation, which is quite sufficient for our purpose. Furthermore, we have assumed that the system possess inversion symmetry so that in particular $\chi(2) \equiv 0$; for noncentrosymmetric systems, the expression of $\chi(3)$ is slightly more lengthy and the second order susceptibility is :

$$\chi_{xxx}^{(2)} = \left(\frac{6e^3}{hV} \right) \frac{i}{2\pi} \int_{B.Z} \left[S_{cv} S_{cv} \left(\Omega_{vv} - \Omega_{cc} \right) - \frac{1}{2} \left(S_{vc} \frac{\partial S_{cv}}{\partial k} - \frac{\partial S_{vc}}{\partial k} S_{cv} \right) \right] dk \tag{3}$$

where Ω_{vv} and Ω_{cc} are interband transition dipole matrix elements. One may formally define also the polarizabilities $\alpha^{(n)}$ for such infinitely extended systems by the relation :

$$\chi^{(n)} = \frac{\alpha^{(n)}}{v} \tag{4}$$

where v is the repeart unit-cell volume : for localized systems of extension much smaller than the optical wavelength the above expressions for $\alpha^{(n)}$ reduce to the well known ones :

$$\alpha^{(1)} \equiv \alpha = \sum_n{}' \frac{\mu_{gn}\mu_{ng}}{E_{ng}} \tag{5}$$

$$\alpha^{(2)} \equiv \beta = 3 \left(\sum_{nn'}{}' \frac{\mu_{gn}\mu_{nn'}\mu_{n'g}}{E_{ng}E_{n'g}} - \mu_{gg} \sum_n{}' \frac{\mu_{gn}\mu_{ng}}{E_{ng}^2} \right) \tag{6}$$

$$\alpha^{(3)} \equiv \gamma = 4 \left(\sum_{n,n',n''}{}' \frac{\mu_{gn}\mu_{nn'}\mu_{n'n''}\mu_{n''g}}{E_{ng}E_{n'g}E_{n''g}} - \sum{}' \frac{\mu_{gn}\mu_{ng}}{E_{ng}} \sum_{n'}{}' \frac{\mu_{gn'}\mu_{n'g}}{E_{n'g}} \right) \tag{7}$$

where $\sum_m{}'$ means that terms with m = g will be excluded, m labels the electronic states of the molecule (g is the ground state) and we limit ourselves to the x components ; μ is the electronic dipole moment operator. The calculation of these expressions is quite involved, but useful information about size effects and scaling laws can be obtained with the help of the Unsöld approximation [2]. There is a correspondance between the different terms in the two sets of expressions (1) to (3) and (5) to (7) respectively.

As can be seen from (1) to (3) the susceptibilities are expressed as integrals over the Brillouin zone. The main contribution to the integrals is expected [4,5] to come from some few nonoverlapping ciritical regions in the joint density of states. These are, points E_0, lines E_1 and surfaces E_2 depending on the spatial extension of the electronic density distribution and are defined by the condition [6] :

$$\nabla_k \omega_{cv}(k) = 0 \tag{8}$$

or equivalently $v_{gv} = v_{gc}$ where $v_{gi} = \nabla_k \varepsilon_i$ is the electron group velocity in band i. This is indeed the case in $\chi(1)$ and is also the case for odd-order susceptibilities $\chi(2n+1)$. This directly establishes a relation between the optical susceptibilities $\chi(2n+1)$ and the topology and dimensionality of the joint density of states. Furthermore it expresses $\chi(2n+1)$ in terms of the values Ω_{vc} and ω_{vc} at these critical regions. This functional dependence is precisely [5] the origin of the scaling laws that will be summarized below for the odd order susceptibilities ; in the case of the even order susceptibilities as will become evident later the critical points do not play such a proeminent role and a different approach is needed.

Before we move to derive the scaling laws using the above expressions we wish to address our attention to another important feature in the expressions of the nonlinear susceptibilities. As can be seen from (2), for instance, the magnitude and sign of the nonlinear optical susceptibilities $\chi(n)$ is determined by the competition of two terms an intraband one, the first term in the integrand in (2), that arises from field mixing of Bloch band states within a band and an interband one, the second term in the integrand in (2) that arises from mixing of states across the gap with wave vector conservation ; in contrast, the linear susceptibility $\chi(1)$ only involves interband terms and this has crucial qualitative and quantitative implications. This competition can also be interpreted as displacements of opposite signs of the effective band gaps at the critical points when an electric field is applied. The interband band term can be identified as the analogue of the Stark shift for atomic levels and consists in a repulsion between valence and conduction states at each k and in particular at the critical regions ; this leads to an increase of the effective energy gap there or a negative contribution to $\chi(3)$ as can also be inferred from the negative sign in front of this term in the integrand in (2) The intraband term on the other hand, which is also responsible for the Franz-Keldysh effect, consists in a repulsion of the states within a single band which results in a net repulsion of the critical points from all other states in the same band ; this leads to a decrease of the effective energy gap there or a positive contribution to $\chi(3)$. It is quite evident that the intraband term will be dominant whenever the bands vary strongly with k (wide bands) which will be the case for very delocalized systems or strong overlap between wave functions of neighboring units in the periodic structure ; on the other hand this contribution will be negligeable for highly localized systems (very narrow bands) and the interband term becomes then dominant. Accordingly the sign of $\chi(3)$ gives a crucial indication of the degree of electron delocalization.

3. ODD ORDER SUSCEPTIBILITES IN 1D-2D AND 3D-SYSTEMS

As previously stated the essential contributions to the integrals in (1) to (3) come from some few non overlapping critical regions in the joint density of states the so-called van Hove singularities where (and only there) an infinite density of states accumulates [6]. Their features, in particular their dimensionality and relative positions, are shaped [4] by the topology of the k-space. For a three dimensional (3D) semicondutor these are most commonly a point a line and a surface located at E_0, E_1 and E_2 in the Brillouin zone with $E_2 < E_1 < E_0$; for a two dimensional (2D) semiconductor these are a point and a line located at E_0 and E_1 respectively with $E_1 < E_0$ and for a one-dimensional (1D) semiconductor there is only a point located at E_0.

For an one-dimensional semiconductor one can show [7] on quite general topological grounds that this infinite accumulation of states takes place at a point E_0 located at the edge of the Brillouin zone, $ka = \pi$, where a is the unit cell length, where in addition the transition dipole moment $|\Omega_{cv}|$ attains its maximum value [7]:

$$|\Omega_{cv}(\pi)| = L_d = aN_d \tag{9}$$

and the energy gap is smallest

$$\hbar\omega_{cv}(\pi) = E_0$$

The coincidence of these three features at a single point is an essential characteristic of the 1D systems and is reminiscent of a metallic type behavior [7]. One can show [5] that for $N_d \gg 1$

$$\chi^{(2n-1)} = \chi_0^{(2n-1)}\left(\frac{E_F}{E_0}\right)^{4n-2} \approx N_d^{4n-2} \tag{10}$$

where E_F is the Fermi energy of electron distribution and $N_d = E_F/E_0$ the delocalization parameter defined by (9) : in particular [7,8]:

$$\chi^{(1)} = \chi_0^{(1)}\left(\frac{E_F}{E_0}\right)^2 \approx N_d^2 \tag{11}$$

$$\chi^{(3)} = \chi_0^{(3)}\left(\frac{E_F}{E_0}\right)^6 \approx N_d^6 \approx \left[\chi^{(1)}\right]^3 \tag{12}$$

These relations seem to possess a very wide validity [7]

For the other types of critical regions using certain simplifications in the model density of states, one finds [5]:

$$\chi^{(2n-1)} \cong P_1 / E_1^{3n-1} \tag{13}$$

for the E_1 contribution and in particular [4]:

$$\chi^{(1)} \approx P_1^2 / E_1^2$$

$$\chi^{(3)} \approx P_1^2 / E_1^5$$

similarly [5]:

$$\chi^{(2n-1)} \approx P_2^2 / E_1^{2n+1/2} \tag{14}$$

for the E_2 -contribution and in particular [4]:

$$\chi^{(1)} \approx P_2^2 / E_1^{5/2}$$

$$\chi^{(3)} \approx P_2^2 / E_1^{9/2}$$

The scaling laws for the E_1^- and E_2^- critical regions are subject to uncertainties as they were derived from a very approximate model for the density of states .

4. CORRELATION AND BAND STATE FILLING EFFECTS

The important point we wish to stress here is that the contributions from the different critical regions are <u>additive</u> ; furthermore the energies E_0, E_1 and E_2 can be determined experimentally. Hence one has a very convenient way of evaluating $\chi^{(2n-1)}$ for a given compound or comparing different compounds without resorting to complicated quantum mechanical calculations. In particular, one can follow the implications of any changes of the valence electron density distribution of chemical or other origin.

Clearly to the extend that one wishes cristals with a large transparency region the most favorable case is the one with only the E_0 contribution ; otherwise stated one-dimensional systems, other things being equal, possess the highest odd order nonlinearity per valence electron in one direction but clearly not in the other directions ; on the other hand $\chi(1)$ does not show the same enhancement as $\chi(3)$ neither does $\chi(2)$. Thus the striking differences [8,7] between the inorganic semiconductors like Ge and GaAs on one hand and the polydiacetylenes on the other find a simple and unified explanation qualitatively as well as quantitatively.

Clearly the scaling laws given above were derived within the one electron approach where electron correlation is altogether neglected ; when the latter is taken into account the expression of Ω_{cv} and $\chi(n)$ become exceedingly complex to allow any analytical or even numerical treatment. However we may remark that the many body effects become relevant mostly when the light frequency ω approaches or coincides with one of the critical point energies, E_0, E_1 or E_2 ; well below all these energies these effects can be taken into account by renormalizing the one electron parameters but the exponents may not be severely affected.

In deriving expressions (3), we have not taken properly into account the band filling effect [9,10,11] related to the exclusion of band states that have been victually occupied. This effect can have a dramatic impact on the magnitude and sign of $\chi(3)$. It can be most easily seen and estimated in the case of the third order susceptibility $\chi(3)(\omega,-\omega,\omega)$ related to the optical Kerr effect. This is defined through the relation :

$$\delta\varepsilon = 12\pi \, \chi(3)(\omega,-\omega,\omega) \, E_\omega^2$$

where $\delta\varepsilon$ is the dielectric constant change induced by a macroscopic electric field of frequency ω. This change $\delta\varepsilon$ can be related by a Kramers-Kronig relation to the photoinduced change of absorption $\delta\alpha(\omega')$ linear in the beam intensity. In $\delta\alpha(\omega')$ because of the Pauli exclusion principle states that have already been populated by light induced transitions are not any longer accessible to further occupation ; this essentially amounts to a photoinduced increase of the effective gap. Through the Kramers-Kronig relation this band filling effect also affects $\delta\varepsilon$ and $\chi(3)(\omega,-\omega,\omega)$ and can have important implications on their magnitude and sign but presumably not on the scaling law.

5. EVEN ORDER SUSCEPTIBILITIES

There is an essential difference between odd and even susceptibilities : the later vanish identically for centrosymmetric materials while the former does not. This implies that the charge asymmetry within a unit cell plays the essential role : it is the accumulation of such asymmetric units pointing predominantly in one direction that leads to a $\chi(2n) \neq 0$. It is therefore instructive to limit our attention to an 1D semiconductor.

Because of this charge asymmetry within a unit cell the matrix elements are in general complex quantities. Careful analysis of (3) reveals that contributions to the integrand only come from regions where Ω_{cv} is complex while it gets no contribution from region in the B.Z. where Ω_{cv} is real or pure imaginary [7].The later precisely occurs at the edge of the B.Z. of an one-dimensional system namely at the critical point E_0 where the electron states are also most delocalized. Thus in contrast to $\chi(3)$, $\chi(2)$ does not take full advantage of the infinite densities of states there and the highly delocalized character of these states. In particular with the critical point contribution being suppressed in $\chi(2)$, integration over the whole B.Z. is now required to extract the behavior of $\chi(2)$. This implies that $\chi(2)$ is sensitive to local properties of the electron density and more precisely to the asymmetry of the valence electron distribution within a unit cell which may not be compatible with delocalization over several unit cells.

Millers rule which states that [12]:

$$\chi_{ijk}^{(2)} = \delta_{ijk} \chi_{ii}^{(1)} \chi_{jj}^{(1)} \chi_{kk}^{(1)} \qquad (15)$$

where δ_{ijk} was found [13] to be roughly proportional to the dipole moment of the electron density distribution within a unit cell ; this indicates that delocalization and asymmetry effects can be roughly factorized but this requires more careful analysis. Thus in order to increase $\chi(2)$ one must optimize the parameters that determine the charge asymmetry and extension within a unit cell.

6. ZERO DIMENSIONALITY SYSTEMS. NON RESONANT NONLINEARITIES

Under the term zero-dimensionality systems we understand material systems which are obtained from infinitely extended periodic systems by reducing their extension in all directions to a size much smaller than the wave length. These can be isolated molecular systems like finite polymer chains, metal or semiconductor microcrystallites. In comparing such systems embedded in a transparent dielectric (glass, water...) with corresponding infinite ones previously discussed one has to take into account two major effects the quantum confinement effect which results in a spectral redistribution of the quantum states and the dielectric confinement effect which takes into account the local field amplitude redistribution because of the interface between the finite size system and its surrounding medium ; clearly both effects crucially depend on the boundary conditions and this greatly complicates the extraction of meaningful relations between their nonlinear properties and other physical parameters.

The dielectric confinement effect to a good approximation, can be taken into account by the effective medium approach for composites [14]. The quantum confinement on the other hand needs careful consideration for each case : but certain trends can be extracted by simple models

Thus for linear conjugated carbon chains without bond alternation one obtains [7] :

$$\alpha^{(2n-1)} \approx L^{2n+1} e^{2n} / a\beta^{2n-1} \qquad (16)$$

valid for large $N = L/a$, where L is the half-length of the chain, a is the interatomic distance and β is the resonance (hopping) energy between neighboring atoms. In particular, one has for $n = 1$ and $n = 3$ respectively :

$$\alpha = L^3 e^2 / a\beta$$

$$\gamma = L^5 e^4 / a\beta^3$$

the L^5 dependence of γ was initially derived [15] using the free electron model to describe the electronic states.

For conjugated carbon chains with bond alternation analogous relations can only be extracted through numerical calculation of the quantum mechanical expressions (5) (6) and (7). If we denote β_1 and β_2 the two resonance energies there and we introduce the delocalization parameter :

$$N_d = \frac{\beta_1 + \beta_2}{\beta_1 - \beta_2} \qquad (17)$$

the same as in (9), one can show [16] that within the Hückel approximation the analogue of the tight binding approximation used for infinite 1D systems, for a chain of 2N atoms and such that $N_d < N$

$$\gamma_{2N} \cong N_d^6 \qquad (18)$$

which is the same as the one previously derived for our infinite 1D semiconductor (ex. bond alternated infinite chain). This is a very important result since it indicates the validity range of the scaling law (10) for actual 1D-systems when defects are present along the chains which interrupt its periodicity ; as long as the average distance between two neighboring defects is larger than the delocalization length L_d defined by (9) and N_d by (17) the scaling law does not break down.

7. COMPOSITES. METAL AND SEMICONDUCTOR PARTICLES

Along with the substantial effort underway to understand and improve the optical nonlinearities of homogeneous materials, crystalline or amorphous, a new class of materials is presently attracting a lot of interest namely the composite materials [17]. These are most commonly formed by homogeneously dispersing small metal or semiconductor microcrystallites in a transparent dielectric, for instance glass, but a great variety of other combinations can be envisionned. In contrast to the homogeneous materials discussed in the previous sections where the attention is mostly concentrated on the optical nonlinearities in their transparency region, in the composites one exploits the enhancement that is brought in by resonances intrinsicly related to the reduction of the dimensions in all three directions of the material with delocalized valence electrons (metal or semiconductor) and its insertion in a different dielectric medium (glass) .

The behavior of the delocalized valence electrons in bulk and semiconductor crystals is described by certain characteristic lengths which correspond to certain averagings ; to this scale of lengths also corresponds a scale of characteristic energies. If the size of the microcrystallite is larger than all these lengths their properties are essentially the same as in the bulk crystal. However as their size is reduced below any of these lengths the corresponding averaging process cannot be used and the details of certain interactions become proeminent. These features are classified into quantum and dielectric confinement effect

a) Dielectric confinement. Metal particles.

This is an effect of classical electromagnetic origin and is related to the redistribution of the electric field strength that results from the confinement of a microcrystallite in dimensions much smaller than the wavelength of the light λ in a dielectric of different dielectric constant. The effect is particularly pronounced in the case of metallic particles [18,19] embedded in small volume concentration in a dielectric . The electric field inside such a particle is enhanced in a certain frequency region the so-called surface plasmon resonance. This enhancement can be easily derived by electrostatics which is valid since $d<<\lambda$, where d is the diameter of the metal inclusion assumed spherical. One obtains that the electric field effectively acting inside such a metallic particle of frequency dependent dielectric constant $\varepsilon_m(\omega)$ surrounded by a transparent dielectric of dielectric constant ε_0 is

$$E_i = \frac{3\varepsilon_0}{\varepsilon_m(\omega) + 2\varepsilon_0} E_0 \equiv f_1 E_0 \tag{19}$$

where E_0 is the externally applied electromagnetic field. We may use the Drude model for $\varepsilon_m(\omega)$ or :

$$\varepsilon_m(\omega) = 1 + \delta\varepsilon_b - \frac{\Omega_p^2}{\omega\left(\omega + \frac{i}{\tau}\right)} = \varepsilon' + i\varepsilon'' \tag{20}$$

where $\delta\varepsilon_b$ is bound electron contribution, Ω_p is the free electron plasma frequency and τ is the electron scattering time in the bulk τ_0 corrected [18] to also include the electron scattering from the walls or :

$$\frac{1}{\tau} = \frac{1}{\tau_0} + \frac{d}{v_F} \tag{21}$$

v_F is the Fermi level electron velocity. One can easily see from (19) than E_i is enhanced close to frequency ω_s such that :

$$\mathrm{Re}\,\varepsilon_m\!\left(\omega_s\right) + 2\varepsilon_0 = 0 \qquad (22)$$

which is the condition for the surface plasmon resonance.

The effective linear dielectric constant of the composite $\tilde{\varepsilon}$, when the volume concentration of the inclusions p<<1, [14,17] :

$$\tilde{\varepsilon} = \varepsilon_0 + 3p\varepsilon_0 \frac{\varepsilon_m - \varepsilon_0}{\varepsilon_m + 2\varepsilon_0} \qquad (23)$$

and the effective optical Kerr effect susceptibility $\chi^{(3)}(\omega,-\omega,\omega)$ is given by [19,20]:

$$\tilde{\delta\varepsilon} = 12\pi\,\tilde{\chi}^{(3)}\,E_0^2 = 12\pi\,pf_1^4\,\chi_m^{(3)}E_0^2 \qquad (24)$$

where $\chi_m^{(3)}\!\left(\omega,-\omega,\omega\right)$ is the optical Kerr effect susceptibility of the metallic particles of diameter d ; in (24) we have neglected the contribution $\chi^{(3)}_0$ from the surrounding medium.

Because of the presence of f_1 in the fourth power in (24) $\tilde{\chi}^{(3)}$ is immensely enhanced when $\omega \cong \omega_s$ the resonance condition (22 ; well outside the width of this resonance, $\Delta\omega \sim \tau^{-1}$, $\tilde{\chi}^{(3)}$ drops down by several orders of magnitude and essentially $\tilde{\chi}^{(3)} \cong \chi^{(3)}_0$. It should be

noted by passing that only the optical Kerr effect susceptibility has fourth power dependence on the resonance since all local field factors f_1 become resonantly enhanced in one and the same frequencies. This effect was proposed and observed by Ricard et al [20]. Thus for gold particles of volume concentration 10^{-6} embedded in glass $\tilde{\chi}^{(3)}/\chi^{(3)}_0 \cong 3.10^3$ at $\omega = \omega s \cong$

0.53μ and in addition the induced anisotropy was measured $\tilde{\chi}^{(3)} / \tilde{\chi}^{(3)} \cong 1/8$. From these values and (24) one can also extract the value $\chi_m^{(3)} \cong 10^{-8}$ esu. The recovery time was

found[20,21] to be short less than 5ps which was the experimental time resolution. These measurements were done [21,22] as a function of the size of the metallic inclusions that allowed to single out the mechanism of the optical nonlinearity $\chi_m^{(3)}$.

There are three mechanisms that contribute [22] to $\chi_m^{(3)}$ and the corresponding contributions

to $\chi^{(3)}$ are all three imaginary :

- intraband [21]which results from resonant transitions at $\omega = \omega_s$ among the confined states of the s-p electrons (in the bulk these are free) : for this contribution one finds that $\mathrm{Im}\chi_m^{(3)} < 0$ and $\chi_m^{(3)} \sim d^{-3}$ a pronounced quantum confinement effect.

- interband [22]which results from resonant transitions at $\omega = \omega_s$ from the bound d-electron states to the confined s-p electron states : for this contribution one finds $Im\chi_m^{(3)} < 0$

and $\chi_m^{(3)}$ size independent

- population redistribution [22] which results from the modification of $\varepsilon_m^{''} = Im\varepsilon_m(\omega)$ because of the temperature rise in the electron system consequent to the absorption at $\omega = \omega_s$; for this contribution one finds that $Im\chi_m^{(3)} > 0$ and size independent.

The analysis of the measured values of $\chi^{(3)}$ on different glass samples containing gold particles of different average diameters in the range 30 Å to 100 Å showed [22] that the dominant mechanism in $\chi^{(3)}$ is the third namely the incoherent population rearrangement ; the second one, the coherent interband contribution is by one order smaller and contributes to $\chi^{(3)}$ while the first one, the coherent intraband contribution is negligeable with respect to the other two. In particular the results showed[22] that was no detectable quantum confinement on the size range studied for the gold particles.

b) Quantum confinement
In the metallic particles the quantum confinement effects are "washed" out by the incoherent hot electron contribution (population rearrangement). In the semiconductor microcrystallite, on the other hand, the quantum confinement effects are expected [23-25] to be pronounced on the nonlinear susceptibilities when the microcrystallite size is close or below the exciton diameter but have not been experimentally evidenced yet.
Up till now the main experimental effort has been concentrated [26-33] on semiconductor doped glasses containing $CdS_{1-x}Se_x$ crystallites which are commercially available ; in this case the particle diameter is larger than any exciton diameter. Extensive studies of linear and nonlinear optical properties in these systems close to their absorption edge showed that these are similar to those in the bulk and no size dependant effects could be observed. In particular the state filling model adequately describes [27,28] all these properties and in particular leads to $\chi^{(3)}/\alpha \cong$ constant independent on frequency. The recovery time of the optical Kerr effect measured [27,29,30] by the four wave mixing technique shows different regimes corresponding to intraband, interband, Auger [32,33] and surface impurity traping [27] processes. The later one can be quenced when the glass is exposed to high fluences and "photodarkening" sets in [27].
For semiconductor particles of diameter close or smaller than the exciton diameter quantum confinement effects will introduce substantial changes as can be expected from the changes that occur on the spectrum. These changes can be simply described and evaluated within a simple framework [34].
A basic assumption in this description is that the electron and hole characteristics are those given by effective mass approximation in a two-band model ; these are in particular their effective masses m_e and m_h, and their radii $a_e = h\varkappa/m_e e^2$ and $a_h = h\varkappa/m_k e^2$ respectively where $\varkappa$ is the semiconductor permittivity. In adddition one assumes that the electrons and holes wave functions do not leak out from the particle and that the surface states are decoupled. With these assumptions one can derive the main changes that occur on the spectrum of size a crystallite as its radius a = d/2 is varied relative to a_e and a_h..

- for $a < a_h < a_e$, the cinetic energy $h^2/\mu a^2$ where $\mu = m_e m_h/(m_h+m_e)$ dominates over the

Coulomb energy $e/\varkappa a$ and at first approximation the electorn only senses the potential of the particle walls which are assumed infinite. The essential consequence is that all exciton features disappear and the continuum of band to band transitions of the bulk are replaced by a series of discrete transitions ; in particular the gap E_g of the bulk increases to :

$$h\omega_1 = E_g + \frac{h^2\pi}{2\mu a^2}$$

also termed 1s-1s transition.

- for $a_h < a < a_e$, because of the higher electron energy, the hole moves into a potential averaged over the electron distribution (adiabatic apporximation) and the hole energy levels are described by those of a 3D isotropic oscillator ; the main consequence is that every interband transition is converted to a series of closely spaced sub-levels.

- for $a_n < a_e \leq a$ the exciton states can be formed since their binding energy is larger than the corresponding energies of the hole and electron. The quantum confinement results in a shift of the exciton line with respect to its position in the bulk. This shift results from the quantification of the exciton translational motion with mass $M = m_e + m_h$ inside the finite particle extent ; this shift is of the order of $h^2\pi/Ma^2$.

Inclusion [35,36] of additional effects (leaking walls, Coulomb potential...) does not alter substantially these features. Actually in a given sample the semiconductor spheres show a distribution of sizes whose analytical form has been derived [37] under certain conditions regarding nucleation and condensation of the crystallites.

Quantum size effects in the linear properties (ex. absorption) has been observed [36,38-40]. The first evidence of such effects in nonlinear properties was recently obtained [41] in hole burning experiments which allow one to assess in the line width of the 1s-1s transition the relative importance of the size distribution (inhomogeneous broadening) and that of the electron-phonon coupling (homogeneous broadening).

8. CONCLUSIONS

In the above discussion we concentrated our attention on some general trends that nonlinear optical susceptibilities possess when delocalization is a dominant feature of the valence electron density distribution. Here by delocalization we understand the electron wave function extension over several repeat units of the periodic structure and should be clearly distinguished from the electron delocalization that occurs within a molecule.

Exploiting certain general topological properties of the electron state distribution in periodic systems, in particular the existence of few non overlapping critical regions, we have shown that the contributions to $\chi^{(2n+1)}$ from these regions satisfy the additivity rule and besides crucially depend on their dimensionality in k-space. The most favorable case is the one-dimensional one. On the other hand we showed that $\chi^{(2n)}$ takes no advantage from this delocalization as the critical points do no contribute to $\chi^{(2)}$. This quantity is dominated by local intramolecular properties which are insensitive to the details of the k-space distribution.

We also have shown that as the size of the extended system is reduced below certain characteristics lengths that describe the bulk properties of the delocalized valence electrons the spectrum of the latter undergo drastic changes as a consequence of the dielectric and quantum confinements. These also drastically modifiy the nonlinear susceptibilities and the dynamics of the photo-induced carriers. These changes can be controlled in the artificially formed composite materials and may lead to the emergence of materials with tunable nonlinearity taillored to satisfy certain constraints.

ACKNOWLEDGEMENTS

The author is very much indebted to D. Ricard, P. Roussignol and F. Hache for numerous and interesting discussions.

REFERENCES
1. See for instance "Nonlinear Optics :Materials and Devices, Eds. C. Flytzanis and J.L.Oudar, Springer Verlag, Heidelberg, 1986
2. C. Flytzanis, in "Treatise of Quantum Electronics, H. Rabin and C.L. Tang Eds, Vol 1, part A, Academic Press, New York, 1975
3. V.N. Genkin and P.M. Mednis, Sov. Phys. JETP 27, 609 (1968) (Zh. Eksp. Teor. Fiz. 54, 1137)
4. M. Cardona and F.H. Pollack, in "Optoelectronic Materials", G.A. Albers ed. Plenum, New York, 1971
5. C. Flytzanis in "Nonlinear Optical Properties of Organic Molecules and Crystals" D. Chemla and J. Zyss Eds., Vol. 2, p. 121, Academic Press, New York, 1987
6. W. Ashcroft and N.D. Mermin, "Solid State Physics",Holt Saunders, Tokyo, 1981
7. G.P. Agrawal, C. Cojan and C. Flytzanis,Phys. Rev. B17, 776 (1978)
8. C. Sauteret, J.P. Hermann, R. Frey, F. Pradère, J. Ducuing, R.R. Chance and R.H. Barghman, Phys. Rev. Lett. 36, 956 (1976)
9. See for instance, H. Haug and S. Schmitt-Rink, J. Opt. Soc. Amer. B2, 1135 (1985)
10. See for instance B.S. Wherett and N.A. Higgins, Proc. R. Soc. (London) A379, 67 (1982)
11. S. Schmitt-Rink, D. Miller and D. Chemla, Phys. Rev. B37, 941 (1987)
12. R.C. Miller, Appl. Phys. Lett. 50, 17 (1964)
13. C. Flytzanis and J. Ducuing, Phys. Rev. 178, 1218 (1969)
14. J.C. Maxwell-Garnett, Philos. Trans. Roy. Soc. (London) 203, 385 (1904) ; 205, 237 (1906)
15. K. Rustagi and J. Ducuing, Opt. Comm. 10, 258 (1972)
16. G.P. Agrawal and C. Flytzanis, Chem. Phys. Lett.,44 366 (1976)
17. C. Flytzanis, F. Hache, D. Ricard and Ph. Roussignol in "The Physics and Fabrication of Microstructures and Microdevices", M.J. Kelly and C. Weisbuch Eds., Springer Verlag, Heidelberg, p. 331 (1986)
18. See for instance J.A.A.J. Perenboom, P. Wyder and F. Meier, Physics Reports, 78, 173-292 (1981)
19. Contribution of D. Ricard in ref. 1
20. D. Ricard, Ph. Roussignol and C. Flytzanis, Opt. Lett. 10, 511 (1985)
21. F. Hache, D. Ricard and C. Flytzanis, J. Opt. Soc. Am. B3, 1647 (1986)
22. F. Hache, D. Ricard, C. Flytzanis and V. Kreibig (to appear)
23. S.Schmitt-Rink, D.A.B. Miller and D.S. Chemla, Phys. Rev. B35, 8113 (1987)
24. T. Takagahara and H. Hanamura, Phys. Rev. Lett. 56, 2533 (1986)
25. T. Takagahara, Phys. Rev. B36, 9293 (1987)
26. R.K. Jain and R.C. Lind, J. Opt. Soc. Am. 73, 647 (1983)
27. P. Roussignol, D. Ricard, J. Lukasik and C. Flytzanis, J. Opt. Soc. Am. B4, 5 (1987)
28. Ph. Roussignol, D. Ricard and C. Flytzanis, Appl. Phys. A44, 285 (1987)
29. S.S. Yao, C. Karaguleff, A. Gabel, R. Fortenberry, C.T. Seaton and G.I. Stegerman, Appl. Phys. Lett. 46, 801, (1985)
30. G.R. Olbright, N. Peyghambarian, S.W. Kock and L. Banyai, Opt. Lett. 12, 413 (1987)
31. M.C. Nuss, W. Zinth and W. Kaiser, Appl. Phys. Lett. 49, 1717 (1987)
32. F. de Rougemont, R. Frey, P. Roussignol, D. Ricard and C. Flytzanis Appl. Phys. Lett.50, 1619 (1987)
33. Ph. Roussignol, M. Kull, D. Ricard, F. de Rougemont, R. Frey, C. Flytzanis Appl. Phys. Lett. 51, 1982 (1987)
34. A.L. Efros and A.I. Efros, Sov. Phys. Sem. 16, 772 (1982)
35. L.E. Brus, J. Chem. Phys. 80 4403 (1984)
36. L.E. Brus, I.E.E.E. J. Quant. Electr. QE22, 1909 (1986)
37. I.M. Lifshitz and V.V Slezov, Sov. JETP 35, 331 (1959)
38. A.I. Ekimov, A.L. Efros and A.A. Onushenko, Solid St. Comm. 56, 921 (1985)

39. N F. Borelli, D.W. Hall, H.J. Holland and D.W. Smith, J. Appl. Phys. <u>61</u>, 5399 (1987)
40. A I. Ekimov and A.A. Onushenko, JETP Lett. <u>40</u>, 1136 (1987)
41. Ph. Roussignol, D. Ricard, C. Flytzanis and N. Neuroth (to be published)

THEORETICAL DESIGN OF ORGANIC MOLECULES AND POLYMERS FOR OPTOELECTRONICS.

J. DELHALLE, M. DORY, J.G. FRIPIAT and J.M. ANDRE

Laboratoire de Chimie Théorique Appliquée, Facultés Universitaires Notre-Dame de la Paix
Rue de Bruxelles, 61 B-5000 Namur, Belgium.

1. INTRODUCTION

Materials which exhibit high nonlinear responses are currently subject of large research activities. The interest in organics [1-5] not only lies in their enhanced NLO responses over a wide frequency range and ultrafast response times, but also in the inherent adaptability of their molecular structures, the possibility of film forming and processing, and higher laser damage thresholds. In particular, conjugated oligomer and polymer chains are currently considered as very promising for devices based on third-order nonlinear effects. To be useful as materials, these compounds must combine, in addition to high electric susceptibilities, other properties such as proper organisation at the molecular level with possible symmetry constraints, chemical stability, etc. Moreover it is expected that the compounds used to form materials will depend on the particular application and therefore it is likely that there will be a continuous need for designing new molecules for optoelectronics.

The first and minimal condition for organic molecules to be of potential interest for optoelectronics and nonlinear optics is certainly to have high microscopic electric responses. Electric responses can be related to the basic expression describing the change in molecular dipole moment **p** upon interaction with an oscillating external electric field [6,7].

$$p_i = p_{oi} + \sum_i \alpha_{ij} E_j + \sum_{jk} \beta_{ijk} E_j E_k + \sum_{jkl} \gamma_{ijkl} E_j E_k E_l + ... \tag{1}$$

Here p_i is the ith component of the total dipole moment **p** (permanent + induced contributions), E_j is the jth component of the applied electric field, α is the linear polarisability, β the first hyperpolarisability and γ the second hyperpolarisability. Large values of these coefficients are thus required conditions for high electric responses.

For organics, high electric responses have invariably been obtained in systems containing delocalised π-electrons. Out of possible conjugated structures, many can readily be discarded on the basis of simple considerations (number of π-electrons, chain length, etc.), but more quantitative assessments require direct quantum mechanical calculations of the coefficients. For instance, factors such as the geometry and the resulting electron density distribution which turn out to have a significant influence on these properties are hardly predicted from rules of thumb.

The basic framework to compute α, β and γ is by now established [8,9], and it is in principle possible to compute these quantities for model compounds to very high accuracy using correlated wave functions [10,11]. However the computational difficulties are so enormous that such accurate calculations are only possible for very small systems. Using *ab initio* coupled Hartree-Fock procedures [12,13] it is nevertheless possible to compute these properties consistently and with a reasonable accuracy.

In this contribution we report *ab initio* calculations of the static electric polarisability as a way to evaluate the relative merits of conjugated systems which could be incorporated into active main chain polymers. By voluntarily limiting the study to the linear response coefficient, we implicitly assume that trends (enhancement or attenuation) obtained for the mean polarisability $<\alpha> (= \{ \alpha_{xx} + \alpha_{yy} + \alpha_{zz} \}/3)$ will be similar for the average second

J. Messier et al. (eds.), Nonlinear Optical Effects in Organic Polymers, 13–28.
© 1989 by Kluwer Academic Publishers.

hyperpolarisability , $<\gamma>$ $(= \{\gamma_{xxxx} + \gamma_{yyyy} + \gamma_{zzzz} + \gamma_{xxyy} + \gamma_{xxzz} + \gamma_{yyzz}\}/5)$. The main reason to limit our calculations to $<\alpha>$ lies in the difficulty to obtain accurate values for $<\gamma>$ which require quite diffuse basis functions and thus prohibitive computation times. Test calculations [14,15] on polyene chains tend to indicate that these assumptions are indeed reasonable. In any event, the results on $<\alpha>$ must be taken with caution when inferring trends for $<\gamma>$.

2. SOME METHODOLOGICAL QUESTIONS

All the computations reported in this work have been made with the GAUSSIAN 82 series of programs [16] that perform Hartree-Fock Roothaan (HFR) calculations at an *ab initio* level. The HF model is the most rigorous single-particle scheme, and calculations based on this model characterize the independent motion of a single electron in the electrostatic field of the nuclei and in the average Coulomb and exchange fields of the other electrons of the system. That level of the quantum mechanical theory results in the traditional molecular-orbital (MO) language. The total wave function is an antisymmetrised product of the occupied one-electron orbitals φ_i that are self-consistent eigenfunctions of the one-electron Fock operator h_0. In the linear combination of atomic orbitals (LCAO or Roothaan) scheme, the one-electron orbitals φ_i are expanded in a linear combination of N atomic basis functions χ_p,

$$\varphi_i = \sum_p \chi_p C_{pi}, \qquad p = 1, N, \qquad (2)$$

and, in the framework of the restricted Hartree-Fock Roothaan (RHFR) method, the one-electron Hamiltonian is represented by its matrix elements h_{pq} between the functions of the basis :

$$h_{pq} = <p|h_0|q> = <p| -\frac{\nabla^2}{2}|q> - \sum_A Z_A <p|\frac{1}{|r-R_A|}|q>$$

$$+ \sum_i \sum_{rs} C_{ri}C_{si} [2(pq|rs) - (pr|qs)] \qquad (3)$$

In Eq.(3) the summation over i runs over the occupied one-electron states, and from left to right are the one-electron kinetic, nuclear attraction, and two-electron integrals, respectively.

It is our opinion that it is more important to ensure consistent predictions on model molecules than having *ad hoc* agreement with measurements where additional factors such as bulk effects, temperature dependence, etc. exist, but are not accounted for in the theoretical models. Since both the equilibrium geometry and the electric polarisability of the model compounds are needed, we have preferred an *ab initio* approach to the cheaper semiempirical methods, the later being parametrized to reproduce specific properties of interest, but often at the expense of loosing control on the others. On the contrary *ab initio* methods are known to predict several ground-state properties with satisfactory accuracy and reasonable confidence; their main disadvantage lies in the computational cost.

The optimized structures have been determined by searching the minimum for the total energy with respect to all geometrical parameters; the search is based on the Fletcher-Powell procedure [17] which minimises the total energy by cancelling the forces on the nuclei by using analytical procedures.

The polarisability components of the molecules in their computed equilibrium geometry have been calculated with the so-called finite-field method [12] proposed by Cohen and Roothaan. The principle of the method is to add to the Hamiltonian a term $-\mathbf{p}.\mathbf{E}$ that describes the interaction energy between the external electric field $\mathbf{E}$ and the dipole moment $\mathbf{p}$ of the molecule. Accordingly, the Fock operator h_0 is modified as follows,

$$h(\mathbf{r}) = h_0(\mathbf{r}) + e(\mathbf{r}.\mathbf{E}) \tag{4}$$

and the field-dependent dipole moment $\mathbf{p(E)}$ is evaluated for different values of the electric field $\mathbf{E}$, ± 0.001, ± 0.02 au (1 au of elecric field $= 5.1423 \times 10^{11}$ Vm^{-1}) in three directions of the space. The polarisability components are obtained by numerical differentiation of $\mathbf{p(E)}$:

$$\alpha_{ij} = \left[\frac{\partial p_i(E_j)}{\partial E_j} \right]_{E=0} \approx \frac{p_i(E_j) - p_i(-E_j)}{2E_j}, \qquad i,j = x,y,z, \tag{5}$$

In the limit of zero field, this approach is equivalent to the analytical coupled HF method. It is the best method to compute second-order energy corrections in one-particle schemes. The average electronic interactions in the presence of the electric field are fully taken into account, which is not the case in perturbational methods based on the sum over the HF one-electron states. Indeed, if HF single determinants give reasonable ground-state descriptions at zero field, they may not have the flexibility to describe the true ground state in the presence of a perturbation. The advantage of the finite-field approach is its inclusion of self-consistent orbital relaxation from the presence of the perturbing electric field.

An important aspect of *ab initio* calculations is the choice of the basis set which provides the aforementioned flexibility. It determines both the quality of the results and their cost. Because of the large size of the systems treated and the need for results of comparable accuracy, the minimal STO-3G basis [18] has been used for all molecules considered in this paper. In spite of its limitation the STO-3G basis predicts molecular structures [19] reasonably well in the sense that errors on bond distances and angles are relatively constant for a wide variety of molecular structures and thus corresponds to the conditions for theoretical design of organic molecules and polymers for optoelectronics. Similarly, good qualitative estimates for the polarisability of conjugated hydrocarbons [12,20-22] can be obtained from STO-3G calculations. It is on the basis of these observations that we have chosen STO-3G results to illustrate possible contributions of Quantum Chemistry to optoelectronics.

3. ILLUSTRATIVE APPLICATIONS

To illustrate the use of finite-field *ab initio* quantum mechanical calculations in the context of organics as materials for optoelectronics, we simulate some of the questions that can arise in the design of active main chain molecules (oligomers) and polymers. In the followings, distances are expressed in Å and polarisabilities in atomic units (1 au of polarisability is equal to 1.6488×10^{-41} C^2 m^{2}J^{-1}).

3.1. Assessment of the monomeric units.

In addition to problems concerning synthesis, stability, yield, etc., the question arises of evaluating the intrinsic merits of prospective monomeric entities for conjugated polymer structures. This evaluation cannot be isolated from the general aspects of the problem. For example, one may decide to favour the organisation at the molecular level (order, packing, etc.), or pay attention to the stability of the resulting material, or both. This often imposes to compromise between the wish of having very effective molecular constituents (*e.g.* highly polarisable monomeric templates) and the need to meet the aforementioned constraints. In spite of their usefulness, rules of thumb can be misleading. For example, contrary to a widespread belief the polarisability of conjugated structures does not always scale linearly with the number of π-electrons they incorporate. Structural aspects have a strong incidence on the polarisability patterns. Quantum mechanical calculations of the type presented in the previous section can provide relatively fast and economic ways of performing reliable assessments. Furthermore, and this is one of the merits of *ab initio* calculations, it is possible to consider with an acceptable margin for failure quite unfamiliar structures and eventually point their interest to researchers in the optoelectronics field.

The polarisability of four molecules, benzene (1), 1,3,5-*trans*-hexatriene (2), vinylacetylene (3) and butatriene (4), all containing 6 π-electrons are compared in Table 1.

Table 1. Longitudinal polarisability α_{zz} of 6 π-electron molecules. Polarisability in au and Λ in Å.

Molecule	α_{zz}	Λ	$(\alpha_{zz}).\Lambda^{-1}$
benzene (1)	45.26	2.78	[16.28]
1,3,5-*trans*-hexatriene (2)	104.13	6.13	16.99
vinylacetylene (3)	44.65	3.50-3.94	12.76-11.33
butatriene (4)	74.23	3.85	19.29

Calculations have been performed at the STO-3G level [21,23] and all geometrical parameters have been optimized at that level of the theory. The corresponding structures are represented in Fig. 1, with selected geometrical parameters. The molecular planes coincide with the (x,z)-plane, the z-axis being along the direction of maximal extension of the molecules and the y-axis perpendicular to the molecular plane. As indicated in the figure, Λ is the vector distance (in Å) between the most distant carbons, it provides a qualitative (admittedly arbitrary) measure of the molecules' extension.

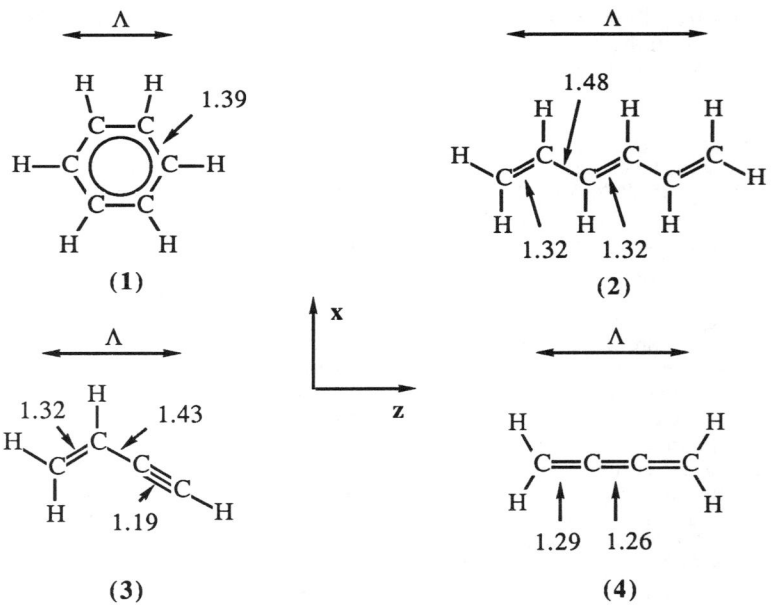

FIGURE 1. Schematic representation of molecules (1) to (4). Relevant distances (in Å) are indicated.

Structures (3) and (4) are isomers and therefore the comparison of their polarisabilities is not biased by different numbers of hydrogens. In the case of benzene, because of its bidimensional character, Λ is not a fully appropriate measure of the molecule's extension.

Values in Table 1 show important differences in the polarisability values of the various structures (1) to (4) in spite of the fact that they contain the same number of π-electrons. First to notice is that vinylacetylene and benzene, which has good chemical stability when incorporated in polymer backbones and also tends to favour the packing, are the less polarisable of the four structures. Hexatriene is quite polarisable due to its delocalised π-electron distribution over a larger molecular dimension. Vinylacetylene, the template monomeric structure found in polydiacetylenes, is much less polarisable than (2). This is due to a strong confinement of the four π-electrons within the C≡C triple bond. Finally butatriene (4), also known as a cumulenic structure, shows an appreciable polarisability value. Such differences for systems having 6 π-electrons stress that it is not possible to analyse polarisability trends without knowing the compounds' structures. In particular the extension of the conjugated backbone over which the π-electrons are delocalised is an important information to know. In absence of experimental data on the structure, quantum chemistry methods can predict geometrical structures with quite acceptable confidence.

If α_{zz} is scaled according to the molecular extension as in the rightmost column in Table 1, some reordering among the systems occurs : butatriene turns out to be better than hexatriene. Thus, if preparation of polycumulenes could be made such that the resulting system would pack densely and be chemically stable, it is very likely that the resulting materials would be among the best polymers for optoelectronics [23,24].

Inspection of the molecular structures in Fig. 1 suggests that the more homogeneous the molecular structure and electron distribution and the more polarisable will be the system [21] . This seems to be generally true and could be used as a convenient rule. However, to obtain quantitative information on the importance of polarisability modifications with respect to geometry changes, direct calculations are nevertheless necessary.

3.2. Polarisability of conjugated chains and the influence of the structure on conjugation.

We have just seen that the more extended a conjugated structure, the larger its polarisability. A lot of activity is devoted to engineer conjugated chain molecules and polymers with best possible delocalisation. For example, in polydiacetylenes the one-dimensional delocalisation that results from the polymerisation of diacetylenes produces a dramatic enhancement of the optical nonlinearities of these compounds. The TCDU (R = - $(CH_2)_4OCONH\phi$) polymer [25] shows an effect 6×10^2 times that of the TCDU monomer. The size-dependency of the electronic polarisability can be illustrated for the polyene (polyacetylene) series. For regular (metallic) polyene chains, both free-electron [26] and Hückel [27] theories predict a dependence of the longitudinal , α_{zz}, proportional to Λ^3, where Λ is the chain extension. Accordingly, the longitudinal polarisability should grow as the second power of the chain length and thus diverge in the limit of an infinite chain. However, free-electron and simple Hückel models do not take into account Coulombic interactions explicitly. The effect of size and bond alternation on polyenes containing increasing numbers of carbons has been investigated at the *ab initio* level by studying the evolution of the longitudinal polarisability per vinylic moiety, α_{zz}/n [28,29]. As seen in Fig. 2, α_{zz}/n increases with the number of double bonds, but eventually saturates which indicates that a limit of linear increase has been reached. The value at which α_{zz}/n levels off and the number of double bonds needed to reach the plateau depends very much on the degree of alternation. Here, the alternation degree controls the efficiency of the π-electron delocalisation along the chain. Thus everything else being similar, an ideal strategy would not only be to connect π-electron rich moieties but also to exert some control on the resulting structure to ensure efficient delocalisation. Now we consider two study cases of the influence of oligomerisation on the resulting polarisability in connection with the delocalisation ability of the system.

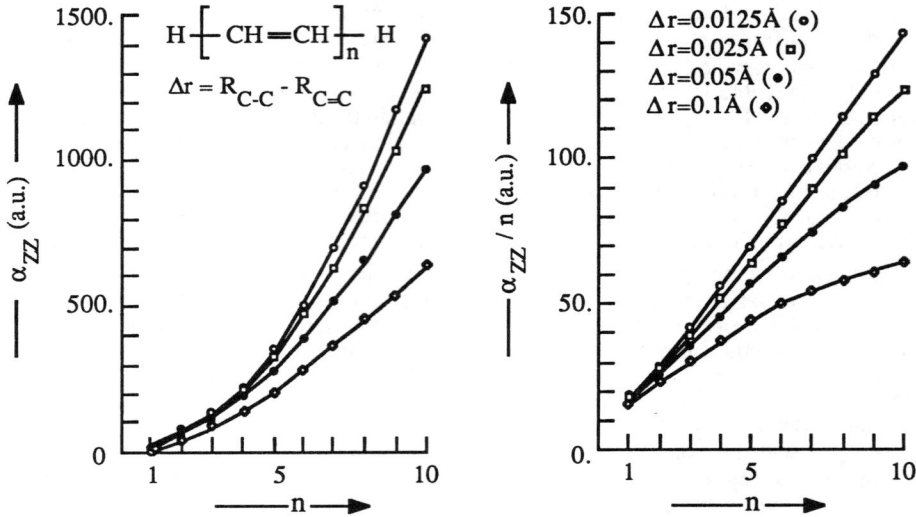

FIGURE 2. Longitudinal polarisability α_{zz} (total [left] and per C_2H_2 unit [right]) of polyenes as a function of the number n of units.

The first case compares the influence of the oligomerisation on the longitudinal polarisability of 1,5-hexadiene-3-yne (5) and 1,5,9-decatriene-3,7-diyne (6) with that of 1,3,5-*trans*-hexatriene (7) and 1,3,5,7,9-*trans*-decapentaene (8), respectively. The structures are shown in Fig. 3a, the corresponding polarisability data [21,23] are given in Table 2.

Table 2. Average $\langle\alpha\rangle$ and longitudinal polarisability α_{zz} (in au) for molecules (5) to (10).

Molecule	α_{zz}	$\langle\alpha\rangle$
1,5-hexadiene-3-yne (5)	44.65	-
1,5,9-decatriene-3,7-diyne (6)	85.28	-
1,3,5-*trans*-hexatriene (7)	104.13	-
1,3,5,7,9-*trans*-decapentaene (8)	217.25	-
phenylethylene (9)	76.39	47.53
3,6-dimethylene-1,4-cyclohexadiene (10)	125.41	55.81

Going from (5) to (6) means adding 10 more π-electrons with the $-C\equiv C-CH=CH-C\equiv C-$ moiety between two vinylic groups, while going from (7) to (8) amounts to insert three double bonds carrying a total of 6 π-electrons between the two vinylic moieties. The net gain in polarisability is substantially more important for the polyenic backbone, obviously because of a more efficient delocalisation of the π-electrons in this geometrical framework. Again the triple bonds, notwhistanding the fact that they are π-electron richer than the double bonds, are geometrically more concentrated and form a less efficient delocalised structure.

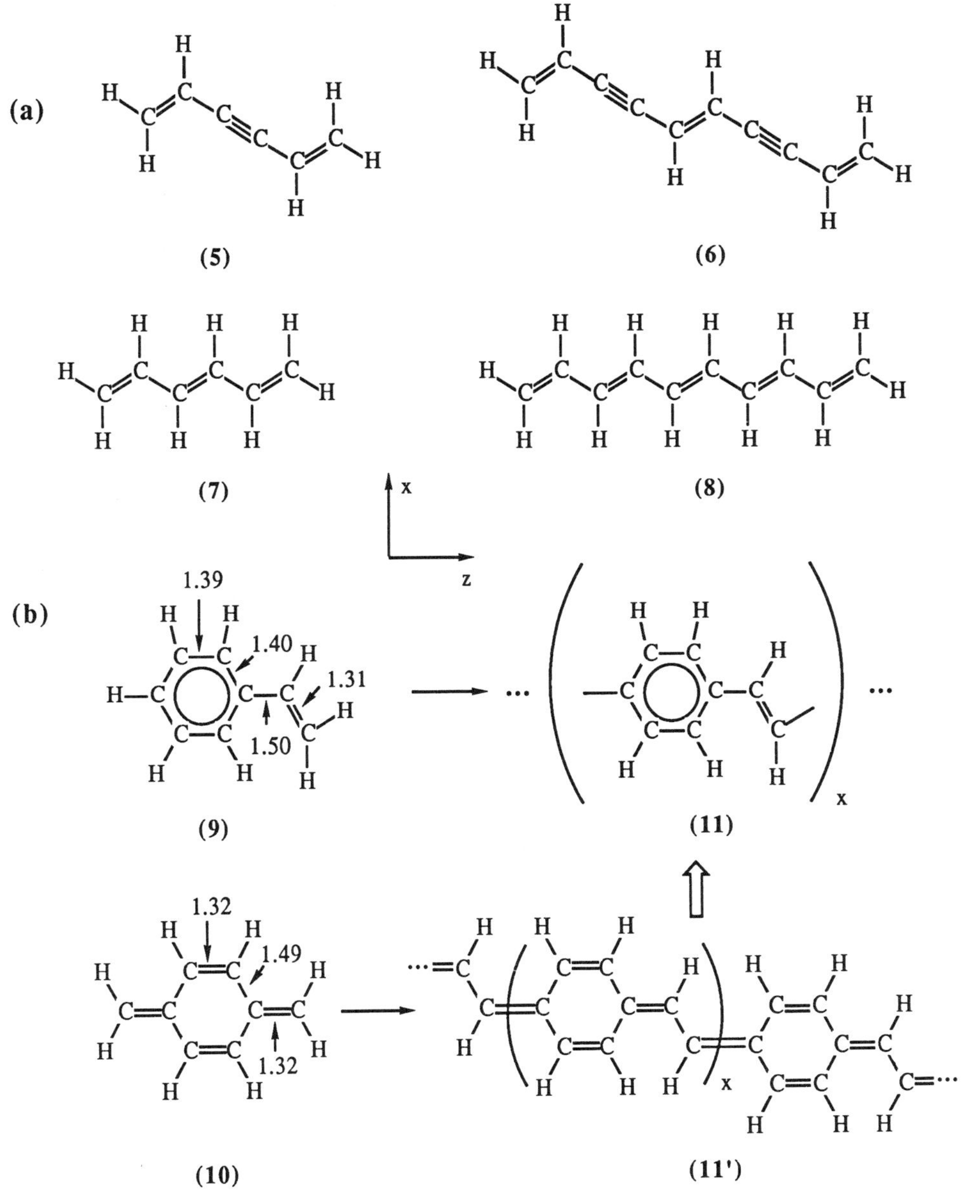

19

FIGURE 3. Geometrical representation of compounds (5) to (11).

In the second case we consider phenylethylene (or styrene) (9) and its quinoidic isomer
3,6-dimethylene-1,4-cyclohexadiene (10) represented in Fig. 3b. Polarisability data [30] are

given in Table 2. It is accepted nowadays that, due to its aromatic character, benzene disrupts the conjugation and thus, as seen above, has an adverse influence on the overall polarisability of the system. One way to go around could thus be to form the quinoidic isomer to remove the aromatic character and have a more efficient delocalisation. Quinoidic structures like (10) are not common and connecting such units to form polymers is probably difficult. Therefore it is useful to know beforehand the extent of the polarisability enhancement and rate the gains versus synthesis problems. The results are summarized in Table 2. Quite surprisingly, the polarisabilities of both isomers are not drastically different and one can anticipate that there will be no advantage in forming the hypothetical polymeric structure (11'), Fig. 3b. First because the polarisability gain is small, and second because, as the number of units in the polymer increases, (11') will gradually transform into the energetically more favourable isomeric polyparaphenylenevinylene (11), Fig. 3b. This can easily be conjectured from the total energy obtained from *ab initio* theoretical calculations. Phenylethylene (9) is predicted to be more stable than 3,6-dimethylene-1,4-cyclohexadiene (10) by about 104.5 kJ.mole^{-1} (computed at the 6-31G level).

3.3. Heteroatomic functions in conjugated carbon backbones.

In the preparation of new materials it can be valuable, but also difficult to either introduce or remove specific linkages and/or terminal groups in a conjugated backbone. Thus it is useful to evaluate the influence on the polarisability of heteroatomic moieties in otherwise highly polarisable conjugated hydrocarbon chains.

In the followings we first analyse the effect on the longitudinal polarisability of substituting -CH= parts in 1,3,5,7-*trans*-octatetraene by -N= moieties. This choice is dictated by the concern to identify effects pertaining to the nature of the functional groups while keeping as much as possible unchanged other factors such as the length and conformation, the total number electrons and the number of π-electrons. Furthermore, the π-electron channel whose length is measured by Λ, the distance between the most distant atoms of the second period, is quite comparable for all compounds even though significant geometry readjustments can take place locally in the chains.

Table 3. Longitudinal polarisability α_{zz} (in au) of 8 π-electron molecules and Λ (in Å).

Molecule	α_{zz}	Λ	$(\alpha_{zz})\Lambda^{-1}$
1,3,5,7-*trans*-octatetraene (12)	140.44	8.59	16.35
1-aza 1,3,5,7-*trans*-octatetraene (13)	134.86	8.53	15.81
4-aza 1,3,5,7-*trans*-octatetraene (14)	137.29	8.43	16.29
1,2 diaza-1,3,5,7-*trans*-octatetraene (15)	137.31	8.37	16.41
3,4-diaza 1,3,5,7-*trans*-octatetraene (16)	136.46	8.30	16.44
1,3-diaza 1,3,5,7-*trans*-octatetraene (17)	131.08	8.46	15.49

The compounds studied are the 1,3,5,7-*trans*-octatetraene (12), 1-aza 1,3,5,7-*trans*-octatetraene (13), 4-aza 1,3,5,7-*trans*-octatetraene (14), 1,2 diaza-1,3,5,7-*trans*-octatetraene (15), 3,4-diaza 1,3,5,7-*trans*-octatetraene (16) and 1,8-diaza 1,3,5,7-*trans*-octatetraene (17). Full geometry optimisation at the STO-3G level has been made for compounds (12) to (17); the corresponding molecular structures are shown in Fig. 4. The values for the longitudinal polarisability α_{zz} and the length of π-channel [22] are given in Table 3.

Out of the six compounds, octatetraene is the most polarisable structure which is also in agreement with the previous results indicating that chains characterised by higher homogeneity in their geometrical structure and π-electron distribution are more polarisable than the comparatively less regular structures. It is worth noticing that systems such as (14) and (16)

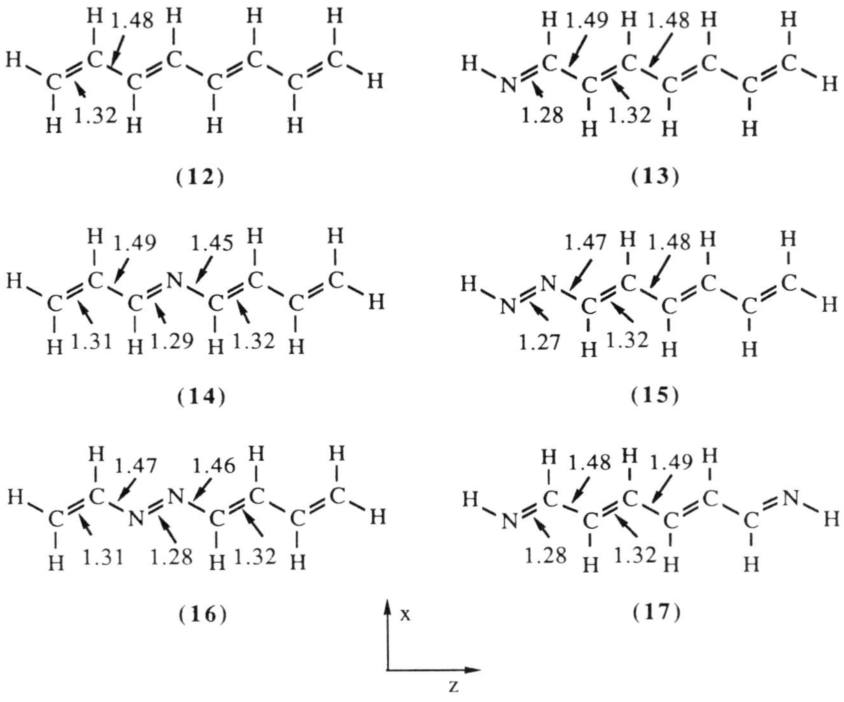

FIGURE 4. Structure of chain molecules (12 to (17) with relevant bond lengths (in Å).

which include heteroatomic functions in the interior of the chain have slightly higher polarisability per chain length than those having the heteroatomic moieties located at the end(s) of the chain, *i.e.* (13) and (15) respectively. (14) and (16) undergo noticeable structural modifications in the vicinity of the heteroatomic moiety in the direction of a more efficient delocalisation.

Another example is the incorporation of a peptide or amide bond, -NH-CO- into a polyenic backbone. As shall be discussed in section 3.4., peptide bonds can be used as chemical tools to organise active molecules at the molecular level . Therefore it is interesting to assess their influence on the resulting polarisability due to their presence in conjugated chains. In Fig. 5 are shown two molecules, N-vinylacrylamide (18) and N-butadienyl formamide (19), containing two double bonds and an amide group. In the first system (18) the double bonds are separated by the -NH-CO- linkage while in (19) the two double bonds are linked together. In Table 4 are given the longitudinal and the mean polarisability, respectively α_{zz}

and $<\alpha>$; Λ, the distance between the most distant carbon atoms, is also given for the sake of comparison. These results have been obtained at the STO-3G level on the fully optimised

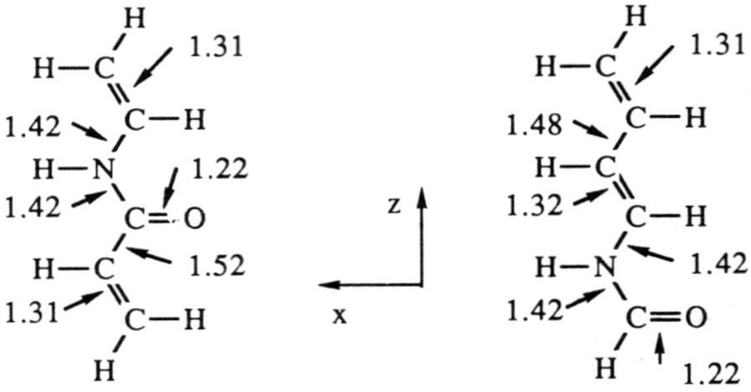

(18)   (19)

FIGURE 5. Structure of the two isomeric conjugated molecules (18) and (19) incorporating and amide group in their backbone, bond distances in Å.

geometries constrained to be planar, see Fig. 5. The results for 1,3,5-*trans*-hexatriene (1) have been added to facilitate the comparison.

Table 4. Average $<\alpha>$ and logitudinal polarisability α_{zz} (in au) for molecules (2), (18) and (19). Λ in Å.

Molecule	α_{zz}	$<\alpha>$	Λ
N-vinylacrylamide (18)	64.29	36.26	6.04
N-butadienyl formamide (19)	76.36	41.25	6.15
1,3,5-*trans*-hexatriene (2)	104.13	43.93	6.13

The amide group, -NH-CO-, even with its 4 π-electrons, is less efficient than a double bond for the overall conjugation. This is further supported by the total energy results which predict isomer (19) to be slightly more stable than (18) , by approximately 4 kJ.mole[-1]. The results are in agreement with the classical view of a peptide bond where the C-N bond has a high degree of double bond character resulting from the delocalisation of the nitrogen lone pair into the carbonyl group. However this high degree of double bond character prevents effective conjugation with connected C=C double bonds and to some extent spoils the delocalisation. Thus peptide bonds disrupt the conjugation when incorporated in a conjugated hydrocarbon chain and should be avoided except if they can serve other purposes which lead to a positive trade off .

3.4. Hydrogen bonded systems.
    In the molecular engineering of new materials for optoelectronics, controlling the molecular organisation, increasing the density of electroactive species as well as stiffening the

material turn out to be very important [31]. Various approaches aiming at the control of the molecular organisation exist and are actively pursued : Langmuir-Blodgett film deposition, topochemical solid-state polymerization, liquid crystal formation, etc. In the case of liquid crystalline systems, chain ordering by hydrogen bond formation has recently been demonstrated for an organic system containing a carboxylic group [32]. Similarly, the property of hydrogen bonding to impose specific patterns to the molecular organisation has been used elegantly to engineer molecular ferromagnets [33]. These two examples suggest a more systematic use of hydrogen bond properties to exert some control on the molecular assembly of electroactive conjugated organic chains. As already mentioned, free-electron and Hückel phenomenological models predict a Λ^3 dependence for the polarisability if Λ is the length of the conjugation path [26,27]. In long chains, however, conformational freedom can result in defects and twists which disrupt the conjugation and thus spoil the expected benefit of the Λ^3 behaviour. Here we would like to consider the possibility of using hydrogen bonds to prevent and/or minimize the occurrence of these undesirable effects, and more specifically illustrate the principles for a study of the intrinsic polarisability properties of some relevant hydrogen bonded patterns. Before considering the polarisability questions, it might be useful to indicate how the amide linkage could be used both directly in the backbone and as terminal group to improve the packing (increased order and/or density).

   Nylons owe part of their mechanical properties to hydrogen bonds which, among other things, induce better organisation at the molecular level. Fig. 6 provides a schematic representation of the hydrogen-bonded sheets with amide groups in nylon 6 as inferred from crystal structure determination [34], scheme (20). Provided monomer preparation is not exceedingly difficult, and stability as well as reactivity do not turn out unfavourable, it could be conceivable to form by polycondensation unsaturated analogs of nylon 6 as shown in scheme (21). If chemistry permits, longer chains could be incorporated between the amide groups with the benefit of locking the double bond sequence in an ordered way and thereby minimize the occurence of kinks and twists in the conjugated path.

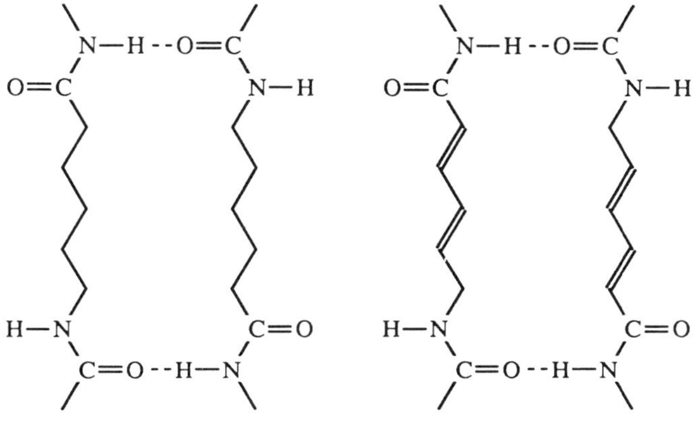

(20)   (21)

FIGURE 6. Schematic representation of the hydrogen bond pattern in nylon 6 (20) and in the hypothetical conjugated analog (21).

24

Hydrogen bonds could also be used to increase the crystal density and force the conjugated backbone to align. For instance, it is known that muconic acid (22) [35] forms hydrogen bonds in the crystal; these bonds force the molecules to fully align owing to the cyclic hydrogen-bonded pairs (Fig. 7) between two carboxylic groups. In contrast, dimethyl-*trans,trans*-muconate (23) [36] - with a structure very close to muconic acid but without the possibility of forming cyclic hydrogen-bonded pairs - exhibits a quite different organisation in the crystal, see also Fig. 7.

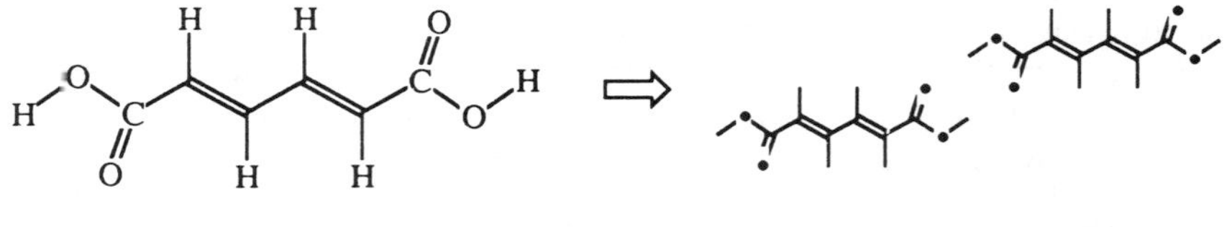

(22)

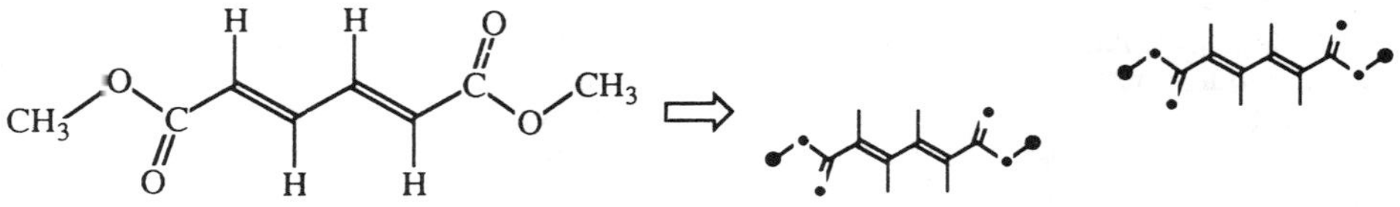

(23)

FIGURE 7. Packing mode for muconic acid (22) and dimethyl-*trans,trans*-muconate (23).

There are many more examples of hydrogen bonding patterns [37,38] which could be used to exert some control on the molecular organisation of the active species.

By *ab initio* quantum mechanical calculations one can study the effect of forming intermolecular hydrogen bonds on the polarisability of dimeric systems. It must be stressed here that properties of hydrogen-bonded molecules are generally difficult to obtain compared to those of more strongly bonded systems. Accordingly, the results presented in this section should be used with caution.

The first illustrative case we consider is to compare the longitudinal polarisability of the dimer (26) to the sum of the separate constituents (24) and (25). As before, the calculations have been made at the STO-3G level and the equilibrium structures obtained by full geometry optimisation on the systems; planarity was the only constraint imposed. These structures are schematically represented in Fig. 8, and as usual the most relevant structural parameters are also given.

Table 5 Longitudinal α_{zz} and average $<\alpha>$ polarisability (in au) for (24) to (26), Λ in Å.

Molecule	α_{zz}	$<\alpha>$	Λ
acrylamide (24)	34.50	22.02	3.68
N-vinyl formamide (25)	38.46	23.47	3.71
dimer (26)	72.18	46.96	-

The hydrogen bond between acrylamide (24) and N-vinyl formamide (25) has limited incidence on the resulting polarisability of the dimer compared to that of the isolated molecules. This can also be seen from the structure of the constituent molecules in the dimer compared to the separate molecules. Within the present level of calculation, no significant changes occur in the C-N, C-C, C=C and C=O bonds and thus no chance to enhance the delocalisation along the chain.

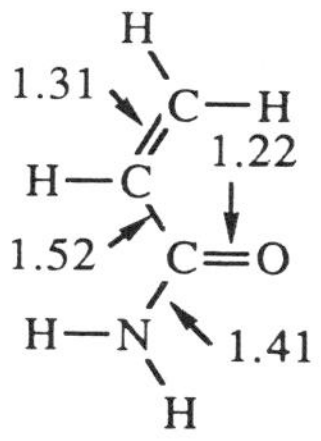

(24)

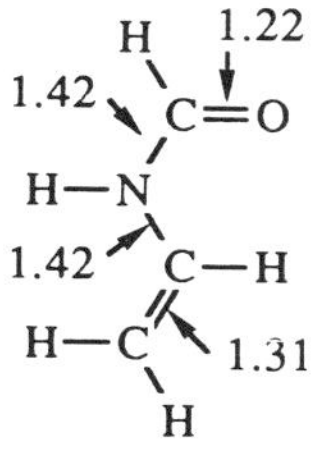

(25)

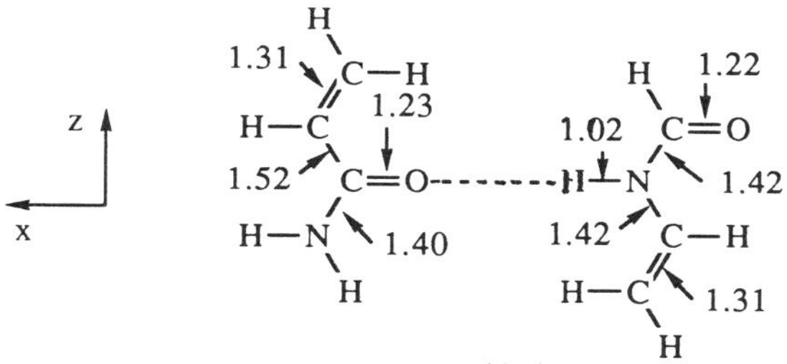

(26)

FIGURE 8. Geometrical arrangement of (24) and (25) and of the corresponding dimer (26).

The second case is a comparison of the polarisability of cyclic [39] and linear hydrogen bonded dimers of formamide. The STO-3G optimized structures are shown in Fig. 9. Table 6 contains the polarisability data.

Table 6. Longitudinal α_{zz} and average $<\alpha>$ polarisability (in au) for structures (27) to (28), Λ in Å.

Molecule	α_{zz}	$<\alpha>$	Λ
Cyclic formamide dimer (27)	30.00	23.27	3.85
Linear formamide dimer (28)	38.24	22.40	6.70
Formamide monomer	-	10.48	-

Here again, there is litttle to be expected from the hydrogen bond formation at the polarisability level. The average polarisability <α> is barely more than twice that of the monomer. Larger basis sets have been considered in the case of the cyclic dimers and these trends are confirmed [39]. Thus the types of hydrogen bonded structures considered in this paper do not seem to be beneficial from the sole point of view of the polarisability.Nevertheless they are potentially useful for structural organisation at the molecular level. Furthermore, it must be stressed that the realm of hydrogen bonding is very large and other situations can be just the opposite. Work is in progress along these lines.

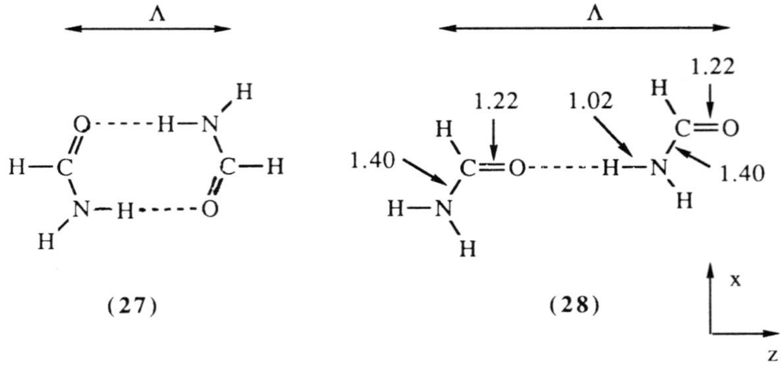

FIGURE 9. Structure of the cyclic (27) and linear (28) dimers of formamide.

Other works along similar lines have been made, e.g study of the doping influence on the polarisability [28]. Now in progress is the analysis of the influence of the order (disorder) of sequences (alternating, block, random) of two or more monomers linked in conjugated chain systems [40].

4. CONCLUDING REMARKS

As hopefully suggested in this contribution, Quantum Chemistry can provide some help to the field of optoelectronics. However, to be really fruitful in the integrated effort of molecular engineering for molecular electronics, theoretical approaches must be directly combined with other expertises and skills from the other fields involved: synthesis, physical characterisation and material engineering. It should be stressed that Quantum Chemistry can contribute a lot more than just assess electric response coefficients. Other properties, which are sometimes as important as polarisability and hyperpolarisabilities themselves (stability , reactivity, structure, molecular organisation, etc.) can be investigated with reasonable confidence.

If the presently available quantum mechanical tools are already capable of both interpreting and predicting the polarisability behaviour of conjugated organic backbones, it remains, as far as the *ab initio* level is concerned, to develop more appropriate techniques to deal accurately with molecules containing at least twelve atoms from the first row of the periodic table. Among the developments that should take place one can mention : improvements of the wave functions to reach a comparable quality in the prediction of all components of the polarisability and hyperpolarisability tensors, correlation corrections to deal with frequency dependence, and treatments of infinite systems [41].

ACKNOWLEDGEMENTS
The authors are very grateful to Dr. J. Messier for allowing them to present their work at this NATO workshop.They also want to thank Dr. J.L. Brédas for helpful discussions on some of the topics developed in this paper. They acknowledge with appreciation the support of this work under the ESPRIT-EEC contract n°443 on Molecular Engineering for Optoelectronics. Finally they acknowledge the National Fund for Scientific Research (Belgium), IBM Belgium and the Facultés Universitaires Notre-Dame de la Paix (FNDP) for the use of the Namur Scientific Computing Facility.

REFERENCES

1. Williams DJ (ed): Nonlinear Optical Properties of Organic and Polymeric Materials.Washington DC : American Chemical Society, 1983, ACS Symposium Series 233.
2. Williams DJ, Angew. Chem. Int. Ed. Engl., 690 (1984).
3. Zyss J, J.Mol. Electronics, 1, 25 (1985).
4. Singer KD, Lalama SJ, Sohn JE, SPIE, Integrated Optical Circuit Engineering II, 578, 130 (1985).
5. Chemla DS, Zyss J (eds): Nonlinear Optical Properties of Organic Molecules and Crystals. New York: Academic Press, 1987. vols. 1 and 2.
6. Bloembergen N : Nonlinear Optics. New York: Benjamin, 1965.
7. Shen YR: The principles of Nonlinear Optics. New York: Wiley, 1984.
8. Bogaard MP, Orr BJ, International Review of Science, Physical Chemistry Series 2, vol. 2, Buckingham AD (ed). London: Butterworths, 1975.
9. Elliott DS, Ward JF, Mol. Phys. 51, 45 (1984).
10. Bishop DM, Pipin J, Phys. Rev. A36, 2171 (1987).
11. Bishop DM, Lam B, Phys. Rev. A37, 464 (1988).
12. Cohen HD, Roothaan CJ, J. Chem. Phys. 43, S34 (1965).
13. Zyss J, Berthier G, J. Chem. Phys. 77, 3635 (1982).
14. André JM, Barbier C, Bodart V, Delhalle J, in Nonlinear Optical Properties of Organic Molecules and Crystals. Chemla DS, Zyss J (eds). New York: Academic Press, 1987. vol. 2, pp. 137-158.
15. Hurst GJB, Dupuis M, Clementi E: *Ab initio* analytic polarizability, first and second hyperpolarizabilities of large conjugated organic molecules : Applications to polyenes C_4H_6 to $C_{22}H_{24}$. J. Chem. Phys., in press.
16. Binkley JS, Frisch MJ, Defrees DJ, Raghavachari K, Whiteside RA, Schlegel HB,Fluder EM, Pople JA : Gaussian 82 Program (Carnegie Mellon University,Pittsburgh,1981).
17. Schlegel B, J. Comput. Chem. 3, 214 (1982).
18. Hehre WJ, Stewart RF, Pople JA, J. Chem. Phys. 51, 2657 (1969).
19. Hehre WJ, Radom L, v.R. Schleyer P, Pople JA : *Ab initio* Molecular Orbital Theory. New York : Wiley, 1986.
20. Chablo A, Hinchliffe A. Chem. Phys. Lett. 72, 149 (1980).
21. Bodart VP, Delhalle J, André JM, Zyss J, Can. J. Chem. 63, 1631 (1985).
22. Younang E, Delhalle J, André JM, New J. Chem. 11, 404 (1987).
23. Bodart VP, Delhalle J, Dory M, Fripiat JG, André JM, J. Opt. Soc. Am. B4, 1047 (1987).
24. Delhalle J, Bodart VP, Dory M, André JM, Zyss J, Int. J. Quantum Chem. Symp. 19, 313 (1986).
25. Sauteret C, Hermann JP, Frey R, Pradere F, Ducuing J, Baughman RH, Chance RR, Phys. Rev. Lett. 36, 956 (1976).
26. Rustagi KC, Ducuing J, Opt. Commun. 10, 258 (1974).

27. Davies PL, Trans. Faraday Soc. <u>47</u>, 789 (1952).
28. Bodart VP, Delhalle J, André JM, Zyss J, in Polydiacetylenes : Synthesis, Structure and Electronic Properties. Bloor D, Chance RR (eds), Dordrecht : Martinus Nijhoff , 1985. pp. 125-133.
29. André JM, in Large Finite Systems, Jortner J, Pullman A, Pullman B (eds), The Jerusalem Symposia on Quantum Chemistry and Biochemistry, vol. 20. Dordrecht : Reidel, 1987. pp. 277-288.
30. Dory M, Bodart VP, Delhalle J, André JM, Brédas JL, in Nonlinear Optical Properties of Polymers. Heeger AJ, Orenstein J, Ulrich DR (eds). Pittsburgh: Materials Research Society, 1988. vol. 109. pp. 239-250.
31. Roberts GG, Adv. Phys. <u>34</u>, 475 (1985).
32. Faranjpe AS, Kelkar VK, Mol. Cryst. Liq. Cryst. <u>102</u>, 289 (1984).
33. Pei Y, Verdaguer M, Kahn O, Sletten J, Renard JP, J. Am. Chem. Soc. <u>108</u>, 7428 (1986).
34. Parker JP, Lindenmeyer PH, J. Appl. Polym. Sc. <u>21</u>, 821 (1977).
35. Bernstein J, Leiserowitz L, Isr. J. Chem. <u>10</u>, 601 (1972).
36. Filippakis SE, Leiserowitz L, J. Chem. Soc. B, 290 (1967).
37. Leiserowitz L, Acta Cryst. <u>B32</u>, 775 (1976).
38. Leiserowitz L, Nader F, Acta Cryst <u>B33</u>, 2719 (1977).
39. Dory M, Delhalle J, Fripiat JG, André JM, Int. J. Quantum Chem. Quantum Biology Symp. <u>14</u>, 85 (1987).
40. Hennico G, Delhalle J, André JM, to be published.
41. C. Barbier, Chem. Phys. Lett. <u>142</u>, 53 (1987).

STRUCTURAL RELAXATION AND NONLINEAR ZERO-POINT FLUCTUATIONS AS THE ORIGIN OF THE ANISOTROPIC THIRD-ORDER NONLINEAR OPTICAL SUSCEPTIBILITY IN TRANS-(CH)x

M. SINCLAIR[#], D. MOSES, K. AKAGI[*] and A. J. HEEGER

Department of Physics and Institute for Polymers and Organic Solids
University of California, Santa Barbara, CA 93106

1. INTRODUCTION

Although organic polymers have great potential for eventual application in nonlinear optical elements, an understanding of the mechanism (or mechanisms) underlying their nonlinear susceptibilities is necessary before the design and synthesis flexibility afforded by organic chemistry can be applied to the development of new and better materials. Detailed experimental studies of prototype systems must therefore be performed in order to guide a parallel theory effort aimed at a general understanding of the nonlinear optical properties of organic polymer materials.

The promise of conducting polymers as fast response nonlinear optical materials has been recently emphasized[1,2]. Polymers such as polyacetylene, polythiophene and the soluble and processible poly(3-alkylthienylenes) contain a high density of π-electrons, and they are known to exhibit photoinduced absorption and photoinduced bleaching, indicating major shifts of oscillator strength upon photoexcitation[2,7]. For polyacetylene, these nonlinear effects have been studied in detail in the picosecond[5a,b] and sub-picosecond[5c] time regime and correlated with the photoproduction of charge carriers through fast photoconductivity measurements[6]. The data have demonstrated ultra-fast response with nonlinear shifts in oscillator strength occurring at times of the order of 10^{-13} seconds. These resonant nonlinear optical properties are intrinsic; they originate from the nonlinearity of the self-localized photoexcitations[7] which characterize this class of polymers: solitons, polarons and bipolarons[4].

In any material where photoexcitation results in shifts of oscillator strength (as is the case in conducting polymers), the optical properties will be highly nonlinear. The magnitude of the resonant $\chi^{(3)}$ can be estimated from the magnitude and frequency dependence of the photoinduced absorption and bleaching. For example, as a result of the shift in oscillator strength subsequent to photoexcitation, the complex index of refraction is intensity dependent

$$n(\omega) = n_0(\omega) + n_2(\omega,\omega_p)\, I(\omega_p) \tag{1}$$

where the second term describes the nonlinear response at frequency ω due to an intense pump at pump frequency ω_p. Under pumping conditions which are resonant with the π-π^* transition of polyacetylene ($\hbar\omega_p$ = 2.0 eV), the existing data yield an estimate[2,7] for n_2(1.4eV, 2.0eV) $\approx 10^{-4}$ (MW/cm^2)$^{-1}$. This large value for n_2 implies a correspondingly large value for Im$\chi^{(3)}$ through the relation

J. Messier et al. (eds.), Nonlinear Optical Effects in Organic Polymers, 29–45.
© *1989 by Kluwer Academic Publishers.*

$$n_2 = 4\pi^2/c\varepsilon \; \chi^{(3)} \tag{2}$$

where ε is the dielectric constant at the probe frequency (ω). Using the above value for n_2, we obtain $\mathrm{Im}\chi^{(3)}(-\omega_2 = \omega_1-\omega_1-\omega_2) = 5 \times 10^{-8}$ esu, an impressive value even under resonant conditions. From a Kramers-Kronig analysis of the photoinduced absorption data, one concludes that the real parts of n_2 and $\chi^{(3)}$ are correspondingly large. Based upon these observations, experimental studies of third harmonic generation in polyacetylene and related conducting polymers were initiated in order to explore directly the third order susceptibility under nominally nonresonant conditions where the pump is well below the principal interband (π-π^*) transition.

In this paper, we present the results of a series of third harmonic generation (THG) measurements which probe the magnitude and the origin of the third order nonlinear optical susceptibility of polyacetylene. By performing reflection THG measurements relative to a silicon standard, we have unambiguously determined the magnitude of the third order susceptibility associated with tripling the fundamental of the Nd:YAG laser to be $\chi_{||}^{(3)}(3\omega;\omega,\omega,\omega) = (4 \pm 2) \times 10^{-10}$ esu, where $\chi_{||}^{(3)}$ indicates that component of the third order susceptibility tensor with all indices parallel to the chain direction. We have successfully measured the anisotropic THG in oriented thin films and demonstrated that all the nonlinearity is associated with the π-electron polarizability along the conjugated chain. The magnitude and anisotropy are directly compared with the results obtained from single crystals of polydiacetylene-(toluene-sulfonate), PDA-TS, measured simultaneously and in the same apparatus. When pumping at 1.06 μm (1.17 eV), the third harmonic power generated (on reflection) from a cleaved single crystal of PDA-TS is about a factor of two greater than that from an oriented film of trans-(CH)$_x$. Finally, we have measured THG in cis-rich and trans isomers of the same sample. The measured response of the cis-rich samples scales with the residual trans content of the sample; $\chi_{||}^{(3)}$ of the trans isomer is 15-20 times larger than that of the cis isomer. This symmetry specific aspect of $\chi_{||}^{(3)}$ implies an underlying mechanism which is sensitive to the existence of a degenerate ground state, as in trans-(CH)$_x$. The large $\chi_{||}^{(3)}$ when pumped sub-gap is therefore consistent with the virtual generation of nonlinear solitons as a mechanism for the nonlinear optical susceptibility of polyacetylene. This mechanism is developed and discussed in detail; we conclude that the nonlinear zero point fluctuations of the ground state lead to an important mechanism for nonlinear optics, particularly in polymers with a degenerate ground state.

2. EXPERIMENTAL METHODS

Since trans-(CH)$_x$ is strongly absorbing at the third harmonic (355 nm) of the Nd:YAG pump laser, the technique developed by Burns and Bloembergen[8] for measurement of third harmonic generation in absorbing media is most applicable. This method relies on measurement of the backward moving wave at 3ω radiated by the nonlinear polarization. When the sample thickness is large compared to the absorption depth at 3ω, the amplitude of the reflected signal is independent of the sample thickness, and, for the case where the fundamental wave is normally incident on a "thick" sample, the amplitude of the reflected third harmonic wave amplitude (E_R) is given by

$$|E_R(3\omega)|^2 = |\chi^{(3)}(3\omega)|^2 F(n_0,N_1,N_3)|E_i(\omega)|^6 \tag{3}$$

where $E_i(\omega)$ is the incident field amplitude, n_0 is the refractive index of the medium from which the pump is incident, N_1 and N_3 are the complex indices of refraction at ω and 3ω, respectively, and

$$F_{(n_0 N_1 N_3)} = \frac{1024\pi^2 n_0^6}{|n_0 + N_1|^6 |n_0 + N_3|^2 |N_1 + N_3|^2} \tag{4}$$

The principal difficulty in determining $\chi^{(3)}$ from Eqs. 3 and 4 is that of determining $E_R(3\omega)$ and $E_i(\omega)$ accurately. This is overcome by measuring the third harmonic generated by the sample relative to that generated by a reference sample with known $\chi^{(3)}$. In this way, uncertainties associated with determining the pump intensity at the sample and absolute detector calibration are avoided. If P_s and P_r are the third harmonic powers generated by the sample and the reference under identical conditions, then one can determine the unknown $\chi^{(3)}$ using the following relation

$$\frac{P_s}{P_r} = \frac{|\chi s^{(3)}|^2 Fs}{|\chi r^{(3)}|^2 Fr} \tag{5}$$

for the ratio.

The experimental arrangement used to measure the third harmonic intensity is shown in Fig. 1. It consists of a mode-locked Nd:YAG laser and fiber-grating pulse compressor which produce an 82 MHz train of pulses with autocorrelation FWHM of 4.5 ps and ~5 KW peak power. A half-wave plate and polarizing cube

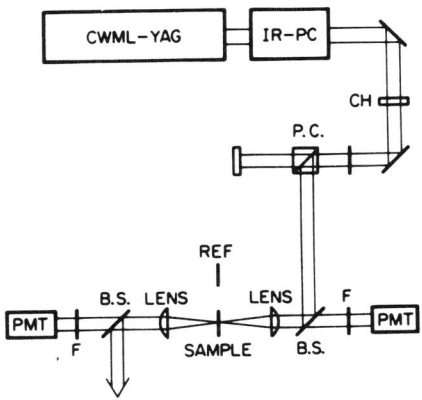

FIGURE 1. Schematic diagram of the experimental arrangement used to measure the third harmonic intensity.

were used as a variable attenuator, and the beam was directed onto the sample by means of a dichroic filter which was chosen to have high reflectivity at $\lambda = 1.06$ μm and high transmissitivity at $\lambda = 355$ nm. The fundamental beam was focused to a spot size of ~ 30 μm at the sample by a converging lens which also served to recollimate the reflected third harmonic. After recollimation, the third harmonic passes through the dichroic filter and the notch filter to reach the photomultiplier tube. A recollimating lens and dichroic filter were also mounted after the sample so that third harmonic generation could also be measured in the transmission mode. The fundamental beam was chopped, and lock-in detection was used to measure the output of the photomultiplier tube. For the polarization dependence measurements, an additional half-wave plate was inserted between the dichroic filter and the focusing lens to allow the polarization of the fundamental beam to be rotated at the sample.

The sample was attached to the cold finger of a Helitran cryostat; the vacuum shroud of the cryostat was mounted on an XYZ translation stage to facilitate focusing and to allow the samples to be moved in and out of the beam. A small piece of intrinsic silicon (cut with a (111) surface) was mounted immediately adjacent to the polymer sample to serve as reference. Relative measurements were taken by first measuring the third harmonic signal reflected from the sample and then (without changing any of the parameters) translating the cryostat to measure the signal from the silicon reference.

The polyacetylene samples used in the anisotropy measurements were films of aligned trans-$(CH)_x$ grown on glass by the novel technique in which the catalyst for the polymerization reaction was suspended in a liquid crystal solvent[9,10]. The polymerization reaction was carried out with the liquid crystal (and catalyst) oriented in the ~1 tesla field of an electromagnet. Film thickness varied from ~ 0.1 μm (for transmission measurements) to ~ 1 μm (for reflection measurements). Studies of the anisotropy of the infrared absorption of polyacetylene samples prepared in this way[9] indicate that the polymer chains are aligned with a typical mosaic spread of $\approx \pm 15$ - 20°. The comparative cis-trans measurements were carried out using non-oriented material prepared on glass substrates using the standard Shirakawa method. All reflection measurements were taken with the fundamental beam incident on the sample from the glass side ($n_0 = 1.5$) in order to take advantage of the superior surface quality on that side.

3. MAGNITUDE AND ANISOTROPY OF $\chi^{(3)}$ FOR TRANS-$(CH)_x$ AND FOR PDA-TS: A DIRECT COMPARISON

The third harmonic power reflected from the (glass side) surface of the oriented trans-$(CH)_x$ samples (when the polarization of the fundamental beam is parallel to the alignment direction of the sample) is approximately 2000 times larger than the power reflected from the surface of intrinsic silicon under identical conditions. Assuming that the optical properties of the oriented samples are those determined for oriented Durham trans-polyacetylene[11], the appropriate optical constants are as follows:

for silicon[12]

$$N_1 = \sqrt{\varepsilon_1} = 3.55 \text{ and } N_3 = \sqrt{\varepsilon_3} = 5.55 + i3.04;$$

for trans-$(CH)_x$ [13]

$N_1 = \sqrt{\varepsilon_1} = 3.8$ and $N_3 = \sqrt{\varepsilon_3} = 0.7 + i1.4$.

Using these values and Eq. 4 yields $F_{(CH)x} = 3.4 \times 10^{-2}$ for <u>trans</u>-$(CH)_x$ and $F_{Si} = 2.4 \times 10^{-4}$ for silicon. Finally, with the help of Eq. 5 we calculate

$$\chi^{(3)} = [P_{(CH)x}/P_{Si}]^{1/2} \, [F_{Si}/F_{(CH)x}]^{1/2} \chi_{Si}^{(3)} \tag{6}$$

or

$$\chi_{\parallel}^{(3)} \approx 4\chi_{Si}^{(3)} = (4 \pm 2) \times 10^{-10} \text{ esu} \tag{7}$$

where we have used $\chi_{Si}^{(3)} \approx 10^{-10}$ esu.

We have estimated the error bars in eqn. 7 based on the repeatability of the result from measurement to measurement. The main source of error in this determination of $\chi_{\parallel}^{(3)}$ is sample inhomogeneity (i.e., the magnitude of the reflected third harmonic power varied somewhat from place to place on the sample and from sample to sample). This can be reduced by the development of higher quality, more uniform samples. We note, again, that we have used values for the complex indices or refraction taken from Kramers-Kronig analysis of reflectance data from oriented Durham <u>trans</u>-polyacetylene[8]. This uncertainty can be removed by developing methods of measuring the linear optical constants of the actual samples used.

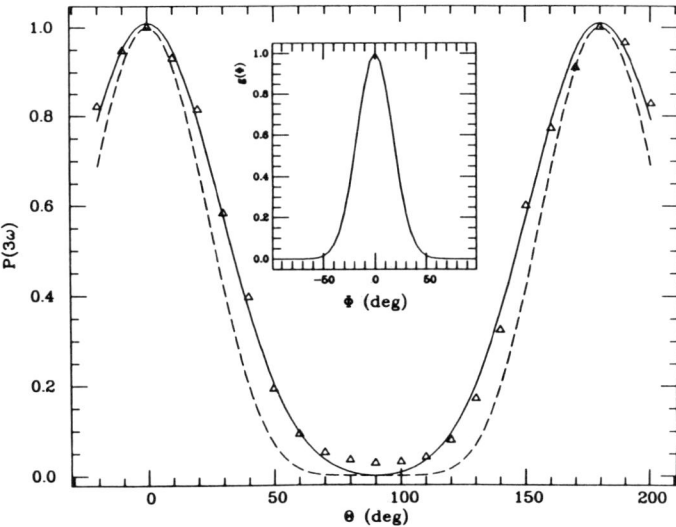

FIGURE 2. The polarization dependence of third harmonic generation from an aligned sample. The dashed curve shows the type of behavior expected for a perfectly aligned one-dimensional system ($\cos^6\Theta$); the solid line represents the best fit of the functional form of Eq. 9 to the data. The best fit corresponds to a chain orientation distribution, $g(\phi)$ (as shown in the inset) with FWHM of 40°.

Figure 2 displays the results of the polarization dependence of third harmonic generation from an aligned sample. In this measurement (performed in the transmission mode), all experimental parameters were held constant except the direction of the polarization of the incident beam. The triangles in Fig. 2 indicate the (normalized) measured power as a function of the angle between the chain direction and the polarization of the fundamental beam. When the fundamental field is polarized parallel to the chain direction, the magnitude of the third harmonic signal is approximately 30 times larger than when the fundamental field polarization and chain directions are orthogonal.

If we assume perfect alignment and that the only significant component of the third order susceptibility tensor is $\chi_{\parallel}^{(3)}$, then the third harmonic power is expected to vary as

$$P(3\omega) \; \alpha \; |\chi_{\parallel}^{(3)} E_{\parallel}^{3}|^2 \; \alpha \; |\chi_{\parallel}^{(3)} \cos^3\Theta|^2 |E|^6 \tag{8}$$

where $E_{\parallel}$ is the component of the fundamental field along the chain direction, and Θ is the angle between the polarization and chain directions. Note that the nonlinear polarization is parallel to $E_{\parallel}$. The dashed curve in Fig. 2 shows the type of behavior expected for a perfectly aligned one-dimensional system.

For comparison, we have measured the magnitude and the polarization dependence of the third harmonic power from a single crystal of PDA-TS. The magnitude was obtained in a relative measurement in which the polydiacetylene third harmonic power was directly compared to that from a trans-$(CH)_x$ samples (both oriented and non-oriented) which were in turn referenced to silicon.

The third harmonic power reflected from the PDA-TS sample varied considerably as a function of position on the sample surface. These variations were attributed to surface roughness of the PDA polymer crystal. Since the reflection technique depends critically on surface quality, the spots yielding the highest third harmonic (and hence the best surface quality) were used. The reflected third harmonic power was reproducible from "good" spot to "good" spot. For polyacetylene, the response was somewhat more uniform. We found that the third harmonic power generated on reflection from single crystal PDA-TS was about a factor of two larger than that from oriented trans-polyacetylene. In order to complete the comparison between the nonlinear susceptibilities of these two materials, accurate values of their linear optical constants are required. If we take $N_1 \approx 2$ for PDA-TS and assume that the two materials have the same N_3, then $\chi_{\parallel}^{(3)}$ of trans-$(CH)_x$ is somewhat larger than that of PDA-TS. More precise measurements of these linear optical constants are currently under way.

In evaluating the comparison between these two conjugated polymers, one must note that this measurement tends to favor PDA-TS since the "good" spots on a cleaved single crystal surface should be of much higher surface quality than the surface of a (fibrillar) polyacetylene film. Moreover, with the incident beam at 1.06 μm (1.17 eV), the third harmonic response of trans-$(CH)_x$ is at a minimum between strong resonances[3] in $\chi^{(3)}$. As a result, $\chi_{\parallel}^{(3)}$ for trans-$(CH)_x$ is substantially larger than that of PDA-TS for wavelengths deeper in the infrared.

The polarization dependence of the PDA-TS results are shown in Fig. 3. The third harmonic power from the polydiacetylene crystal accurately follows the $\cos^6\Theta$ relation[14]. We conclude, therefore, that it is likely that the deviations from the $\cos^6\Theta$ dependence in Fig. 2 are the result of imperfect chain alignment in partially oriented trans-$(CH)_x$ samples.

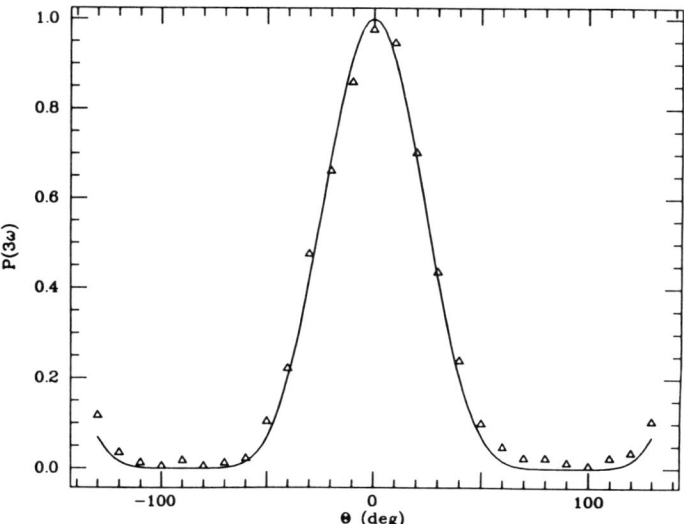

FIGURE 3. The polarization dependence of the THG from PDA-TS; the third harmonic power from the polydiacetylene crystal accurately follows the $\cos^6\Theta$ relation.

A more realistic treatment of the polarization dependence of the third harmonic generation from the polyacetylene samples should therefore allow for some misalignment of the polymer chains. To account for this, we have fit the data to a model which incorporates a gaussian distribution of chain orientations which is centered on the overall alignment direction. Specifically, we have assumed that the total response of the sample can be written as a sum of terms which represent chains which are oriented at an angle Φ with respect to the alignment direction, i.e.,

$$P(3\omega) \propto \left| \int_{-\pi/2}^{\pi/2} \cos^3(\Theta - \Phi)\exp{-(\Phi/\Phi_0)^2}d\Phi \right|^2 \qquad (9)$$

where the width of the distribution is used as the fitting parameter. The solid line in Fig. 2 represents the best fit of the functional form of Eq. 9 to the data. The value of Φ_0 which achieves best fit corresponds to a chain distribution with FWHM of 40°, in good agreement with the value inferred from analysis of the anisotropy of the infrared absorption and from examination of electron micrographs of the fibrillar alignment[9].

4. SYMMETRY SPECIFIC ORIGIN OF $\chi^{(3)}$: COMPARATIVE MEASUREMENTS OF $\chi^{(3)}$ IN CIS AND TRANS-POLYACETYLENE

Polyacetylene is unique in that it can be prepared in two different forms: cis-polyacetylene and trans-polyacetylene; the two isomers are shown in Fig. 4. The existence of two different isomers with different symmetry allows one to explore the specific origin of observed phenomena. Trans-$(CH)_x$ has a two-fold degenerate ground state and can support solitons as the fundamental nonlinear excitations[15]. In cis-$(CH)_x$, this ground state degeneracy has been lifted so that for the cis isomer the important nonlinear excitations are polarons and bipolarons. Thus, for example, subsequent to resonant (interband) photoexcitation, the shifts in oscillator strength will be quite different in the two cases. Moreover, the implied changes in the nonlinear optical properties (bigger shifts in oscillator strength imply larger $\chi^{(3)}$) due to this fundamental change in polymer symmetry can be probed on the same physical sample; conversion from cis- to trans-$(CH)_x$ can be accomplished simply by heating the sample to $\approx 150°$ C for about 1/2 hour.

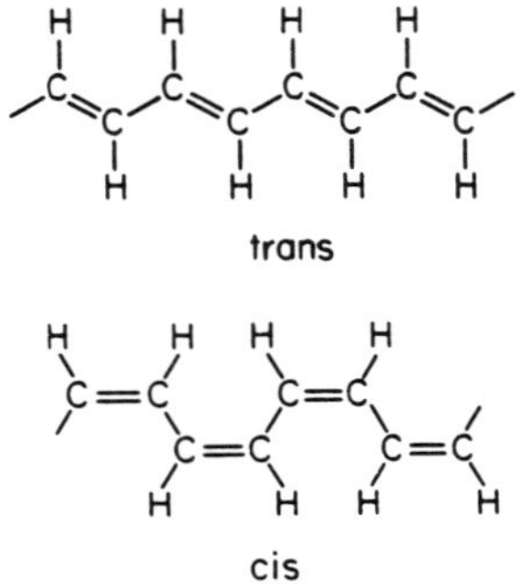

FIGURE 4. Chemical structure diagrams of the two different isomers of polyacetylene.

Films of cis-$(CH)_x$ were synthesized using the Shirakawa method. By carrying out the polymerization and subsequent washing etc. at -78° C, nearly 100% cis-$(CH)_x$ can be obtained. The cis-$(CH)_x$ samples were prepared as thick films (several microns in thickness) on pre-cut glass substrates made to fit into the sample holder on the cold-finger of the cryostat used for the nonlinear optical measurements.

Since partial conversion to the trans-$(CH)_x$ isomer is unavoidable when the temperature of the polyacetylene film is raised, our initial experiments were designed to minimize the time period at room temperature. By coordinating the sample preparation with pre-cooling of the cryostat, we were able to reduce the room temperature transfer time to approximately 5 minutes. After mounting the sample in the cryostat, the system was immediately cooled.

A precisely parallel experiment was carried out to characterize the cis-trans content by infrared absorption. The sample transfer time into the cryostat mounted on the FTIR instrument was essentially identical to that required for the transfer into the THG measurement system. From the integrated intensities of the characteristic cis-$(CH)_x$ and trans-$(CH)_x$ infrared absorption bands, we found that after transfer the polyacetylene films were approximately 15 - 20% trans-$(CH)_x$.

The methodology of the initial experiment was therefore as follows:

    (1) Synthesis of cis-$(CH)_x$;

    (2) Transfer of the sample into the measurement cryostate (15 - 20% conversion during the transfer);

    (3) Measurement of P(3ω) relative to the Si standard;

    (4) Remove the polyacetylene film and isomerize at 150° C for 1/2 hour;

    (5) Replace the same sample into the cryostat and measure P(3ω) relative to the Si standard.

This process was carried out for two independently prepared samples with the same result. We found that the third harmonic power scales with the residual trans-$(CH)_x$ content in the polyacetylene films; for a sample which is 15 - 20% trans, the measured $\chi^{(3)}$ is approximately 15 - 20% of the $\chi^{(3)}$ of the fully isomerized sample. Note that since P(3ω) is proportional to $|\chi^{(3)}|^2$, the measured third harmonic power from the "cis"-$(CH)_x$ samples was down from that of trans-$(CH)_x$ was down by a factor of approximately 40. The results are consistent with negligible contribution to P(3ω) from the cis-$(CH)_x$ portion of the sample. After including the measurement uncertainties in P(3ω) and in the determination of the cis-trans content, we conclude that $\chi^{(3)}(\omega,\omega,\omega)$ for cis-$(CH)_x$ is at most one-tenth of that of trans-$(CH)_x$.

To improve this upper limit, cis-$(CH)_x$ samples were prepared as thick (several microns) films in a specially constructed glass cell that allowed measurements without raising the temperature of the sample above -78°C. After completing the THG measurements on cis-$(CH)_x$, the sample was isomerized in the same cell, and the same sample was remeasured as trans-$(CH)_x$. This process was carried out for two independently prepared samples with the same result; P(3ω) from cis-$(CH)_x$ was between 250 and 500 times smaller than that from trans-$(CH)_x$, with the range due to variations in the sample surface. Since P(3ω) is proportional to $|\chi^{(3)}|^2$, the third order susceptibility of cis-$(CH)_x$ is smaller than that of trans-$(CH)_x$ by a factor of (conservatively) 15-20.

If the mechanism for the nonlinear optical response is related to virtual production of the nonlinear excitations of the polymer as argued in the Introduction, conversion from cis-$(CH)_x$ to trans-$(CH)_x$ should have a major effect. The experimental results presented in the preceding paragraphs are consistent with this hypothesis.

5. DISCUSSION OF THE MECHANISM: $\chi^{(3)}$ FROM VIRTUAL SOLITONS ENABLED BY QUANTUM ZERO-POINT FLUCTUATIONS OF THE POLYMER CHAIN

The anisotropy of the THG for both trans-$(CH)_x$ and PDA-TS demonstrates that the nonlinearity is entirely associated with the nonlinear polarizability of the π-electrons in the conjugated polymer backbone. The nearly identical magnitude of the third harmonic response in these two materials is quite remarkable in the context of traditional explanations[16] in which the mechanism for nonlinear optical response is nonlinear polarizability of the delocalized π-electrons within a rigid lattice (and a rigid band structure). In this point of view, the third order susceptibility would be strongly dependent on the magnitude of the single particle energy gap (~sixth power)[16]. This is certainly not the case for the experimental results described in the preceding sections.

The onset of absorption in trans-$(CH)_x$ is well below that of PDA-TS (<1.5 eV compared with ≈2eV). More importantly, the absorption edge at 2eV in PDA-TS is generally accepted to be due to a neutral exciton[17]. The interband transition, as indicated by the onset of band photoconductivity[18a] and by electroreflectance measurements[18b], is at approximately 2.4 eV. In contrast, for trans -(CH)$_x$ the onset of absorption at approximately 1.5 eV is due to an interband transition broadened on the leading edge by dynamical effects (see below) and by disorder. In trans -(CH)$_x$, the onset of photoconductivity coincides with the onset of absorption demonstrating that there is no evidence of any bound exciton[6]. Moreover, in trans -(CH)$_x$ the signatures of photogeneration of charged solitons have the same excitation profile[19a] as that of the photoconductivity, and both follow the low energy absorption tail[19b]. The sub-gap absorption below 2 eV in trans -(CH)$_x$ has been shown to be consistent with a mechanism in which the absorption tail is caused by anharmonic quantum fluctuations of the lattice[20]. In cis -(CH)$_x$, the absorption tail is supressed relative to that of trans -(CH)$_x$ consistent with the supression of anharmonic quantum fluctuations of the lattice due to the lifting of the ground state degeneracy. Although the absorption edge of cis-(CH)$_x$ is close to that of PDA-TS, the third harmonic response of the latter is at least an order of magnitude greater. Thus, the close agreement between the THG in trans -(CH)$_x$ and PDA-TS must simply be considered accidental; the nonlinear mechanisms in the two cases are different, and in neither case is the conceptually simple nonlinearity arising in third-order perturbation theory from the rigid band structure[16] the appropriate mechanism.

The data of Figs. 2 and 3 demonstrate that the nonlinear susceptibility is polarized entirely along the polymer backbone. Thus, in comparing $\chi_{\parallel}^{(3)}$ for cis- and trans-(CH)$_x$, one should consider the one-dimensional gaps for intrachain π-π^* excitation; for cis- and trans-(CH)$_x$, these are nearly equal. For trans -(CH)$_x$, $E_g(1d) \approx 1.7$ eV has been estimated from resonant Raman scattering[21] and 1.9 eV from a fit of the absorption tail arising from anharmonic quantum fluctuations of the lattice[20]. For cis -(CH)$_x$ the 1d gap is about 2 eV. Therefore, the major difference in $\chi^{(3)}$ for the two isomers of polyacetylene appears to be larger than can be accounted for in the context of rigid band theory (the E_g^6 depenence would predict a difference of only a factor of two), and it implies a mechanism which is sensitive to the existence of a degenerate ground state.

In attempts directed toward a deeper understanding of the mechanisms for nonlinear optical response of conjugated polymers, it is important to develop the connection between the **nonresonant** nonlinear response for pumping well below the absorption edge to the **resonant** nonlinear response for pumping directly into the principal absorption band. For example, in the polydiacetylene case, Greene et al[22] have analyzed the resonant nonlinear response in terms of phase-space filling by one-dimensional excitons. In a beautiful argument, they showed how the concept of exciton-polaritons could be used to generalize this mechanism to the nonresonant regime well below the exciton absorption edge.

For trans -(CH)$_x$, we have shown that for **resonant** pumping, the shifts in oscillator strength due to the photogeneration of charged solitons lead to relatively large changes in the optical constants. In the following, we generalize this idea to the nonresonant regime where the nonlinear response is due to virtual soliton-antisoliton pairs enabled by nonlinear quantum lattice fluctuations. Following Yu, Matsuoka and Su[20a], we include fluctuations in the staggered order parameter $\phi_n = (-1)^n u_n$ which are described by the following one-parameter family of configurations (see Figure 5a):

$$\phi(x=na) = \phi_0\{1 - tanh(2x_0/\xi)[tanh(x-x_0/\xi) - tanh(x+x_0/\xi)]\} \qquad (10)$$

These configurations can be viewed as soliton-antisoliton pairs where x_0 represents roughly half the separation of the soliton and antisoliton, ξ is the coherence length, and $\underline{a}$ is the lattice constant. Yu et al show that within this class of configurations, the total energy is of the form $E(s) = E_p(s) + (M/2)(ds/dt)^2$ where $E_p(s)$ is shown as curve (a) in Figure 5b, and s is related to x_0 by a coordinate transformation $x_0(s)$. They then treat $E_p(s)$ as an effective potential such that

$$H_0(s)= E_p(s) + (M/2)(ds/dt)^2 \qquad (11)$$

is an effective Hamiltonian, and solve for the allowed bound states. They find the ground state at $E_G = 0.05$ eV above the classical ground state with wave function $\phi_0(s)$ describing the nonlinear ground state fluctuations. In addition, they have shown that there are five vibrational excited states of the s-configuration ; the

(a)

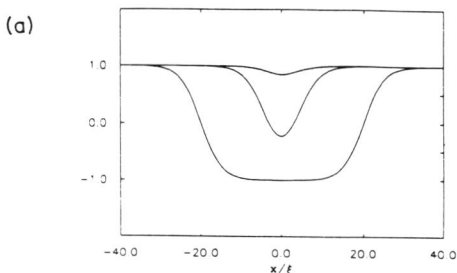

(b)

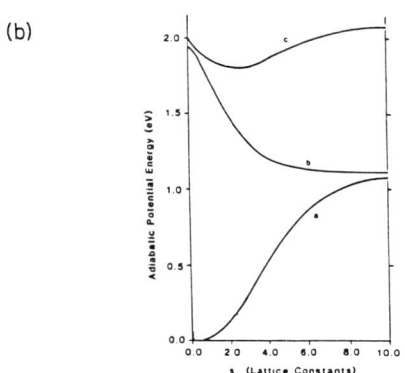

FIGURE 5. a) Sketch of the fluctuations in the staggered order parameter described by the one-parameter family of configurations: $\phi(x = na) = \phi_0\{1 - tanh(2x_0/\xi)[tanh(x-x_0/\xi) - tanh(x+x_0/\xi)]\}$. b) Adiabatic potential curves for the electronic ground state (a) and first (b) and second (c) excited states as a function of s for the configuration of Fig. 5a (from ref. 20a).

highest one at an energy $\approx 1.8 eV$ above the ground state and a series of four others approximately equally spaced below 1 eV.

For each value of s there is a soliton-antisoliton pair configuration; for small s the two gap states are close to the band edges, while for large s the two gap states are those of widely separated solitons and appear at mid-gap. Thus for each s, there is a change in the absorption coefficient $\delta\alpha = (2\omega/c)\delta k$, where (n+ik) is the complex index of refraction, $\varepsilon = (n+ik)^2$. For a given s, let $\delta n_s(\omega,\omega_p)$ be the change in n at a frequency ω due to pumping a configuration s at ω_p. For example, if ω_p is above the gap, then one makes real solitons (delayed in time by $\sim 10^{-13}$ seconds) by the Su-Schrieffer mechanism[23]. Thus for ω_p above the gap, δn_s time evolves to $\delta n_\infty(\omega)$ which comes from the shift in oscillator strength from the interband transition into the mid-gap transition. Since the time resolved spectroscopy has been thoroughly studied for pumping at $\hbar\omega_p > E_g$, one knows the complete time evolution of $\delta\alpha(\omega)$ and of $\delta n(\omega)$ subsequent to pumping into the interband transition.

If the pump frequency is in the region between 2Δ and $(4/\pi)\Delta$, then one pumps the specific nonlinear configuration s of equation (10) which is resonant with the pump; the excited s-configuration then evolves (in time) following down the excited state energy curve $E_{p'}(s)$ (see Figure 5) toward a well-separated charged soliton pair each with a state at mid-gap. Again, one expects a corresponding δn_s which also evolves with time after the photon is absorbed from that characteristic of the originally pumped s(t=0) evolving to a charged soliton pair each with a state at mid-gap. Pumping at $\hbar\omega_p = 4\Delta/\pi$ directly generates a free soliton pair with a corresponding $\delta n_\infty(\omega)$; when pumping at the soliton pair creation energy, there is no time delay nor any time evolution since the free soliton pair is created directly (although with vanishingly small probability). The above are all resonant processes; they involve a real absorption of photons to form either electron-hole pairs $(\hbar\omega_p > E_g)$ or excited s-configurations $(\hbar\omega_p < E_g)$ which time-evolve to separated charged soliton pairs. Thus, these resonant processes lead to highly nonlinear optical phenomena with characteristic time evolution and with corresponding changes in the complex index of refraction.

What about truly virtual processes for $\hbar\omega_p$ below the $(4/\pi)\Delta$ threshhold? We argue that

$$<\delta n(\omega,\omega_p)> = \Sigma_s(\text{probability of finding a charged configuration s})\delta n_s(\omega).$$

Thus,

$$<\delta n(\omega,\omega_p)> \approx \Sigma_s|\phi_0(s)|^2 \, [\, |Q|^2/ E_s \,] \, \delta n_s(\omega) \tag{12}$$

where $\phi_0(s)$ is the ground state wavefunction for the s-configuration. The last equality holds if $\hbar\omega_p < E_s$. In eqns. 9 and 10, $|Q|^2 = e^2 f_s^2 |E|^2$ is the square of the dipole matrix element betwen the ground state and the (classical) electronic excited state of the s-configuration at $E_{p'}(s)$; $|Q|$ is therefore proportional to the square of the oscillating electric field (i.e. the intensity of the pump) at $\omega_p < E_s$ where E_s is the appropriate energy denominator (E_s decreases from 2Δ to $4\Delta/\pi$ as s increases from 0 to ∞). Thus, from eqn. 12,

$$n(\omega) = n_0(\omega) + \Delta n(\omega,\omega_p)I(\omega_p)$$

(where $n_0(\omega)$ is the linear index) and from the standard phenomenology, $\Delta n(\omega,\omega_p) = (4\pi^2/c\varepsilon)\chi^{(3)}(\omega,\omega_p,-\omega_p)$ (see eqn. 2) . Equation 12 predicts a nonlinear index (which arises directly from the nonlinear zero-point fluctuations) in response to pumping at frequencies well below the energy gap.

Note that if the ground state of the system is nondegenerate, $E_p(s)$ increases initially much more steeply with s and becomes linear for large s (soliton confinement). This suppresses the nonlinear ground state fluctuations and suppresses the magnitude of $<\delta n(\omega,\omega_p)>$. The predicted suppression of the nonlinear response is in agreement with our experiments comparing cis and trans-$(CH)_x$.

A parallel analysis can be made of the contribution of the virtual solitons (s-configurations) to the third order nonlinearity responsible for THG. In this case, the standard third order perturbation theory on the excited states of the s-configurations yields[24];

$$\chi_{||}^{(3)} = N_0 e^4 \Sigma_{nml}(f_{gn}f_{nm}f_{ml}f_{lg}) \times$$
$$[(1/(E_{ng}-3\omega)(E_{mg}-2\omega)(E_{lg}-\omega) + 1/(E_{ng}+\omega)(E_{mg}-2\omega)(E_{lg}-\omega)$$
$$+ 1/(E_{ng}+\omega)(E_{mg}+2\omega)(E_{lg}-\omega) + 1/(E_{ng}+\omega)(E_{mg}+2\omega)(E_{lg}+3\omega)] \qquad (13)$$

where $E_{ng}=(E_{s,n} - E_{s,g})$, $E_{s,g}$ is the ground state energy, $E_{s,\alpha}$ (α=n,m,l) are the energies of excited states of the s-configurations, $f_{\alpha\beta}$ are the dipole matrix elements. In eqn. 13, $N_0\approx n_0\rho_c$ is the average number of soliton-antisoliton pairs on a chain of length L, and ρ_c is the density of chains per unit area. Monte Carlo simulations[25] have shown that the reduction of the dimerization order parameter due to zero point motion is about 15%. Since this is also approximately the reduction in average order parameter in a polyacetylene ring of size 4ξ (just large enough for an $S\bar{S}$ pair), we estimate $n_0 = 1/4\xi$.

In the sum, n and l denote excited states with symmetry opposite to g, and m is an excited state with the same symmetry as g. Yu, Matsuoka and Su[20a] have shown that there are five vibrational excited states of the s-configuration with the same parity as the ground state which contribute to $\chi_{||}^{(3)}$; one at ≈ 1.8eV and a series of four others approximately equally spaced below 1 eV. These excited states are in product form, $\psi(s)\zeta(s)$ where $\psi(s)$ is a vibrational eigenstate and $\zeta(s)$ is the many-electron wavefunction associated with the classical configuration s. Thus, the matrix elements factorize into an electronic dipole matrix element and an overlap integral, as in the usual Born-Oppenheimer approximation. The lattice overlap integral would be zero except for the nonlinear gound state fluctuations.

To estimate the magnitude of $\chi_{||}^{(3)}$ we consider the term in the full sum of eqn. 13 where the matrix elements go from $g \to E_p'(s) \to g \to E_p'(s) \to g$; i.e we ignore the various vibrational states of the s-configuration and consider only this single contribution ($\chi_{||}^{(3)}|_0$):

$$\chi_{||}^{(3)}|_0 = N_0 \Sigma_{ss'}|\phi_0(s)|^2|\phi_0(s')|^2(|f_s|^2|f_{s'}|^2/2\omega) \times$$
$$[-(1/(E_s-3\omega)(E_{s'}-\omega) + 1/(E_s+\omega)(E_{s'}+3\omega)]. \qquad (14)$$

For $3\omega < E_s$

$$\chi_{||}^{(3)}|_0 \approx -4N_0 \Sigma_{ss'}|\phi_0(s)|^2|\phi_0(s')|^2|f_s|^2|f_{s'}|^2/E_s^2 E_{s'} \qquad (15)$$

where f_s is the matrix element between the ground state with wave function $\phi_0(s)$ and the excited state with energy E_s relative to the ground state (see eqn 12). Since

$$\alpha = N_0 \Sigma_s |\phi_0(s)|^2 \alpha_s \qquad (16)$$

is the linear term in the polarizability,

$$\chi_\parallel^{(3)}|_0 \approx -(4\alpha)\Sigma_s |\phi_0(s)|^2 (\alpha_s/E_s) \qquad (17)$$

This term has a simple semi-classical origin; it arises from the modulation of the linear polarizability of the configuration s by the oscillating electric field well below resonance.

To estimate the magnitude of $\chi^{(3)}|_0$, we note that the low frequency dielectric constant[13] of trans-$(CH)_x$ is $\varepsilon_\parallel \sim 15$. Since the fraction of the total oscillator strength in the sub-gap absorption tail is approximately 20% of the total, and since that oscillator strength has an effective gap of about 1.75 ev (whereas the mean energy of the interband transition is at about 2.5 eV), we estimate $\alpha \sim 1/2$. From eqn. 17,

$$\chi_\parallel^{(3)}|_0 \sim (8\xi/a)(N\rho_c)^{-1}\alpha^2/E_0 \qquad (18)$$

where E_0 is the typical energy in the tail (~ 1.75 eV), and $N\rho_c$ is the density of carbon atoms. This is of the correct order of magnitude; i.e. $\chi_\parallel^{(3)}|_0 \sim (1-5) \times 10^{-10}$ esu. We note, furthermore that the energies of the two- and three-photon resonances implicit in eqn 13 are in good agreement with the resonant THG response (at 0.9 eV and for $\hbar\omega < 0.5$eV) observed by Kajzer et al[3].

Detailed numerical calculations of both terms contributing to $\chi^{(3)}$ have been carried out[26]; the magnitude of the nonlinear response, the observed two-photon resonance, and the sensitivity to the lifting of the ground state degeneracy are all in agreement with experiment.

6. CONCLUSION

We have determined the magnitude of the third order susceptibility associated with tripling the fundamental of the Nd:YAG laser (1.06 μm, 1.17 eV) to be $\chi_\parallel^{(3)}(3\omega;\omega,\omega,\omega) = (4 \pm 2) \times 10^{-10}$ esu, where $\chi_\parallel^{(3)}$ refers to that component of the third order susceptibility tensor with all indices parallel to the chain direction. By measuring anisotropic third harmonic generation in oriented films, we have shown that this component dominates. The magnitude and anisotropy were directly compared with results obtained from single crystals of polydiacetylene-(toluene-sulfonate) measured in the same apparatus. The third harmonic power generated by PDA-TS was found to be about a factor of two greater than that of trans-$(CH)_x$ (for pumping at 1.06 μm). The anisotropy of the THG for both trans-$(CH)_x$ and PDA-TS demonstrates that the nonlinear optical properties are entirely associated with the nonlinear polarizability of the π-electrons in the conjugated polymer backbone.

The close agreement between the THG in trans-$(CH)_x$ and PDA-TS was inferred to be simply accidental; the nonlinear mechanisms in the two cases are different, and in neither case is the conceptually simple nonlinearity arising from the rigid band structure in third-order perturbation theory the appropriate mechanism.

Third harmonic generation was measured in both cis-rich and trans isomers of the same sample (before and after thermal isomerization). The measured response of the cis-rich samples was found to scale with the residual trans content, indicating that $\chi_{||}^{(3)}$ of the trans isomer is at least an order of magnitude larger than that of the cis isomer.

Since $\chi^{(3)}(\omega=\omega_1+\omega_2+\omega_3)$ is a function of three independent variables, the measured $\chi^{(3)}(3\omega=\omega+\omega+\omega)$ represents a single point on a complex three-dimensional surface. The larger $\chi^{(3)}$ values implied by the resonant photoinduced absorption measurements are located in another region of this parameter space. These two regions are not unrelated, however, since one knows that the nonresonant values of the third order suscetibility will be determined by the nature of the excited states which dominate the resonant properties.

The symmetry specific aspect of $\chi_{||}^{(3)}$ is an important result and implies a mechanism which is sensitive to the existence of a degenerate ground state, as in trans-$(CH)_x$. This experimental fact is consistent with a mechanism for the third order nonlinear optical susceptibility of polyacetylene which involves the generation of virtual nonlinear solitons. This mechanism was explored in considerable detail, and the connection was made to the nonlinear quantum zero point fluctuations of the polyene chain. In the absence of quantum fluctuations of the lattice, nonlinear excitations such as solitons do not affect nonresonant processes. These configurations involve lattice distortions around the photogenerated carriers; since there are no direct matrix elements between the ground state and the soliton excited states, they could only be produced by photoexcitation and subsequent decay. Inclusion of quantum fluctuations allows matrix elements directly connecting the ground state and the nonlinear excited states and thereby enables virtual solitons to affect nonresonant processes.

Expressions were derived for $\chi_{||}^{(3)}(\omega,\omega_p,-\omega_p)$ (which leads to an intensity dependent complex index of refraction), and for $\chi_{||}^{(3)}(\omega,\omega,\omega)$ (which leads to third harmonic generation). In both cases, the magnitude of the virtual soliton terms are large enough to make important contibutions to $\chi_{||}^{(3)}$. We conclude, therefore, that the nonlinear zero point fluctuations of the ground state lead to an important mechanism for nonlinear optical properties, particularly in polymers with a degenerate ground state. Lifting the ground state degeneracy suppresses the nonlinear response, in agreement with our experimental results. In addition, the usual interband transitions (i.e. electron-hole pairs)[16] can be expected to influence $\chi_{||}^{(3)}$. An important goal of future work will be to sort out the relative importance of the two processes in order to understand the large nonlinear susceptibility of trans-polyacetylene and in order to guide the development of new and better materials.

ACKNOWLEDGMENTS: We thank Dr. W.-P. Su for sending a preprint of his work prior to publication and for important discussions concerning the role of solitons in $\chi^{(3)}$. We are grateful to D. McBranch for help with measurements and for comments on the proposed mechanism. We thank Prof. H. Shirakawa for helping with arrangements for Dr. Akagi's visit to Santa Barbara. This research was supported by the Office of Naval Research through N00014-86-K-0514.

44

REFERENCES
1. (a) Nonlinear Optical Properties of Organic and Polymeric Materials, Ed. by
    D.J. Williams, American Chemical Society Symposium, Series 233 (Amer.
    Chem. Soc., Wash, D.C. ,1983).
    (b) Nonlinear Optical Properties of Polymers, Ed. by A.J. Heeger, J.
    Orenstein and D. Ulrich (Materials Research Society, Pittsburgh, 1988).
2. A. J. Heeger, D. Moses and M. Sinclair, Syn. Mtls. 15, 95 (1986).
3. F. Kalzar, S. Etemad, G.L. Baker and J. Messier, Syn. Mtls. 17, 563 (1987);
    Sol. St. Commun. 63, 1113 (1987).
4. (a) J. Orenstein, in Handbook of Conducting Polymers, Ed. by T.A.
    Skotheim, (Marcel Dekker, New York and Basel, 1986), Vol. 2.
    (b) A. J. Heeger,Polymer Journal, 17, 201 (1985).
5. (a) Z. Vardeny, J. Strait, D. Moses, T.-C. Chung and A.J. Heeger, Phys.
    Rev. Lett. 49, 1657 (1982).
    (b) C. V. Shank, R. Yen,R. L. Fork, J. Orenstein and G. L. Baker, Phys. Rev.
    Lett. 49, 1660 (1982).
    (c) C. V. Shank, Science, 219, 1027 (March 4, 1984).
6. (a) M. Sinclair, D. Moses and A.J. Heeger, Sol. State. Commun. 59, 343
    (1986).
    (b) M. Sinclair, D. Moses and A.J. Heeger, Phys. Rev.B36, 4296 (1987).
7. A.J. Heeger, D. Moses and M. Sinclair, Syn. Mtls. 17, 343(1987).
8. W.K. Burns and N. Bloembergen, Phys. Rev. B4, 3437 (1971).
9. (a) K. Akagi, S. Katayama, H. Shirakawa, K. Araya, A. Mukoh and T.
    Narahara, Syn. Mtls. 17, 241 (1987).
    (b) K. Araya, A. Mukoh and T. Narahara, K. Akagi, H. Shirakawa, K. Araya,
    A. Mukoh and T. Narahara, Syn. Mtls. 17, 247(1987).
10. M. Aldissi, J. Polym. Sci., Polym. Lett. Ed. 23, 167 (1985).
11. W. J. Feast, in Handbook of Conducting Polymers, Ed. by T.A. Skotheim,
    (Marcel Dekker, New York and Basel, 1986) Vol. 1.
12. See references listed in ref. 8.
13. These values were obtained from Kramers-Kronig analysis of the
    reflectance   from highly oriented Durham polyacetylene; G. Leising (private
    communication).
14. C. Sauteret, J.-P. Hermann, R. Frey, F. Pradere, J. Ducuing, R.H.
    Baughman and R.R. Chance, Phys. Rev. Lett. 36, 956 (1976).
15. (a) S. Etemad, A.J. Heeger and A. G. MacDiarmid, Ann. Rev. Phys. Chem.
    33, 443 (1982).
    (b) J.R. Schrieffer, Highlights of Condensed Matter Theory, Soc. Italiana de
    Fisica, 89, 300 (1985)
    (c) S. Kivelson, Solitons, Ed. by S. Trullinger and V.L. Pokrovsky (Elsevier
    Science, B.V., 1986).
16. (a) C. Flytzanis, in reference 1a.
    (b) W.K. Wu and S. Kivelson, in reference 1b, p.229.
17. Polydiacetylenes, Ed. by D. Bloor and R.R. Chance, NATO ASI Series, No.
    102 (Martinus Nijhoff Dordrecht,Boston, Lancaster, 1985).
18. (a) K.J. Donovan and E.G. Wilson, Phil. Mag. B, 44, 31 (1981).
    (b) L. Sebastian and G. Weiser, Chem. Physics 62, 447, (1981).
19. (a) G.B. Blanchet, C.R. Fincher and A.J. Heeger Phys. Rev. Lett.51, 2132
    (1983).
    (b) B.R. Weinberger, C.B. Roxlo, S. Etemad, G.L. Baker and J. Orenstein
    Phys. Rev. Lett. 53, 86 (1984).

20.   (a) J. Yu, H. Matsuoka and W.P. Su, Phys. RevB (in press).
      (b) J.P. Sethna and S. Kivelson, Phys. Rev. B26, 3513 (1984).
      (c) A. Auerbach and S. Kivelson, Phys. Rev.B33, 8171 (1986).
      (d) Z.B. Su and L. Yu, Phys. Rev. B27, 5199 (1984).
21.   Z. Vardeny, E. Ehrenfreund, O. Brafman, and B. Horovitz, Phys. Rev. Lett.
      51, 2326 (1983).
22.   B.I. Greene, J. Orenstein, R.R. Millard and L.R. Williams, Phys. Rev. Lett.
      58, 2750 (1987).
23.   W.P. Su and J.R. Schrieffer, Proc. Nat. Acad. Sci. USA 77, 5626 (1980).
24.   R. Loudon, The Quantum Theory of Light (Clarendon Press, Oxford, 1983).
25.   a) W.P. Su, Sol. State Commun, 42, 497 (1982).
      b) E. Fradkin and J.E. Hirsch, Phys. Rev. B27, 1680 (1983).
26.   M. Sinclair, D. Moses, A.J. Heeger, J. Yu, H. Matsuoka and W.P. Su, Phys.
      Rev. Lett. (submitted).

THIRD-ORDER NONLINEAR SUSCEPTIBILITY IN
SEMICONDUCTING POLYMERS

J. MESSIER
Laboratoire de Physique Electronique des Matériaux-DEIN/D.LETI
91191 Gif-sur-Yvette Cedex-FRANCE).

I. INTRODUCTION
In that paper we shall describe the experimental results
obtained in measuring third order nonlinear coefficient, χ^3,
in organic semiconducting polymers and some theoretical calculations.
Since the C. Sauteret's pioneering work [1] in 1974 it has
been well known that the χ^3 can be very high in one dimensional π-electron system as in polydiacetylene. The fast increasing of the χ^3 with the π-electron conjugation length, has
been corroborated by numerous theoretical works [2], especially in one dimensional systems.
The χ^3 coefficients are essential in centrosymmetrical compounds where the second order coefficients equal zero. They
are also important in the noncentrosymmetrical molecules.
Moreover these χ^3 coefficients play a part in some experimental determination of χ^2 coefficients (Electric field induced
second harmonic generation (EFISH) in solution).
Several nonlinear optical phenomena depend on a third-order
process: third harmonic generation (THG), EFISH, Optical Stark
effect, index of refraction variation with the light intensity.
The first three phenomena mainly come from the electronic hyperpolarizability and can be described by similar formalisms.
The last one have several causes: thermal dilatation, elec-
trostrictive compression, creation of excitons, polarons,
etc... by one or two photon absorption and also the electronic
hyperpolarizability. It needs in each case appropriate
treatments.

II. THIRD HARMONIC GENERATION ELECTRIC FIELD INDUCED SECOND
HARMONIC GENERATION

II.1. Experimental
II.1.1. Methods of measurement. The experimental set-up is
shown fig.(1). The laser beam frequency can be varied between
0.8 and 2 µm. The measured sample is either an organic thin
layer deposited on a transparent substrate or a polymers solution in a prismatic cell [3]. The reference sample is amorphous silica (THG) or quartz (EFISH). Both of them are in a
vacuum chamber.
In case of thin layer, the modulus of the $\chi^3(-3\omega;\omega,\omega,\omega)$ or
$\chi^3(-2\omega;\omega,\omega,0)$ coefficients can be determined from experimental
results (cf. fig.(2)).

47

J. Messier et al. (eds.), Nonlinear Optical Effects in Organic Polymers, 47–60.
© *1989 by Kluwer Academic Publishers.*

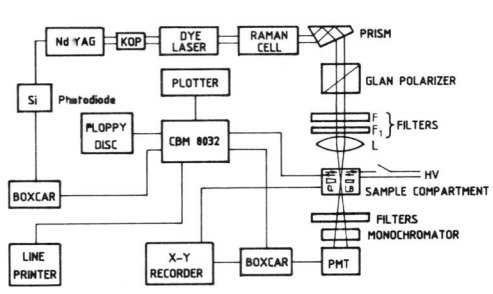

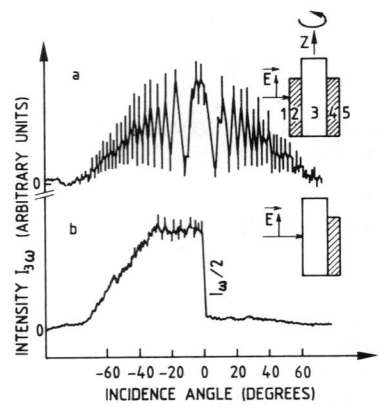

FIGURE 1.

Schematic representation
of experimental arrangement.
The measured sample can be
rotated or translated by a
step-to-step motor and replaced
by a reference sample.

FIGURE 2.

Harmonic light intensity
versus incidence angle
a) Polymer film on both faces of the substrate.
b) Left: polymer film on one
face of the substrate.
Right: substrate only.

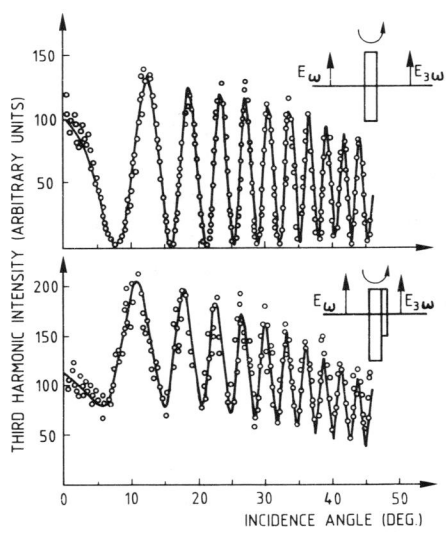

FIGURE 3.

Third harmonic intensity versus incidence angle
a) Rotating silica plate
b) Rotating silica plate + polymer film on a fixed substrate

Theoretically it is also possible to measure the phase by adding a known variable signal generated by a rotating quartz plate to the organic layer signal [4] (fig.(3).). But this method is difficult to perform and now only $|x^3|$ = f(ω) curves are available.
In very thin layer the reabsorption of harmonic is low enough to allow a rather large frequency range of measurements. In the case of a polymer solution where a known constant referen- ce signal is added to the variable signal of the polymer solu- tion $|x^3|$ and the phase can be both simultaneously determined but only in a limited frequency range due to reabsorption of harmonics.
II.1.2. <u>Experimental results</u>. Polydiacetylene thin layers have been made by vacuum evaporation of 4-BCMU, DCH or C_{16-8} mono- mers followed by UV polymerization. The obtained blue form may be converted, except in the case of DCH monomer layers, in a red one by heating at 120°C for 2 hours. The absorption spec- tra of the two forms are shown in fig.(4) and the $x^3_{(\omega)}$ curve in figs.(5) and (6).

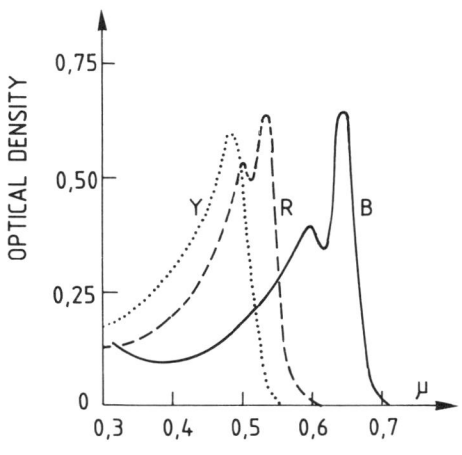

FIGURE 4.

Optical absorption spectra of polydiacetylene thin film (4-BCMU), blue form (B) and red form (R). The yellow form (Y) exists in solution.

EFISH EXPERIMENTS. A DC electric field E_0 (40 kv/cm or less) is applied on a thin layer by two aluminium electrodes. $\vec{E}_0$ is parallel to the light electric field $\vec{E}_\omega$. A second harmonic generation is observed which intensity is proportional to $\left(P^{NL}_{(2\omega)}\right)^2$ where $P^{NL}_{(2\omega)}$ the nonlinear polarizability at 2ω is given by: $P^{NL}_{(2\omega)} = x^3_{(-2\omega;\omega,\omega,0)} E^2_\omega E_0$
The values of $x^3_{(-2\omega;\omega,\omega,0)}$, in the same polydiacetylene layers, are plotted on fig.(7).

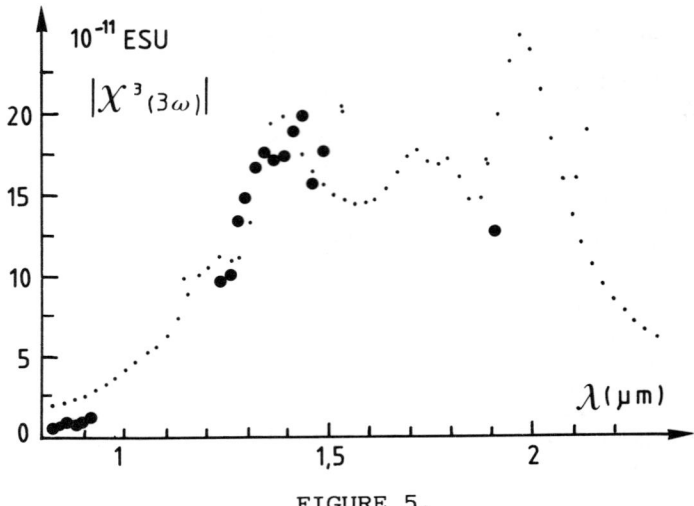

FIGURE 5.

Cubic susceptibility of a polydiacetylene thin film (blue form). Experimental points. Theoretical curves (dotted lines) calculated with the Table I parameters. $\chi^3_{(0)}$ extrapolated value at $\omega = 0$: $7,9.10^{-12}$ esu

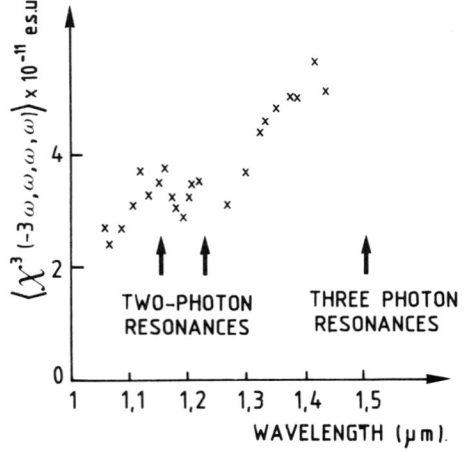

FIGURE 6.

Average cubic susceptibility of 4000 Å thick evaporated C_{16-8} red form PDA film. The two-photon resonances are shifted towards shorter wavelength as compared with the blue forms.

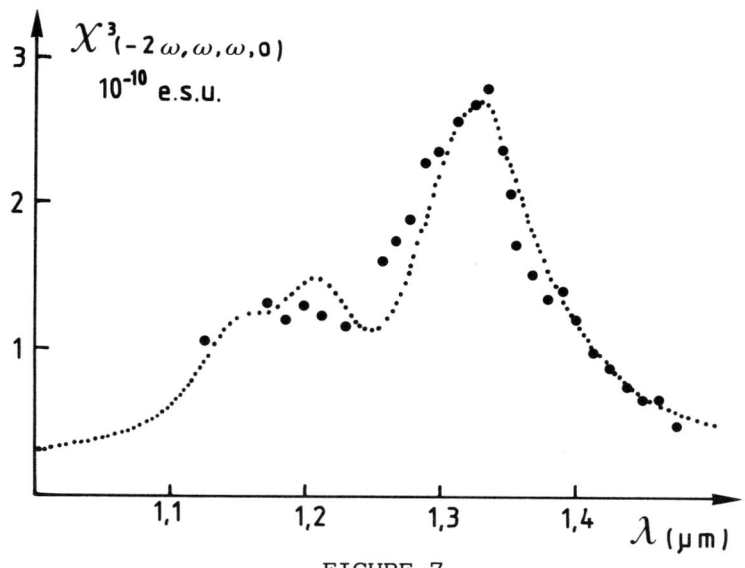

FIGURE 7.

x^3 $(-2\omega;\omega,\omega,0)$ of a thin polydiacetylene (DCH) layer. Experimental points. Dotted line: theoretical curve (formula II)
calculated using Table I parameters.

II.2. Interpretation with a time dependent perturbation model
Resonances from 1,3 to 1,5 μm occur at wavelengths that are
about twice the one of the main absorption transition, and
suggest a two-photon resonance. On the other hand the high
$x^3_{(-3\omega)}$ value and the low $x^3_{(-2\omega)}$ value at 1,907 μm point out a
three photons resonance in this range.
It is the reason why we tried to interpret such results with
by a time dependent perturbation method [5]. This calculations
have been explicited by Ward, particularly in ref. [6] where
difficulties coming from secular terms equal to zero are taken
into account.
For a centrosymmetrical molecule the electronic levels are even
(g) or odd (u) and the third order molecular coefficients
$\Upsilon(-3\omega)$ and $\Upsilon(-2\omega)$ are given, in the case of a three level
model, by:

$$\Upsilon(-3\omega;\omega,\omega,\omega) = \mu^2_{gu}\left[\mu_{ug} \cdot f(\omega) - \mu^2_{gu}g(\omega)\right]\frac{e}{\varepsilon_o} \qquad (I)$$

$$\Upsilon(-2\omega;\omega,\omega,0) = \mu^2_{gu}\left[\mu^2_{ug} \cdot f'(\omega) - \mu^2_{gu}g'(\omega)\right]\frac{e}{\varepsilon_o} \qquad (II)$$

where μ_{gu}, μ_{ug}, are dipolar matrix elements.
F(ω) is a sum of three terms products such as:

$$\frac{1}{E_u \pm \omega \pm i\,\Upsilon_u}\,,\,\frac{1}{E_g \cdot \pm 2\omega \pm i\,\Upsilon_g}\,,\,\frac{1}{E_u \pm 3\omega \pm i\,\Upsilon_u}$$

where $E_u E_g$. are energy differences between the excited states, u, g', and the fundamental one, g. g(ω) have only 1-photon term, f(ω) 2-, 1-, 0-photon terms, g$'_{(\omega)}$ 1-, 0-photon terms. γ_u, γ_g are damping terms.

Theoretical curves calculated from I and II are shown on figs.(5),(7). Used values E_u, $E_{g'}$, γ_u, $\gamma_{g'}$ are indicated in Table I. They have been obtained by fitting the experimental absorption spectra.

m	Eu_m (cm^{-1})	γu_m (cm^{-1})	Eg'_m (cm^{-1})	$\gamma g'_m$	Relative intensity am	Matrix element (mean value per monomer unit)
0	15209	600	14789	720	.42	
1	16610	550	16190	670	.17	$\mu_{g' \cdot u}$ = 9.5 Å
2	17400	550	16980	670	.13	$\mu_{g u}$ = 2.1 Å
3	19000	2100	18580	1800	.28	

Note: m = 0 : zero-zero transition

m = 1,2,3: vibronic side band
$$\chi = \sum_m a_m \chi_m$$

TABLE 1.

To take account of the vibronic side bands, well known in polydiacetylene (cf. fig.(4)) we assume that they come from the same electronic transition as the main absorption band and we insert couples of $E_u/E_{g'}$ electronic levels such as $E_{u_0} - E_{g'_0} = E_{u_1} - E_{g'_1}$ etc...

The E_u, γ_u values have been determined by a least-squares fit to the experimental absorption spectra and, in the same way, $E_{g'}$, $\gamma_{g'}$ values to the χ^3 results.

The I and II formula show that the two-photon resonances are caused, for $\gamma(-3\omega;\omega,\omega,\omega)$, by the term $[E_{g'} - 2\omega - i\gamma_{g'}]^{-1}$ and occur at ω_M with $E_{g'} = 2\omega_M$.

For $\gamma(-2\omega;\omega,\omega,0)$ they are produced mainly by the term $[(E_u - 2\omega - i\gamma_u)(E_{g'} - 2\omega - i\gamma_{g'})(E_u - \omega - i\gamma_u)]^{-1}$. As $E_{g'}$, $E_u < \gamma_{g'}$, γ_u one single resonance is observed between $E_{g'}$ and E_u.

II.3. Third order nonlinear coefficient at zero frequency

Several authors calculate hyperpolarizabilities by "ab initio" computations, with a "finite field" method at a zero frequency [7]. Consequently it is interesting to know what was the relationship between this method and the time dependent perturbation approach. At zero frequency the function $\mathcal{E}_j(E)$, where $\mathcal{E}_j$ is the eigenvalue of an electronic level, can be easily obtained in the case of three levels model for instance by diagonalizing a matrix such as

$$
\begin{vmatrix}
E_g & \vdots & \mu_{ug}.E & \vdots & 0 \\
\dots & \dots & \dots & \dots & \dots \\
\mu_{ug}.E & \vdots & E_u & \vdots & \mu_{gu}E \\
\dots & \dots & \dots & \dots & \dots \\
0 & \vdots & \mu_{gu}E & \vdots & 0 \\
\dots & \dots & \dots & \dots & \dots
\end{vmatrix}
$$

We obtain:

$$
\gamma_{(0)} = -\frac{\partial^4 \mathcal{E}_g}{\partial E^4} = \frac{\mu_{gu}^2}{E_u^2}\left[\frac{\mu_{ug}^2}{E_g} - \frac{\mu_{gu}^2}{E_u}\right] \tag{III}
$$

$$
\gamma_{(0)} = 0 \text{ When } (\mu_{ug}./\mu_{gu})^2 = E_g/E_u \tag{IV}
$$

The same classical result is obtained from I with $\omega \to 0$ since $f_{(0)} = \dfrac{1}{E_0^2 E_g}$, $g_{(0)} = \dfrac{1}{E_\nu^3}$ in the transparency region [8].

Let us notice that a harmonic oscillator defined by an infinite number of g and u equally spaced energy levels obey also IV; but, in this case, all the higher-order nonlinear coefficients equal zero which is not true in a three level model. The relationship between the values of $\gamma(\omega)$ at ω frequency and $\gamma(0)$ is not obvious because the $f(\omega)$, $g(\omega)$ functions depend on E_g, E_u, γ_g, γ_u etc... and may be very different from each other, specially in the near infrared region where two or three photon resonances can occur.

II.4. Computation of third-order nonlinear coefficients by a perturbation method

II.4.1. Centrosymmetrical molecule As a rule the knowledge of all the dipolar transition matrix elements between all the electronic levels is needed. In the most general case the problem is cumbersome. Nevertheless it was interesting to know wether computation was possible in the special case of a regular linear polymer. The results obtained with a (CNDO) self consistant computation method giving all the needed matrix elements between fundamental and excited states are shown below [9]. With a reasonable calculation time (let us say one day) it is possible to determine such elements in a hundred atoms polymer.

In Table II are shown the $\gamma(0)$ coefficients in polythiophene and polydiacetylene with six and twelve monomeric units. The number of products such as $\langle g|x|u\rangle\langle u|x|g'\rangle\langle g'|x|u'\rangle\langle u'|x|g\rangle$ able to contribute appreciably to γ is actually limited and it is generally sufficient to take account of three or four excited u and g states.

On fig.(8) are shown, as a function of the monomer number n, $\gamma_{xxxx}^3(0)$ and $\chi_{xxxx}^3(0) = N\,\gamma_{(0)}^3 L$ where L is a local field factor taken equal to 1 by assuming that the polymer has a cylindrical shape.

Meanwhile several difficulties, now under examination, remain.

In polydiacetylene, for instance, experimentally $E_g - E_u$ = - 0,1eV is found when theoretical calculations give + 0,5eV. It is thought that SCF (CNDO) computations do not take enough into account the electronic correlations [10] which are able to change the relative position of the u and g levels.

Monomer monomeric unit number	excited level n°	sym	energy (eV)	Matrix element (Å)	Matrix element product	γ_0 10^{-48} Si	
Diacetylene	1 :	u	3.34	|0.1 > 3.5|	01210	52	
				|0.3 > 1.0|			
6	2 :	g	3.86	|1.2 >10.8|	01230	13,5	82.2
	3 :	u	4.46	|3.2 >11.6|	03230	13,4	
					03210	3,3	
Diacetylene	1 :	u	3.17	|0.1 > 4.9|	01210	597	
				|0.3 > 1.4|	01230		871
12	2 :	g	3.35	|1.2 > 2.2|	+03210	230	
	3 :	u	3.65	|3.2 > 2.3|	03230	44	
Thiophene	1 :	u	2.32	|0.1 > 3.4|	01210	67	68
	2 :	g	3.09	|0.4 > 0.7|	04240	1	
6	4 :	u	3.7	|1.2 > 8.3|			
				|4.2 > 7.8|			
Thiophene	1 :	u	2.08	|0.1 > 4.9|	01210	1083	
	2 :	g	2.39	|0.3 > 1.4|	03210+	532	
					01230		
12	3 :	u	2.77	|1.2 >16.9|	01410	117	1843
				|1.4 > 1.7|			
	4 :	g	3.13	|3.2 >18.2|	03230	63	
				|3.4 >18 |	03430	48	

Table II.

O = fundamental state. γ_{esu} = 7.16 10^{13} γ_{SI}
The Thiophene-Thiophene distance is arbitrarily fixed to 1.40Å

II.4.2. Noncentrosymmetrical polymer In the most general case the number of large matrix elements becomes large. In some cases nevertheless calculations are still possible. For instance a helicoïdal polythiophene can be described with a four level model and Table III shows the energy levels and matrix element values for several φ angles where φ is the angle between two successive monomer planes. On fig.(9) is shown the main absorption transition energy and the γ value per monomer (γ_M) as a function of φ. An important decrease of the γ_M coefficient is correlated with the polymer distorsion. Let us finally notice that in this noncentrosymmetrical poly-

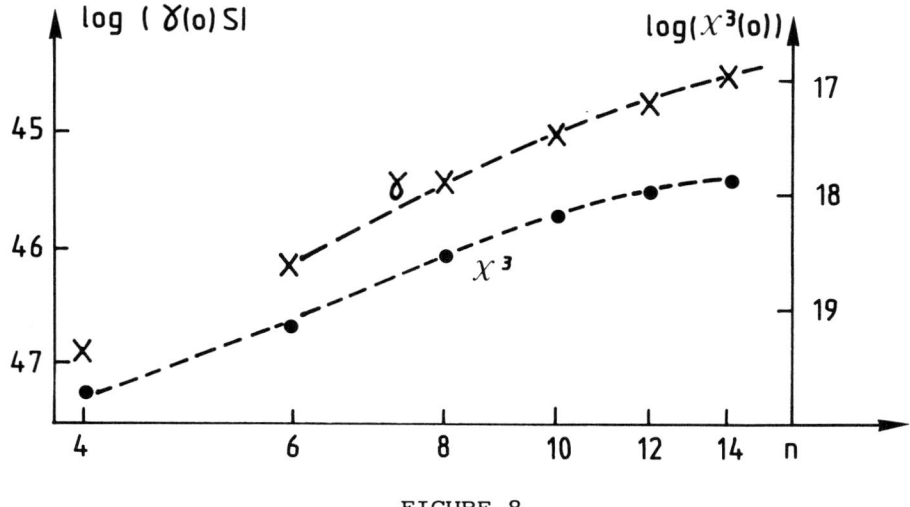

FIGURE 8.

Calculated values of $\gamma^3_{(0)}$ and $\chi^3_{(0)}$ as a function of the monomer number n in polythiophene.

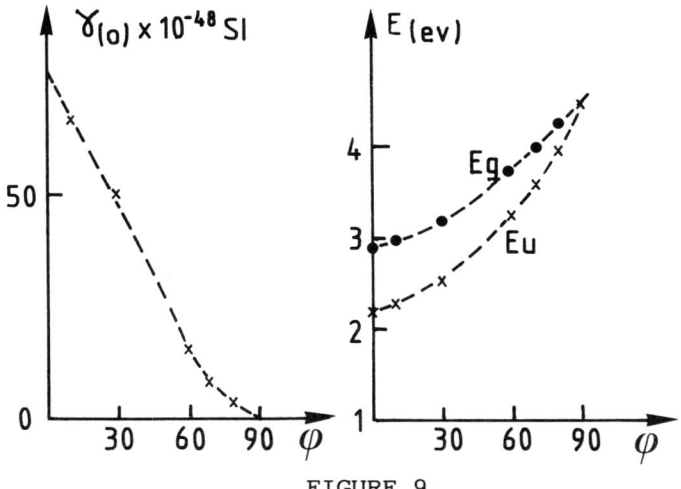

FIGURE 9.

Calculated transition energy and γ_m value as a function of the angle γ between two successive monomer planes.

mers, a two-level model, taking only into account the dipolar moment of excited and fundamental states, is unsuitable. The reason is that matrix elements such as $<g|x|g>, <\lambda|x|\lambda>$ where λ is an excited state are generally small compared with the other one: $<g|x|\lambda>, <\lambda|x|\lambda'>$ etc... We can say as a conclusion of this chapter that the time dependent perturbation calculations give a good approximation of the third order coefficients involved in the THG, EFISH and more generally nondegenerate four waves mixing phenomena. It is related to the fact that the main involved phenomena is only the electronic hyperpolarizability.

φ	excited level n°	sym	energy	Dipolar transition moment (Å)
	:			
0	2 :	u	2.21	|0.2 > 3.4
	4 :	g	2.93	|0.8 > 0.7
	8 :	u	3.59	|2.4 > 8.2
	:			|8.4 > 8.2
	:			
	2 :	u	2.27	|0.2 > 3.3
10	5 :	g	2.98	|0.8 > 0.7
	8 :	u	3.64	|2.5 > 8.1
	:			|8.5 > 8.2
	:			
	2 :	u	2.55	|0.2 > 3
30	4 :	g	3.16	|0.7 > 0.7
	7 :	u	3.77	|2.4 > 8.2
	:			|7.4 > 8.4
	:			
	:			
	2 :	u	3.29	|0.2 > 2.3
	5 :	g	3.73	|0.6 > 0.6
60	6 :	u	4.11	|2.5 > 8
	:			|6.5 > 7.3
	4 :	g	3.13	|3.2 >18.2
	:			|3.4 >18
	:			

Table III.- Cis-Hexathiophene

Excited energy elevel and main dipolar transition moment.
φ : angle between two successive monomer planes.

III. PHASE CONJUGATED WAVES EXPERIMENTS

Several phase conjugated waves experiments have been done in polymer solutions or gels [11].
The phase conjugated wave signal I_c observed in a polydiacetylene gel is shown on figs.(10) and (11)[12] as a function of the time delay Δt between the two writing laser beams P_1 and S and the reading pump P_2.
Near $\Delta t = 0$ a fast response is observed. Its width is very near the one of the two correlated laser beam P_1 and S ($\sim$ 40ps). That means that the time response of these phenomena is shorter than a few ps. For much larger Δt a periodic response appears linked to the acoustic waves generated by the two writing laser beams P_1 and S. On fig.(12) is shown I_c,

normalized with respect to the intensity of a reference CS_2 cell whose signal is proportional to I_p^3. The acoustic response, which is produced at $1.06\mu m$ by a two photon process, is proportional to I_p^5. On the contrary the fast component is proportional to I_p^3 at low intensity and can be described by a constant χ^3 coefficient. But, at higher intensity, due to the creation of excitons by a two-photon process the signal becomes proportional to I_p^5.

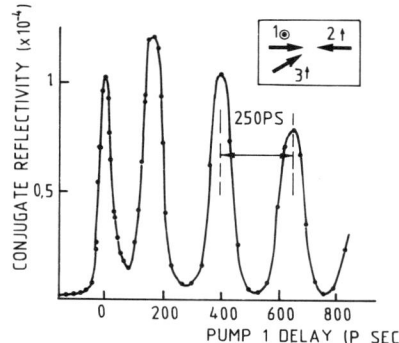

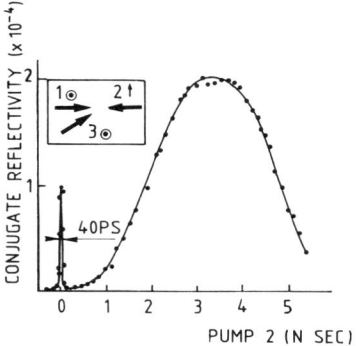

FIGURE 10.   FIGURE 11.

Phase conjugated signal evolution on delaying read pulse (1) in 14% yellow solution. Small spaced grating contribution. Inset depicts polarization of incoming beams.

Phase conjugated signal evolution on delaying read pulse in 14% yellow solution. Large spaced grating contribution. Inset depicts polarization of incoming beams.

Between this fast response and the slower acoustic response a slowly decreasing background is observed [13] at high intensity ($\sim 1GW/cm^2$). It is associated with the slow disappearance of charged species.

Among all these phenomena giving a phase conjugated signal many of them come from the one or two photons light absorption and are not phase correlated with the exciting laser beam. The same formalism as for nondegenerate four waves mixing experiments cannot be used.

Nevertheless F. Charra et al [14] have pointed out that it was possible to exactly solve the degenerate nonlinear response at any high electric field without using perturbative methods. This formalism can be very useful to derive the index variations induced by the electronic hyperpolarizability and also for optical Stark effect.

By using Floquet's theorem the solution of the Schrödinger's equation $i\hbar\frac{\partial\psi}{\partial t} = H(t)\psi(t)$ where $H(t) = H_0 + \mu E\cos\omega t$ can be written: $\psi(t) = \exp(-i\mathcal{E}t)\varphi(t)$ where $\mathcal{E}$ is a "quasi-energy" and $\varphi(t)$

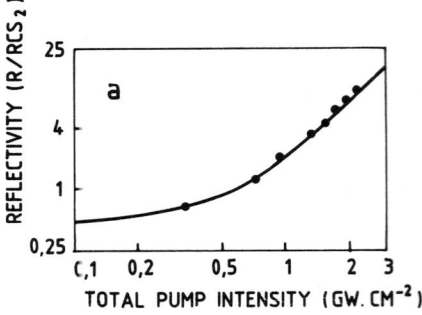

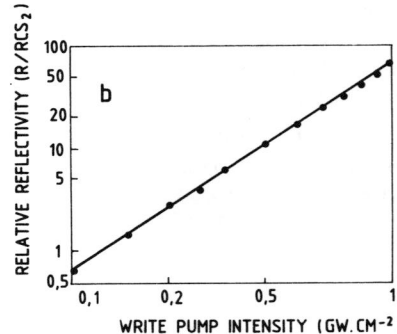

FIGURE 12.a

Relative zero-time conjugate reflectivity in blue gel vs total pump intensity $(I_1 + I_2)$. R=1 for CS_2 and pure chloroben zene gives an horizontal line.

FIGURE 12.b

Relative first acoustic oscillation (t=4.3ns) reflectivity in blue gel vs "write" pump intensity (I_1). At t=0, R=1 for CS_2.

is a periodic three elements column matrix for a three levels model. $\mathcal{E}$ can be obtained by diagonalizing the following infinite matrix

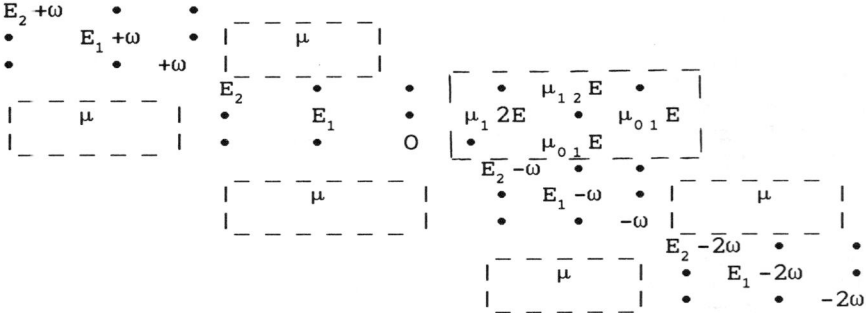

Fortunately this matrix can be truncated as soon as $n\omega$ (n integer) $\gg \mu E$ so that a finite one is obtained.
Fig.(13) shows $\mathcal{E}$ as a function of $I(\sim E^2)$. The α and γ coefficients can be deduced for any E value by the following relation $\alpha(I) = -\dfrac{\partial \mathcal{E}}{\partial I}$, γ being its curvature.
So it is possible to derive the high field refraction index values and also the optical Stark effect which matches experiments [14].

CONCLUSION
Third-order nonlinear coefficients come from a great variety of phenomena. Those arising from nonlinear electronic polarizability can be evaluated reasonably well using time dependent perturbative methods. In the case of degenerate four wave

mixing the nonlinear response can be derived exactly by using the Floquet's theorem for high electric field values. But when the nonlinear response depends on the number of excited species others calculations, taking explicitly into account their finite lifetime, must be considered.

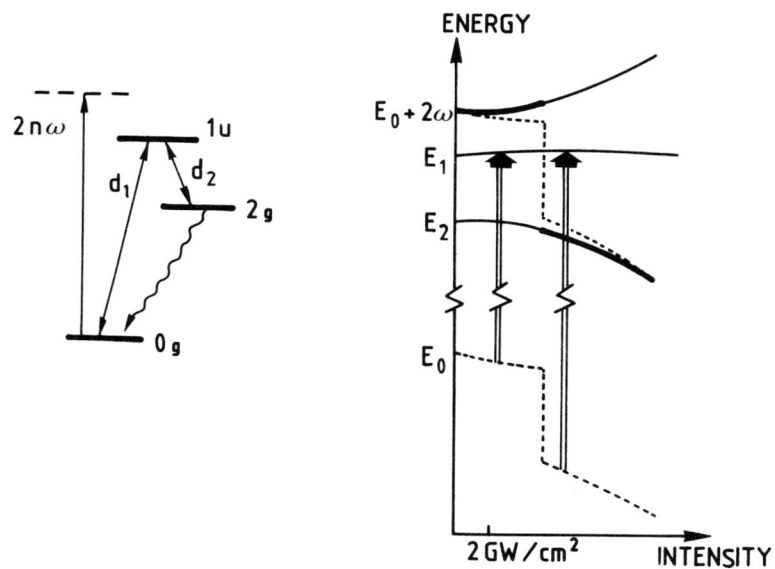

FIGURE 13.

Quasi energy as a function of the light intensity.

REFERENCES

1. C. SAUTERET, J.P. HERMANN, R. FREY, F. PRADERE, R.M. BAUGHMANN and R.R. CHANCE, Phys. Rev. Lett., 36, 5100 (1974).
2. K.C. RUSTAGI and J. DUCUING, Third-Order Polarizability of Conjugated Organic Molecules, Opt. Commun. 10, 258 (1974).
    J. DUCUING, in Nonlinear Spectroscopy, N. Bloemberger ed., North Holland Publ. Comp., Amsterdam 1977, p. 276.
    C. COJAN, G.P. AGRAWAL and C. FLYTZANIS, Optical Properties of One-Dimensional Semiconductor and Conjugated Polymers, Phys. Rev. B15, 909 (1977).
3. F. KAJZAR and J. MESSIER, Rev. Sci. Instrum. 58(11)(Nov. 1987), 2081.
4. F. KAJZAR, J. MESSIER and C. ROSILIO, J. Appl. Phys. 60(9)(1986), 3040.

60

5. P.A. CHOLLET, F. KAJZAR and J. MESSIER, Synth. Metals, 18 (1987), 459.
6. B.J. ORR and J.F. WARD, Mol. Physics, 20(3), 1971, 513. See also J.F. WARD, Rev. of Mod. Phys., 37(1)(1965), 1.
7. J.M. ANDRE, C. BARBIER, V. BODART and J. DELHALLE, Vol. 2. 137, Nonlinear Optical Properties of Organic Molecules and Crystals Quantum Electronics. Academic Press (edited by D.S. Chemla, J. Zyss).
8. C. FLYTZANIS, Synth. Metals, 18 (1987), Vol. 2, p. 121.
9. See also: A.F. GARITO Nonlinear Optical Properties of Polymers: Electron Correlation and Chain Conformation, in Nonlinear Optical Properties of Polymers (Materials Research Society, Boston, 1988), 109, p. 91. P.N. PRASAD. These Proceedings.
10. G.W. HAYDEN and E.J. MELE, Phys. Rev. B, 32, 10(1985), 6527. F. KAJZAR and J. FRIEDEL, Phys. Rev. B, 35, 18(1987), 9514.
11. W.M. DENNIS, W. BLAU and D.J. BRADLEY, Appl. Phys., Lett., 47(3), 1985, 200. W.M. DENNIS, W. BLAU, IEEE Proceedings 133, 1, 1986, 91.
12. J.M. NUNZI and D. GREC, J.A.P. 62(6), 1987, 2198.
13. J.M. NUNZI and F. CHARRA, Proceedings of the Congress on "Organic Materials for Nonlinear Optics". Oxford 28-29 June 1988.
14. F. CHARRA, J.M. NUNZI, Proceedings of the Congress on "Organic Materials for Nonlinear Optics".

RECENT NONLINEAR OPTICAL STUDIES OF MNA AND CONJUGATED LINEAR CHAINS

C. GROSSMAN, J. R. HEFLIN, K.Y. WONG, O. ZAMANI-KHAMIRI, AND
A. F. GARITO

Department of Physics and Laboratory for Research on the Structure of Matter,
University of Pennsylvania, Philadelphia PA 19104

I.   INTRODUCTION

There have been a large number of studies concerning the nonlinear optical properties of organic and polymeric materials in recent years [1]. Compared to the usual inorganic nonlinear optical materials, organic and polymeric materials with conjugated bonding structures possess much larger second and third order electronic nonlinear optical susceptibilites [1,2]. It is now well-established that the macroscopic second order $\chi_{ijk}^{(2)}(-\omega_3;\omega_1,\omega_2)$ and third order $\chi_{ijkl}^{(3)}(-\omega_4;\omega_1,\omega_2,\omega_3)$ nonlinear optical susceptibilities observed for conjugated organic and polymer structures are due to virtual excitations occurring within the π-electron states of individual molecular, or polymeric chain, sites in condensed assemblies and providing macroscopic sources of nonlinear optical response through the corresponding on-site microscopic second order $\beta_{ijk}(-\omega_3;\omega_1,\omega_2)$ and third order $\gamma_{ijkl}(-\omega_4;\omega_1,\omega_2,\omega_3)$ susceptibilities [3 - 7]. Moreover, since the responses originate from on-site electronic virtual excitations, the nonlinear response time is ultrafast of the order of picosecond to femtosecond time scales. Microscopic descriptions of second order and third order π-electron virtual excitation processes are required in order to understand at a fundamental level the physical properties and behavior of $\beta_{ijk}(-\omega_3;\omega_1,\omega_2)$ and $\gamma_{ijkl}(-\omega_4;\omega_1,\omega_2,\omega_3)$. In addition, they provide important insights for developing guidelines for molecular design of new material structures.

A general theoretical method for the calculation of the nonresonant $\beta_{ijk}(-\omega_3;\omega_1,\omega_2)$ and $\gamma_{ijkl}(-\omega_4;\omega_1,\omega_2,\omega_3)$ of organic structures has been successfully developed that is based on a direct summation method of the respective analytical expressions obtained from quantum electrodynamics [7]. The nonresonant susceptibility is purely electronic in origin since the frequencies of the optical input and output are well above molecular vibrational and rotational modes but below electronic resonances. Each susceptibility is evaluated with the excitation

J. Messier et al. (eds.), Nonlinear Optical Effects in Organic Polymers, 61–78.
© *1989 by Kluwer Academic Publishers.*

energies and dipole moments contained in each term calculated using self-consistent field configuration interaction theory. The theory explicitly takes into account electron-electron Coulomb interactions and properly describes electron correlation.

The direct summation method actually proceeds in three stages where, at each stage, theory is compared with experiment. To summarize, in stage I, the molecular basis sets for the configurations are first obtained by an all valence electron self-consistent field molecular orbital (MO) method in the rigid lattice CNDO/S approximation. The required bond lengths and bond angles are obtained from x-ray and electron diffraction data. The basis sets are evaluated for each structure by comparing the calculated one-electron photoelectron spectrum with available experimental emission data.

In stage II, the optical excitation energies and dipole moments contained in each term of the higher order susceptibilities are calculated by configuration interaction theory. With the configuration interaction basis sets from stage I, singly (SCI) and doubly (DCI) excited configurations are included, and the calculated singlet-singlet excitation spectra are compared with experimental optical absorption and emission data.

In stage III, the individual terms contained in the summations of the analytical expression for $\beta_{ijk}(-\omega_3;\omega_1,\omega_2)$ and $\gamma_{ijkl}(-\omega_4;\omega_1,\omega_2,\omega_3)$ are directly evaluated from the singlet-singlet excitation energies and transition dipole moments obtained in stage II. The calculated magnitude, sign, and frequency dependence of $\beta_{ijk}(-\omega_3;\omega_1,\omega_2)$ and $\gamma_{ijkl}(-\omega_4;\omega_1,\omega_2,\omega_3)$ are directly compared with experimental results.

The purpose of this paper is to summarize our recent efforts in the theoretical understanding of the microscopic second and third order responses of organic and polymer structures by reviewing results from two recent studies. The first concerns the theoretical understanding of the macroscopic second order susceptibility of 2-methyl-4-nitroanline (MNA) single crystals (Figures 1 and 2). Although MNA, which is a well known model organic nonlinear optical material, has been under study for many years and the microscopic second order molecular response β_{ijk} is well-understood [4,8], a corresponding understanding at the same level for the crystal phase is still lacking. We approach this problem by both experimental and theoretical studies whose results reveal that the macroscopic crystal response $\chi_{ijk}^{(2)}$ can indeed be described by the sum of responses β_{ijk} from individual units. However, we have found that the individual unit actually consists of an MNA-MNA pair instead of a single MNA site. This is basically a result of the close proximity of pairs of MNA sites within the crystal unit cell with intermolecular coupling apparently taking place.

The second part of this paper concerns the third order susceptibility $\gamma_{ijkl}(-\omega_4;\omega_1,\omega_2,\omega_3)$ for centrosymmetric polyene chains. We begin with a review of recent results for centrosymmetric chains in which it was found that the third order response of these structures is properly described by a theory which takes explicit account of electron correlation. Highly

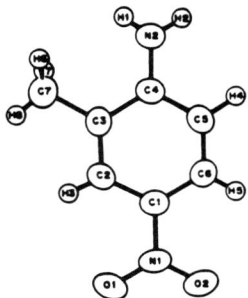

Figure 1   The molecular structure of MNA.

correlated π-electron virtual excitations to high-lying two-photon ^1A_g states turn out to be the dominant contributions to $\gamma_{ijkl}(-\omega_4;\omega_1,\omega_2,\omega_3)$. We will summarize results for $\gamma_{ijkl}(-\omega_4;\omega_1,\omega_2,\omega_3)$ both as a function of structural conformation and of chain length.

II.   MICROSCOPIC ORIGIN OF THE LARGE d_{11} COEFFICIENT OF MNA

In general, the large macroscopic $\chi^{(2)}_{ijk}$ of conjugated organic and polymeric materials are due to highly correlated, π-electron virtual excitations within these structures. The π-electron excitations occur on individual molecular sites that act as macroscopic sources of second order $\chi^{(2)}_{ijk}$ response through the corresponding on-site microscopic susceptibility β_{ijk}. Thus, for organic single crystals, with individual sites orientationally fixed in the lattice, we can relate[9] the macroscopic second harmonic susceptibility $\chi^{(2)}_{ijk}(-2\omega;\omega,\omega)$ to the microscopic susceptibility $\beta_{ijk}(-2\omega;\omega,\omega)$ by the expression

$$\chi^{(2)}_{ijk}(-2\omega;\omega,\omega) = N_u \sum_{s=1}^{p} R^s_{im'} R^s_{jn'} R^s_{ko'} f^{2\omega}_{m'i'} \beta^s_{i'j'k'}(-2\omega;\omega,\omega) f^{\omega}_{j'n'} f^{\omega}_{k'o'} \tag{1}$$

where N_u is the number of unit cells per unit volume, and the summation is over all p molecules in a unit cell. Each molecule indexed by s in the unit cell usually has a different orientation, and thus R^s_{ia} is the rotation matrix from the body axis of the s molecule in the unit cell to the laboratory axis. $\beta^s_{i'j'k'}$ is the microscopic susceptibility tensor component corresponding to the body axis of the s molecule, and f is the local field factor. The macroscopic nonlinear optical susceptibilty of an organic, or polymer, single crystal is thus expressed in terms of the corresponding microscopic susceptibility, and the molecular arrangement within an unit cell which is dictated by the crystal symmetry group.

Single crystal MNA has long been known to possess unusually large second harmonic d_{11} and linear electrooptic r_{11} coefficients due to the parallel alignment of individual dipolar MNA sites along a unique polar axis[10]. Although MNA has been extensively studied, a major outstanding issue has remained the specific relationship between the macroscopic $\chi^{(2)}_{ijk}$ and microscopic β_{ijk} tensor components. The basic approach in recently completed studies in our laboratories was to experimentally determine the frequency dependent second harmonic coefficient d_{11} and concurrently perform theoretical calculations of the same quantity based on $\beta_{ijk}(-2\omega;\omega,\omega)$ and the x-ray determined crystal structure[10]. For the determination of d_{11}, we performed measurements at five fundamental frequencies well below the first electronic reasonance using the wedged Maker fringe method[11]. High optical quality single crystals of MNA were grown from vapor by sublimation of pure powder in which some of the crystals naturally formed as wedge shaped crystals that proved convenient for the SHG measurements. The five different frequencies of light for the dispersion measurement were obtained using stimulated vibrational and rotational Raman scattering from a pressurized H_2 gas cell pumped by a Q-switched Nd:YAG laser. The experimental set up consisted of separate sample and reference arms with the reference arm serving to compensate for any laser power fluctuations. Detailed descriptions of the experimental design are given in Ref. 12.

A theoretical description of the nonlinear optical susceptibility depends on knowledge of the excitation energies, dipole moments, and transition moments of the molecule as well as a local field description that takes into account the crystal structure[13]. The simplest models for a molecular crystal assume that the molecular sites do not interact except for their dipole contribution to the local field. This approximation does not take into account any change in molecular dipole and transition moments due to interactions between nearest neighbor molecules. We found that in order to relate the measured macroscopic d_{ij} components to calculated microscopic β_{ijk} components, we must treat the microscopic unit as a coupled pair of neighboring MNA molecular sites. By choosing a pair of closely oriented MNA molecules as the actual basis unit responsible for the microscopic response, this structure makes up the basis of the crystal lattice as shown in Figure 2. In treating the microscopic unit as a coupled MNA-MNA pair, the local field is calculated for a microscopic polarization at each pair site. The difference between the results of these calculations and those of single MNA sites is that the states of the coupled MNA-MNA pair take explicit account of the mutual dipole fields between neighboring MNA units.

In dc-induced second harmonic generation (DCSHG) experimental studies of MNA molecules, the measured frequency dependent vector part $\beta_x(-2\omega;\omega,\omega)$ of the microscopic susceptibility is given by

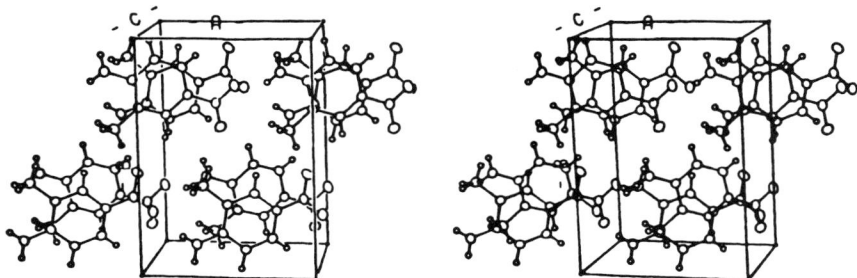

Figure 2. The molecular orientation within the MNA crystal unit cell. Each site of the Bravais lattice has a basis of two closely oriented molecules. (from Ref. 10)

$$\beta_x = \beta_{xxx} + \frac{1}{3}\,(\beta_{xyy} + \beta_{xzz} + 2\beta_{yyx} + 2\beta_{zzx}) \qquad (2)$$

β_x and the gas phase singlet-singlet absorption spectrum were reported earlier [4] (Figure 3). The largest contribution to $\beta_x(-2\omega;\omega,\omega)$ comes from the tensor component $\beta_{xxx}(-2\omega;\omega,\omega)$. The DCSHG results for MNA served as a reference point for theoretically analyzing the frequency dependent principal component $\beta_{xxx,lattice}(-2\omega;\omega,\omega)$ of the microscopic unit for the MNA crystal lattice site. Based on the results of this study, the frequency dependent components $\beta_{xxx}(-2\omega;\omega,\omega)$ and singlet-singlet excitation spectrum for an MNA-MNA pair were theoretically calculated using the complete expression[3]

$$\beta_{xxx} = \frac{-e^3}{2\hbar^2}\left\{\; \sum_{\substack{n\neq m \\ n\neq g \\ m\neq g}} x_{gn}\,x_{nm}\,x_{mg}\,[\;(\omega_{mg}-\omega)^{-1}(\omega_{ng}-2\omega)^{-1} \right.$$

$$+ (\omega_{mg}+\omega)^{-1}(\omega_{ng}+2\omega)^{-1} \;+\; (\omega_{mg}+\omega)^{-1}(\omega_{ng}-\omega)^{-1}\;]$$

$$+ \sum_n [x_{gn}\,x_{gn}\,\Delta x_n\,(\omega_{ng}^2-4\omega^2) + 2x_{gn}x_{gn}\Delta x_n(\omega_{ng}^2+2\omega^2)]$$

$$\left. x\;(\omega_{ng}^2-\omega^2)^{-1}(\omega_{ng}^2-4\omega^2)^{-1}\;\right\} \qquad (3)$$

where $x_{gn} = \langle g|x|n\rangle$, $\Delta x_n = x_{nn}-x_{gg}$, and $\hbar\omega_{ng}$ corresponds to the excitation energy of the state n. The singlet excitation energies and the transition dipole moments x_{nm} are obtained from the self-consistent field configuration interaction theory[7]. The calculations do not take

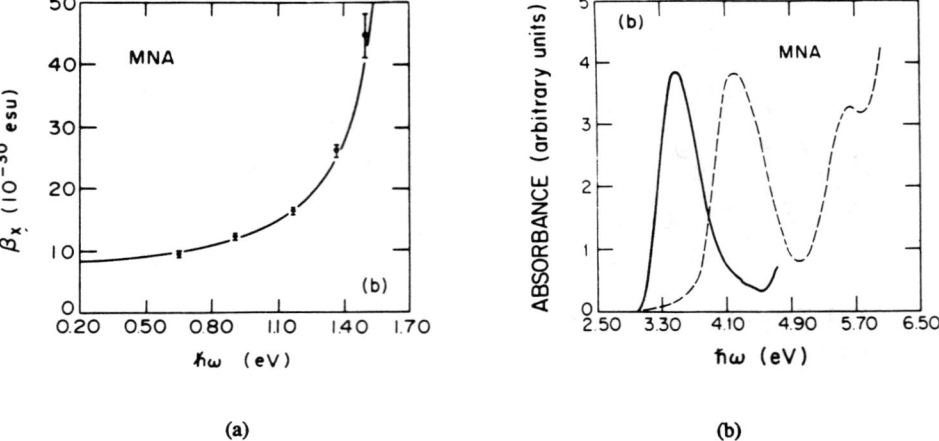

Figure 3. a) The calculated and measured dispersion of β_{ijk} of MNA molecule b) Optical absorption spectra of MNA dissolved in 1,4-dioxane(solid line) and in gas phase(dashed line) (from Ref. 4)

into account the static local field which a molecule experiences in a polar medium. In a molecular crystal such fields are due to the ensemble of ground state dipole moments. To the lowest approximation this field shifts the excited state energies via an electric dipole perturbation. This effect is clearly observed from the difference in the peaks of the gas phase and crystal phase absorption spectra. The lowest energy gas phase absorption peak is centered around 4.22 eV and corresponds to the calculated excited state with energy 4.24 eV. However, in the crystal data the peak is centered around 2.89 eV. This dielectric shift is consistent with a calculation of an electric dipole shift of the first excited state to ground state transition, $-(\mu_{11}-\mu_{gg})E$, where E is the static local field derived from the ground state dipole moment (see the discussion below on local fields). The shift of higher energy excited states was also calculated using the same static field.

Self consistent field configuration interaction calculations were carried out both for the single MNA molecule and the MNA-MNA molecular pair cases in an attempt to elucidate the effect of possible interactions between sites. A comparison of CI excited states would not be particularly illuminating since the single molecule and molecular pair systems are so different, but the net effect on the calculation of β from Eq. (3) illustrates an important difference in the two models. The susceptibility of an independent single MNA site is roughly the same as the interacting MNA-MNA pair unit both far from and near the first electronic resonance. Thus the calculated macroscopic susceptibility based on single independent MNA molecule would be roughly twice as large as that based on the interacting MNA-MNA molecular pair. This is a clearly distinguishable result and the comparison to experiment will show that interactions between MNA sites may be important.

The relation between the microscopic and macroscopic susceptibilities depends on the evaluation of the microscopic local field. We consider two contributions to the local field; one from the bulk uniform polarization, which is written as $\frac{4\pi}{3}$ P, and another from the dipole moments of nearby lattice sites. The latter can be written as $\sum_i L_i m_i$ where the sum is over the nearby lattice sites, i, and m_i is the dipole moments of the ith site. The L_i is the matrix operator that generates the electric field at the site under consideration. The dipole field operator, L_i, only depends on the position vector from the lattice site to the field point and is easily calculated from the lattice constants reported in Ref. 10. The local field is then written as

$$E^{local} = E + \frac{4\pi}{3} P + \sum_i L_i m_i \qquad (4)$$

where E is the applied Lorentz field. Since our microscopic unit is the lattice basis, it is assumed that all units have the same dipole moments and m_i can be removed from the sum. By writing m_i in terms of the microscopic linear optical susceptibility α, defined from $m_i = \alpha E^{local}$, Eq. (4) can be rewritten

$$E^{local} = (I - \sum_i L_i \alpha)^{-1} (E + \frac{4\pi}{3} P) \qquad (5)$$

where I is the unit matrix. From the definiton of the dielectric tensor ε,

$$P = \frac{\varepsilon - 1}{4\pi} E \qquad (6)$$

and the relation $P = Nm = NE^{local}$ (where N is the lattice site density), the Clausius-Mossotti relation, which directly relates α, N, and ε, is derived with a modification that includes the fields of neighboring dipoles,

$$\alpha = [\sum_i L_i + (\frac{\varepsilon + 2I}{3}) (\frac{\varepsilon - I}{4\pi N})^{-1}]^{-1} \qquad (7)$$

The advantage of Eq. (7) is that $\alpha(\omega)$ can be calculated if the dielectric tensor is known at the particular frequency ω. The local field is then calculated from Eq. (5) and the local field factor tensor, f^ω, is defined by

$$E^{local}(\omega) = [I - \sum_i L_i \alpha(\omega)]^{-1} \frac{\varepsilon(\omega) + 2}{3} E(\omega) = f^\omega E(\omega) \qquad (8)$$

The macroscopic SHG tensor is calculated from f^ω and β_{ijk} (see Ref. 13),

$$d_{lmn} = N \, \beta_{ijk} \, f^{2\omega}_{il} \, f^{\omega}_{jm} \, f^{\omega}_{kn} \qquad (9)$$

where d_{lmn} is the equivalent of d_{pq} written in the reduced matrix notation (i.e., $p = l$ and mn reduce to q). The sum of L_i over lattice sites was calculated from the MNA crystal lattice constants and was found to converge, within a radius of 15 sites from the field point, to the value (in units of Å^{-3})

$$\sum_i L_i = \begin{matrix} -2.192\times10^{-3} & 0.0 & -7.584\times10^{-5} \\ 0.0 & 3.190\times10^{-3} & 0.0 \\ -7.584\times10^{-5} & 0.0 & -9.983\times10^{-4} \end{matrix} \qquad (10)$$

The calculation of d^{MNA}_{11} was carried out in four steps. First, $\beta_{xxx}(2\omega;\omega,\omega)$ was calculated from Eq. 3 by the direct summation method[7]. Then $\varepsilon(\omega)$ and $\varepsilon(2\omega)$ were calculated from a Sellmeier dispersion relation[12] and the tensors f^{ω} and $f^{2\omega}$ were calculated from Eqs. (7) and (8). Finally, d^{MNA}_{11} was calculated from Eq. (9). The results of this calculation, with $\hbar\omega$ varying from 0.5 to 1.2 eV, are plotted in Figure 4 along with the average

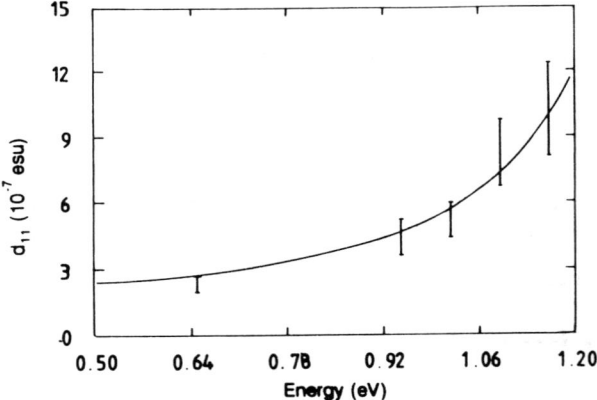

Figure 4. The calculated and measured dispersion of d^{MNA}_{11} plotted as a function of fundamental photon energy. (from Ref. 12)

measured values of d_{11}^{MNA}. The vertical bars are the measured values, and the solid line is calculated from the electronic states of the MNA-MNA pair in the crystal lattice. Within experimental uncertainty, the molecular pair analysis and the experimental results agree satisfactorily. A similar calculation based on a single MNA site was carried out, and we found that the theoretical results differ widely from the experimentally determined values. Thus, within the rigid lattice gas model, the basic microscopic unit appears to be an MNA-MNA coupled pair and not a single MNA site.

We have shown that in the case of an organic molecular crystal, as exemplified by MNA, it may be important to incorporate the mutual interaction of nearest neighbor molecules within a lattice in the calculation of the nonlinear optical susceptibility. This finding is contrary to conventional assumptions that in an organic molecular crystal, the nonlinear optical susceptibility of the macroscopic medium is the direct summation of the nonlinear response of individual molecules, with residual inteactions between molecular sites being taken account by the local field correction. By incorporating nearest neighbor intermolecular interactions through the treatment of coupled pairs of molecules as individual sources of nonlinear response while at the same time taking account of the lattice local field correction and dielectric energy shift, we have found satisfactory agreement between experiment and theory for the crystal second order optical susceptibility. We believe that this analysis should have equal significance in the study of other related systems such as side chain polymers and molecular doped polymers which incorporate dipolar molecular units as pendant side groups or as molecular dopants in isotropic polymers. In these cases, the intermolecular distance may be close enough such that the coupling between neighboring molecular sites cannot be neglected in first order. The analysis we have just summarized should prove useful in theoretical predictions of the macroscopic nonlinear response in terms of the well established SCF-MO-CI procedure that we have been developing for highly accurate descriptions of the microscopic susceptibilities β_{ijk} and γ_{ijkl}.

III. THIRD ORDER OPTICAL PROPERTIES OF CONJUGATED LINEAR CHAINS

The simplest hydrocarbon conjugated linear chains are polyenes in which each carbon site is covalently bonded to its nearest neighbor two carbon atoms and single hydrogen atom (Figure 5). The remaining valence electron of each carbon atom contributes to a delocalized π-electron backbone along the chain having an alternating single-bond/double-bond bonding structure. For description of these systems[15], important evidence for the inadequacy of independent particle theories such as Huckel theory has been provided by the experimental discovery[16] that below the one-photon 1^1B_u π-electronic state exists a two-photon 2^1A_g state in direct contradiction with these simple theories. It was then demonstrated[17 - 21] that a proper account of the electron correlation resulting from the electron-electron interaction was

Figure 5. Schematic diagrams of the molecular structures of (a) *trans*-octatetraene and (b) *cis*-octatetraene.

required in order to obtain the correct ordering of these two states.

We have recently presented[22, 23] a many-electron theory of $\gamma_{ijkl}(-\omega_4;\omega_1,\omega_2,\omega_3)$ for a large range of chain lengths and two structural conformations of polyenes, and our results are in good agreement with available experimental data. Using the direct summation over states expression for $\gamma_{ijkl}(-\omega_4;\omega_1,\omega_2,\omega_3)$ derived from time-dependent perturbation theory[24, 25], we identified the most significant transitions in the virtual excitation process and calculated the dispersion of the third harmonic susceptibility $\gamma_{ijkl}(-3\omega;\omega,\omega,\omega)$ and the dc-induced second harmonic susceptibility $\gamma_{ijkl}(-2\omega;\omega,\omega,0)$ for various chain lengths of the all-*trans* and *cis-transoid* conformations (hereafter referred to as *trans* and *cis*, respectively) of polyenes (Figure 5). The molecular electronic third harmonic susceptibility $\gamma_{ijkl}(-3\omega;\omega,\omega,\omega)$ is defined by the expression

$$p_i^{3\omega} = \gamma_{ijkl}(-3\omega;\omega,\omega,\omega)\ E_j^{\omega}E_k^{\omega}E_l^{\omega} \tag{11}$$

where $p_i^{3\omega}$ is a component of the molecular polarization induced at a frequency of 3ω in response to the cube of an electromagnetic field E^ω oscillating at a frequency of ω. From time-dependent, quantum electrodynamic perturbation theory one obtains the analytic expression

$$\gamma_{ijkl}(-3\omega;\omega,\omega,\omega) = \frac{1}{3!}\left(\frac{e^4}{4\hbar^3}\right) \sum_{n_1 n_2 n_3} \Bigg\{ \frac{P_{jkl}[r_{gn_3}^i\ r_{n_3 n_2}^j\ r_{n_2 n_1}^k\ r_{n_1 g}^l]}{(\omega_{n_3 g}-3\omega)(\omega_{n_2 g}-2\omega)(\omega_{n_1 g}-\omega)}$$

$$+ \frac{P_{jkl}[r_{gn_3}^j\ r_{n_3 n_2}^i\ r_{n_2 n_1}^k\ r_{n_1 g}^l]}{(\omega_{n_3 g}+\omega)(\omega_{n_2 g}-2\omega)(\omega_{n_1 g}-\omega)} \;+\; \frac{P_{jkl}[r_{gn_3}^j\ r_{n_3 n_2}^k\ r_{n_2 n_1}^i\ r_{n_1 g}^l]}{(\omega_{n_3 g}+\omega)(\omega_{n_2 g}+2\omega)(\omega_{n_1 g}-\omega)}$$

$$+ \frac{P_{jkl}[r_{gn_3}^j\ r_{n_3 n_2}^k\ r_{n_2 n_1}^l\ r_{n_1 g}^i]}{(\omega_{n_3 g}+\omega)(\omega_{n_2 g}+2\omega)(\omega_{n_1 g}+3\omega)} \Bigg\} \tag{12}$$

where $r_{n_1n_2}^i$ is the matrix element $<n_1|r^i|n_2>$, $\hbar\omega_{n_1g}$ is the excitation energy of state n_1, and P_{jkl} denotes the sum over all permutations of j, k, and l ensuring that γ_{ijkl} is independent of the ordering of these three indices. The convention has been chosen that the electric fields are represented as $E^\omega \sin(\omega t - k \cdot r)$. A similar expression can be derived for the dc-induced second harmonic generation susceptibility $\gamma_{ijkl}(-2\omega;\omega,\omega,0)$ as well as for the other fundamental third order optical processes. The individual terms of Eq. (12) were directly evaluated from the singlet state excitation energies and transition dipole moments obtained by configuration interaction methods in which all singly (SCI) and doubly (DCI) excited π-electron configurations were included in order to describe electron correlations properly. The values calculated at the SDCI level of the excitation energies for both the one-photon allowed 1^1B_u state and the two-photon allowed 2^1A_g state are in good agreement with experimental values and with previously reported theoretical results[15,17,19].

For all chain lengths and conformations studied, the γ_{xxxx} component of $\gamma_{ijkl}(-3\omega;\omega,\omega,\omega)$ with all fields along the direction of conjugation, is far larger than the others, as expected. This is a direct consequence of the one-dimensional delocalization of the π-electrons along the chain. For the specific case of *trans*-octatetraene (*trans*-OT) with number of carbon sites N=8, the independent tensor components of $\gamma_{ijkl}(-3\omega;\omega,\omega,\omega)$ at a nonresonant fundamental photon energy of 0.65 eV (λ=1.907 μm) are γ_{xxxx}=15.5, γ_{xyyx}=0.6, γ_{yxxy}=0.5, and γ_{yyyy}=0.2x10^{-36} esu.

For centrosymmetric conjugated chains, the π-electron states have definite parity of A_g or B_u, and the one-photon transition moment vanishes between states of like parity. Since the ground state is always ^1A_g, it is evident from Eq. (12) that the π-electron states in a third order process must be connected in the series $g \rightarrow {}^1B_u \rightarrow {}^1A_g \rightarrow {}^1B_u \rightarrow g$. Virtual transitions to both one-photon ^1B_u and two-photon ^1A_g states are necessarily involved. In the summation over intermediate states for *trans*-OT there are two major terms which constitute 70% of γ_{xxxx}. In both of these terms, the only ^1B_u state involved is the dominant low-lying one-photon 1^1B_u π-electron excited state. In addition to its low energy, the importance of this state lies in the large value of the x-component of its transition dipole moment with the ground state of 7.8 D. This is the largest of all transition moments involving the ground state. For one of the two major terms, the intermediate ^1A_g state is the ground state itself; but for the other, it is the 6^1A_g state of OT calculated at 7.2 eV. Since this state has a transition moment with 1^1B_u of 13.2 D, it is much more significant than the 2^1A_g state which has a corresponding transition moment of only 2.8 D.

Figure 6 displays the calculated dispersion curve for $\gamma_{xxxx}(-3\omega;\omega,\omega,\omega)$ of *trans*-OT as a function of the input photon energy. The first resonance located at 1.47 eV (λ = 0.84 μm) and indicated by the vertical dash in the figure, is due to the 3ω resonance of the 1^1B_u state. The second singularity located at 2.08 eV (λ = 0.60 μm), is from the 2ω resonance of the 2^1A_g

72

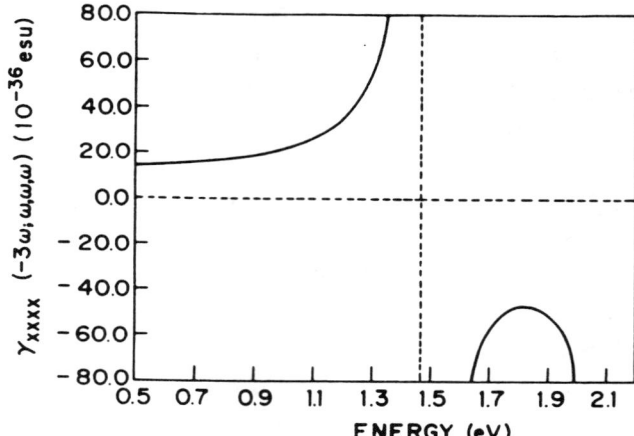

Figure 6. Dispersion of $\gamma_{xxxx}(-3\omega;\omega,\omega,\omega)$ for *trans*-OT. The vertical dash locates the 3ω resonance to the 1^1B_u state. (from Ref. 23)

state. As seen in Eq.(12), the ^1B_u states will have both 3ω and ω resonances in third harmonic generation, whereas the ^1A_g states will have only 2ω resonances. It should also be noted that in real systems, natural broadening of electronic states will prevent divergence at the resonances.

The results for the *cis* conformations are in direct analogy to those for *trans*. For the range of chain lengths we have considered, we find that the transition energies of the 1^1B_u and 2^1A_g states of the *cis* conformations are slightly red-shifted from the values for *trans* by 0.02 to 0.10 eV with the shift monotonically increasing with increased chain length. Just as for *trans*, the dominant tensor component of $\gamma_{ijkl}(-3\omega;\omega,\omega,\omega)$ is γ_{xxxx}, and the most significant virtual transitions and the dispersion are essentially the same as discussed above. The mechanism for $\gamma_{ijkl}(-3\omega;\omega,\omega,\omega)$ is still a symmetry-dictated virtual excitation process involving strongly correlated π-electron states.

The similarity between the two conformations is further emphasized in the transition density matrix contour diagrams which graphically illustrate the electron redistribution upon virtual excitation. The transition density matrix $\rho_{nn'}$ is defined through the expression

$$<\mu_{nn'}> = - e \int r\rho_{nn'}(r)dr \qquad (13)$$

with

$$\rho_{nn'}(r_1) = \int \psi_n^* (r_1, r_2, \ldots r_M) \, \psi_{n'}(r_1, r_2, \ldots r_M) dr_2 \ldots dr_M \qquad (14)$$

where M is the number of valence electrons included in the molecular wavefunction. Contour diagrams for $\rho_{nn'}$ of the ground and 6^1A_g states with the 1^1B_u state for both the *cis* and *trans* forms of OT are compared in Figure 7 where solid and dashed lines correspond to increased and decreased charge density. The (a) *cis* and (b) *trans* virtual $g \rightarrow 1^1B_u$ transition results in a somewhat modulated redistribution of charge with transition moment x-components of 7.9 and 7.8 D, respectively. For the $1^1B_u \rightarrow 6^1A_g$ virtual transition, however, there is a resultant highly separated charge distribution in the (c) *cis* and (d) *trans* conformations. The corresponding transition moments are 12.0 and 13.2 D for the *cis* and *trans* cases, respectively.

Gas phase third order susceptibility measurements of $\gamma_g(-2\omega;\omega,\omega,0)$ for polyenes have been obtained[26] using dc-induced second harmonic generation (DCSHG). At the nonresonant fundamental input of 1.787 eV ($\lambda = 0.694 \, \mu m$), the values for butadiene (BD) with N=4 and hexatriene (HT) with N = 6 are[27] 3.45 ± 0.20 and $11.30 \pm 1.05 \times 10^{-36}$ esu, respectively. Although the BD gas was more than 99% *trans*-BD, the HT was believed to contain as much as 40% *cis*-HT[26,28]. Based on the results of our calculations described above and the appropriate expression for $\gamma_g(-2\omega;\omega,\omega,0)$ with isotropic averaging

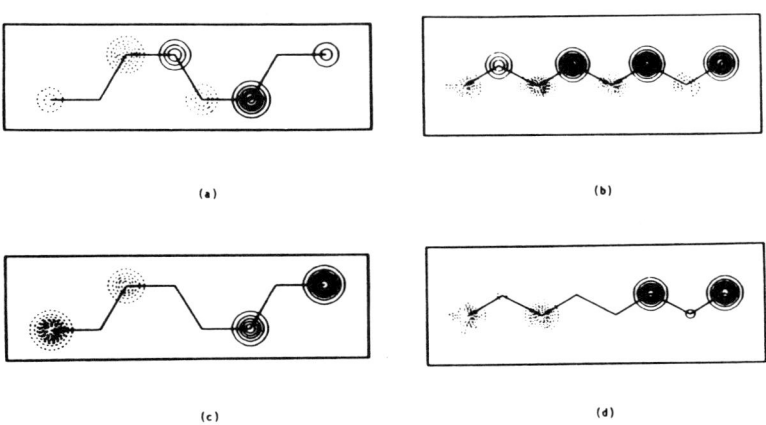

(a)

(b)

(c)

(d)

Figure 7. Transition density matrix contour diagrams for (a) *cis* and (b) *trans* ground states with the 1^1B_u state and (c) *cis* and (d) *trans* 6^1A_g states with the 1^1B_u state of OT. (from Ref. 22 and 23)

$$\gamma_g = \frac{1}{5}[\sum_i \gamma_{iiii} + \frac{1}{3} \sum_{i \neq j} (\gamma_{iijj} + \gamma_{ijij} + \gamma_{ijji})] \tag{15}$$

where the indices i and j represent the Cartesian coordinates x, y, and z, we calculate the π-electron contributions at the same frequency to be 2.1, 11.5, and 9.1 x 10^{-36} esu for BD, *trans*-HT, and *cis*-HT, respectively. Although the σ-electron contribution to γ_g is negligible for longer chains since it increases much more slowly with respect to chain length than the π-electron contribution, it is more signficant for shorter chains and should be included in the cases of BD and HT. After adding in respective σ-contributions of 1.5 and 2.4 x 10^{-36} esu as estimated in Ref. 26, we obtain values for $\gamma_g(-2\omega;\omega,\omega,0)$ of 3.6 and 12.9 x 10^{-36} esu for BD and HT, respectively, in agreement with experiment both in sign and magnitude.

Both the *trans* and *cis* conformations exhibit power law dependences of γ_{xxxx} on the number of carbon sites N with large exponents. For *trans*, the exponent is 5.4 ± 0.2, and for *cis* it is 4.7 ± 0.2. In addition to this smaller exponent for *cis*, for chains with equal numbers of sites, the value of $\gamma_{xxxx}(-3\omega;\omega,\omega,\omega)$ at a fixed frequency is in all cases smaller for the *cis* chain than *trans*. However, when γ_{xxxx} at 0.65 eV is plotted against the actual length of the chain L rather than N, as is done on the lower scale of Figure 8, it becomes clear that the smaller values

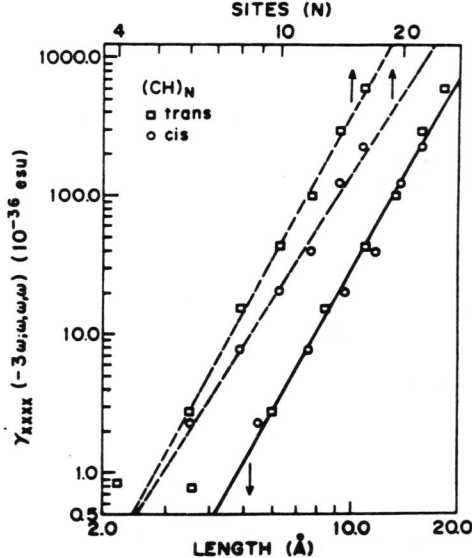

Figure 8. Log-log plot of $\gamma_{xxxx}(-3\omega;\omega,\omega,\omega)$ at 0.65 eV versus the number of carbon sites N (upper axis) and the length L (lower axis) for *cis* and *trans*-polyenes. (from Ref. 22)

for *cis* result simply from the shorter chain length. The length L is here defined as the distance along the x-direction between the two end carbon sites. The calculated values for $\gamma_{xxxx}(-3\omega;\omega,\omega,\omega)$ are thus unified by the general result

$$\gamma_{xxxx} \propto L^{4.6 \pm 0.2} \tag{16}$$

Because of the different geometry, a given *cis* chain is always shorter than its corresponding *trans* form. The difference in $\gamma_{xxxx}(-3\omega;\omega,\omega,\omega)$ values for the two is simply due to this fact, and γ_{xxxx} is, therefore, much more sensitive to the physical length of the chain than the conformation.

The very rapid growth that is observed in the nonresonant $\gamma_{xxxx}(-3\omega;\omega,\omega,\omega)$ with increased chain length is a result of several features common to both conformations that emerge from comparison of polyenes of different length. First, the lowest optical excitation energy decreases proportionally to the inverse of the chain length with a lowering from 5.9 eV in butadiene to 3.7 eV in the case of dodecahexaene (N=12). Second, the magnitudes of transition dipole moments along the chain axis increase steadily with the chain length. Third, while for OT and the shorter chains the nonlinear susceptibility is almost entirely composed of the contributions from only a few states, longer chains have significant contributions from an increasingly larger number of both ^1B_u and ^1A_g states.

From our calculated power law dependence, we can draw several important results for polymers. A typical value of the nonresonant macroscopic third order susceptibility $\chi^{(3)}(-3\omega;\omega,\omega,\omega)$ observed for polymers is 10^{-10} esu[29,30]. For an isotropic distribution of chains considered as independent sources of nonlinear response with a single dominant tensor component $\gamma_{xxxx}(-3\omega;\omega,\omega,\omega)$ we have $\chi^{(3)}=(1/5)N(f^\omega)^3f^{3\omega}\gamma_{xxxx}$, where N is the number density of chains and $f^\omega = (2+n_\omega^2)/3$ is the Lorentz-Lorenz local field factor. Using typical values of N=10^{20} molecules/cm^3 and 1.8 for the refractive index, we derive a γ_{xxxx} of roughly 2×10^{-31} esu. From Eq. (16), this value would correspond to a chain of N≈50 carbon sites, or a length of approximately 60 Å. Since these polymers consist of much longer chains, we infer that γ_{xxxx} must deviate from the power law dependence and begin to saturate at some length shorter than 60 Å.

This major result then suggests that large values of γ, and correspondingly $\chi^{(3)}$, require only chains of intermediate length of order 100 Å (oligomers) and that infinite conjugated polymer chains may not be required in nonlinear optics. These suggestions are especially important in considerations of materials synthesis and processing to obtain high optical quality materials that also possess outstanding secondary properties. Processing is often more tractable for oligomers than for the corresponding infinite chain polymers.

Most recently, we have observed a significant enhancement of γ_{xxxx} due to the lowering of symmetry of a conjugated chain. Heteroatomic substitution of a dicyano acceptor group on one end and a dimethylamino donor group on the other end of a polyene chain results in a noncentrosymmetric structure in which the symmetry selection rules discussed above are lifted. It is found that the low-lying calculated π-electron states can be directly mapped between the centrosymmetric and noncentrosymmetric structures and the analog of the 1^1B_u state experiences a large decrease in excitation energy such that it lies below the analog of the 2^1A_g state. Because of the lowered symmetry, new types of virtual excitation processes are admitted to $\gamma_{ijkl}(-\omega_4;\omega_1,\omega_2,\omega_3)$ which involve the dipole moment difference between the excited and ground states. The result is an enhancement of $\gamma_{xxxx}(-3\omega;\omega,\omega,\omega)$ at 0.65 eV by an order of magnitude for the noncentrosymmetric version of OT, for example, to a value of 173×10^{-36} esu as compared to 15.5×10^{-36} esu for the centrosymmetric chain. Details of these developments will be provided in Ref. 31 and forthcoming communications.

IV.   CONCLUSION

We have reviewed recent developments in the understanding of the origin of the large, ultrafast nonresonant nonlinear optical properties of conjugated organic structures. These studies provide clear evidence of the important role of electron correlation effects in determining the microscopic second order $\beta_{ijk}(-\omega_3;\omega_1,\omega_2)$ and third order $\gamma_{ijkl}(-\omega_4;\omega_1,\omega_2,\omega_3)$ susceptibilities of these unique materials. Furthermore, this microscopic understanding provides basis for the description of the macroscopic responses in crystals and polymers.

Through combined experimental and theoretical studies, we have clarified the origin of the second order nonlinear response of crystal MNA. It is revealed that the macroscopic crystal response can be described by the sum of the independent sources of response from individual units, where each unit is composed of a nearest neighbor, closely coupled pair of MNA molecules. The formalism allows direct calculation of the crystal nonlinear susceptibility through the theoretical framework devised for molecular response.

A theoretical method which takes explicit account of electron correlation was presented for two conformations and several chain lengths of polyenes. The calculated values for $\gamma_g(-2\omega;\omega,\omega,0)$ are in good agreement with available experimental measurements. Details of the mechanism for $\gamma_{ijkl}(-\omega_4;\omega_1,\omega_2,\omega_3)$ were explained in terms of virtual transitions among highly correlated π-electron states, and contour diagrams of transition density matrices $\rho_{nn'}$ provided direct illustration of the most significant virtual transitions. A power law dependence of $\gamma_{xxxx}(-3\omega;\omega,\omega,\omega)$ on the chain length L was found to hold true for both conformations with an exponent of 4.6 ± 0.2.

This research was generously supported by AFOSR and DARPA, F49620-85-C-0105 and NSF/MRL, DMR-85-19059. The calculations were performed on the CRAY X-MP of the Pittsburgh Supercomputing Center.

REFERENCES

1.   See, for example, *Nonlinear Optical Properties of Organic and Polymeric Materials*, edited by D.J. Williams, ACS Symp. Series Vol. 233 (American Chemical Society, Washington, DC, 1983); *Molecular and Polymeric Optoelectronic Materials: Fundamentals and Applications*, edited by G. Khanarian, (Proceedings SPIE, **682**, August 1986); and references therein.

2.   See, for example, A.F. Garito and K.D. Singer, Laser focus **18**, 59(1982).

3.   S.J. Lalama and A.F. Garito, Phys. Rev. A**20**, 1179 (1979).

4.   C.C. Teng and A.F. Garito, Phys. Rev. Lett. **50**, 350 (1983); Phys. Rev. B**28**, 6766 (1983).

5.   A.F. Garito, K.D. Singer and C.C. Teng, in *Nonlinear Optical Properties of Organic and Polymeric Materials*, edited by D.J. Williams, ACS Symp. Series, Vol. 233 (American Chemical Society, Washington, DC, 1983), Chap. 1.

6.   A.F. Garito, Y.M. Cai, H.T. Man and O. Zamani-Khamiri, in *Crystallographically Ordered Polymers*, edited by D.J. Sandman, ACS Symp. Series, Vol. 337 (American Chemical Society, Washington, DC, 1987), Chap. 14.

7.   See, for example, A.F. Garito, K.Y. Wong and O. Zamani-Khamiri, in *Nonlinear Optical and Electroactive Polymers*, edited by D. Ulrich and P. Prasad (Plenum, New York, 1987).

8.   B.F. Levine, C.G. Bethea, C.D. Thurmond, R.T. Lynch, and J.L. Bernstein, J. Appl. Phys., **50**, 2523(1979).

9.   A. F. Garito and K. Y. Wong, Polymer Journal **19**, 51(1987).

10.   G.F. Lipscomb, A.F. Garito, and R.S. Narang, Appl. Phys. lett. **38**, 663(1981); J. Chem. Phys., **75**, 1509(1981)

11.   J. Jerphagnon and S.K. Kurtz, J. Appl. Phys., **41**, 1667(1970).

12.   C. H. Grossman, Ph. D. dissertation, University of Pennsylvania, 1987.

13.   J.A. Armstrong, N. Bloembergen, J. Ducuing, and P.S. Pershan, Phys. Rev.,**127**, 1918(1962).

14.   P. Suppan, J. Chem. Soc. A, 3125(1968).

15.   See, for example, B.S. Hudson, B.E. Kohler, and K. Schulten, in *Excited States*, Vol. 6, edited by E.C. Lim (Academic Press, New York,1982), p. 1 and references therein.

16.   B.S. Hudson and B.E. Kohler, J. Chem. Phys. **59**, 4984 (1973).

17.   K. Schulten, I. Ohmine, and M. Karplus, J. Chem. Phys. **64**, 4422 (1976).

18.   A.A. Ovchinnikov, I.I. Ukrainski, and G.V. Kuentsel, Soviet Phys. Uspekhi **15**, 575 (1973) and references therein.

19.   K. Schulten, U. Dinur and B. Honig, J. Chem. Phys. **73**, 3927 (1980).

20.   Z.G. Soos and S. Ramasesha, Phys. Rev. B**29**, 5410 (1984).

21.   P. Tavan and K. Schulten, Phys. Rev. B**36**, 4337 (1987).

22.   J.R. Heflin, K.Y. Wong, O. Zamani-Khamiri and A.F. Garito, Phys. Rev. B**38**, 1573 (1988).

23.   A. F. Garito, J. R. Heflin, K. Y. Wong, and O. Zamani-Khamiri, in *Nonlinear Optical Properties of Polymers*, edited by A.J. Heeger, D. Ulrich, and J. Orenstein (Mater. Res. Soc. Proc. **109**, Pittsbrugh, PA, 1988),pp. 91-102; Proceedings SPIE **825**, 56(1988); Mol. Cryst. Liq. Cryst. **160**, 37 (1988).

24.   J.F. Ward, Rev. Mod. Phys. **37**, 1 (1965).

25.   B.J. Orr and J.F. Ward, Mol. Phys. **20**, 513 (1971).

26.   J.F. Ward and D.S. Elliott, J. Chem. Phys. **69**, 5438 (1978).

27.   We have converted Ward and Elliot's $\chi(3)$ to our notation by
$$\gamma_g(-2\omega;\omega,\omega,0) = \frac{3}{2} \chi^{(3)}.$$

28.   R.M. Gavin, Jr., S. Risemberg and S.A. Rice, J. Chem. Phys. **58**, 3160 (1983).

29.   C. Sauteret *et .al.*, Phys. Rev. Lett. **36**, 956 (1976).

30.   F. Kajzar and J. Messier, Polymer J. **19**, 275 (1987).

31.   A. F. Garito, J. R. Heflin, K. Y. Wong, and O. Zamani-Khamiri, in *Organic Materials for Nonlinear Optics*, edited by D. J. Ando and D. Bloor (1988).

QUADRATIC NONLINEAR BEHAVIOUR OF VARIOUS LANGMUIR-BLODGETT
MOLECULES.

I. Ledoux, D. Josse and J. Zyss
Centre National d´Etudes des Télécommunications - 196, avenue
Henri Ravera - 92220 - Bagneux - France.

T. McLean, R.A. Hann, P.F. Gordon and S. Allen
ICIplc, Electronics Group - PO Box 11 - The Heath - Runcorn -
Cheshire (UK)

D. Lupo, W. Prass, U. Scheunemann
Hoechst AG, 6230 - Frankfurt - FRG

A. Laschewsky and H. Ringsdorf
Institute of Organic Chemistry, University of Mainz - 6500 -
Mainz - FRG

1 - INTRODUCTION

        The Langmuir-Blodgett (L-B) technique is well known as
a method of building-up ordered arrays of organic molecules,
and especially as a means of imposing a noncentrosymmetric
structure from molecules that crystallize in a   centrosymmetric space group(1-3). L-B films for use in quadratic   nonlinear optics can be ordered in a statistically noncentrosymmetric lattice. The method involves the compression of a   monolayer of the organic molecule, spread on top of a water
surface, into a two-dimensional solid, followed by the
repeated dipping of the substrate to be coated into and   out
of the subphase. Monolayers may be deposited onto the   substrate both on immersion and withdrawal, resulting in a   centrosymmetric arrangement of layers (Y-type deposition),   or
just on immersion (X-type) or withdrawal (Z-type), in   which
case a noncentrosymmetric structure should be obtained.   However, this kind of samples is not very stable and tends to
relax towards disordered structures. An alternative method is
the successive deposition of alternate layers ABAB... where A
and B are two different molecular species stacked following a
Y-type structure, the hydrophobic chain being grafted onto
the acceptor group of the first species A and onto the   donor
group of the second species B (4,5). The quadratic   nonlinear
response, resulting from the addition of the individual   nonlinearities of each moiety, is quite large, with the   additional advantage of a high degree of order and a good stability of the multilayers.
        Nonlinear optical characterization of L-B molecules can
be achieved by measuring the second harmonic (SH) signal
emitted by the L-B film deposited upon a substrate. The   SH
intensity can be exalted by resonance if either or both of
the fundamental and harmonic frequencies are close to   an
absorption band. By virtue of their thinness, L-B films   can
be studied in the frequency domain corresponding to the resonance, for they do not introduce a strong attenuation of the
signal. In addition, SH generation measurements on monolayers
can afford informations on the values of hyperpolarizabilities β of some molecules that cannot be measured by   alterna-

J. Messier et al. (eds.), Nonlinear Optical Effects in Organic Polymers, 79–91.
© *1989 by Kluwer Academic Publishers.*

tive methods such as Electric-Field-Induced-Second-Harmonic-Generation (EFISHG), a technique limited to non-ionic and rather highly soluble compounds. Therefore, the SH signal emitted by a monolayer can be used to estimate the value of β for a wider range of molecular moieties than in the case of the EFISHG technique.

Second harmonic generation (SHG) from monolayers of highly polarizable dyes was reported for the first time by Aktsipetrov et al.(6), the corresponding $\chi^{(2)}$ being of the order of that of LiNbO$_3$. SHG has been evidenced in monolayers of merocyanine (7), hemicyanine (8), amidonitrostilbene (5) and azobenzene derivatives (9). Only a few molecules have been studied up to now, and it would be of interest to investigate new compounds that combine good film quality with large nonlinear susceptibilities.

In this paper we report some β measurements on various families of nonlinear molecules: azobenzene derivatives, polyenes, stilbazium salts and phenylhydrazone derivatives. In addition, we study the behaviour of a monolayer when it is separated from the substrate by a variable number of inactive layers, in order to investigate the structure and the local order of monolayers. The influence of a protective overcoat in some special cases has also been observed.

2 - OPTICAL MEASUREMENTS

2-1 . Linear measurements.

The linear optical behaviour of L-B samples is studied by using a Pye-Unicam SP8-200 absorption spectrometer and a SOPRA ellipsometer working in the UV-visible range. The spectral dispersion of LB films in the absorption coefficient and refractive indices is available for optical energies ranging from 1.5 to 5 eV. Some results are shown in Figure 1 for Y-type pure layers of a diazostilbene derivative, 107474.

2-2 . Nonlinear optical assessments.

SHG measurements are performed using a home-made Q-switched Nd^{3+}:YAG laser operating at 1.06μm. The pulse duration is 10 ns, the incident IR energy ranges from 0.5 to 5 mJ for a repetition rate of 10 Hz. The incident beam can be polarized either horizontally, in the plane of incidence (p-configuration) or perpendicular to it (s-configuration). The sample is set on a rotation stage monitored by a stepping motor so as to vary the incidence angle θ of the fundamental beam the rotation axis being vertical and perpendicular to the laser beam. SHG signals are recorded at θ values ranging from - 50° to +50°. The transmitted harmonic signal is detected by a photomultiplier, sampled and averaged. When working at large integration scales (10 s), the noise level does not exceed 0.1 mV, and the minimum nonlinear susceptibility value of LB films that can be detected is of the order of 2.0×10^{-9} e.s.u. The measurements are calibrated against the SH Maker fringes emitted by a 2 mm-thick quartz plate. In addition, we used a reference consisting of the SH signal $I_R^{2\omega}$ produced by a nonlinear organic powder of NPP irradiated by an additional fundamental beam, sampled and averaged by the same lock-in amplifier as the SH signal $I_{LB}^{2\omega}$ emitted by the

LB films. The signal $I_{LB}^{2\omega}$ is divided by the reference $I_R^{2\omega}$ and recorded. This set-up considerably reduces the variations of the SH signal owing to laser fluctuations and therefore improves the rapidity and the sensitivity of the measurements. In this work we focused only on SHG from p-polarized fundamental beams.

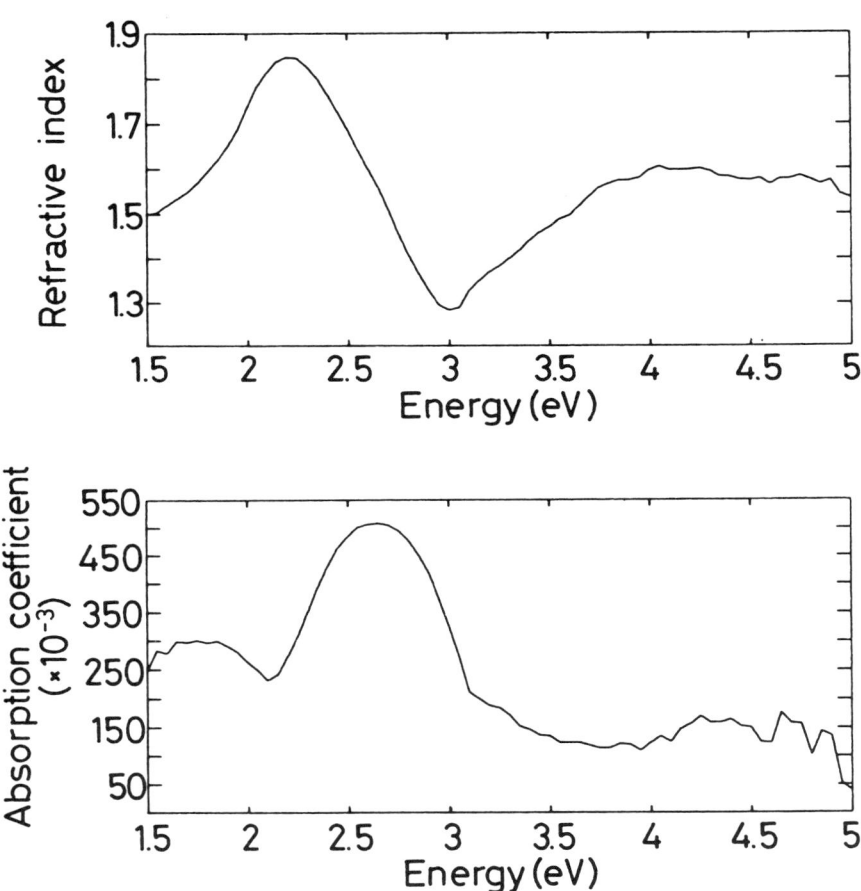

FIGURE 1: *Refractive indices and absorption coefficients of 41 layers of 107474.*

3 - NONLINEAR OPTICAL RESPONSE FROM MONOLAYERS

The dependence of the emitted intensity with respect to the incidence angle shows a well-contrasted fringing pattern resulting from interference between the harmonic beams produced by the front and back layers (Figure 2). This pattern originates from dispersion of the refractive index of the substrate $n_{2\omega}^{G} - n_{\omega}^{G}$, which introduces a dephasing factor

82

between harmonic waves generated at the front and back sides, respectively, of the substrate (Figure 3) (10). n $_\omega$ and n $_\omega$ are the fundamental and harmonic refractive indices of the glass substrate, respectively. The dephasing factor varies with the optical path in the substrate as the plate is rotated. The interfringe $\delta\theta$ is given by:

$$\delta\theta = \lambda /[2L \quad \cos\theta \ (tg\theta^G_\omega - tg\theta^G_{2\omega})]$$

where L is the thickness of the substrate. θ^G_ω (resp. $\theta^G_{2\omega}$) is the angle of incidence of the fundamental (resp. harmonic) beam inside the glass substrate, and λ is the fundamental wavelength. A quite good agreement is found between calculated and experimental values of $\delta\theta$ for the samples studied here.

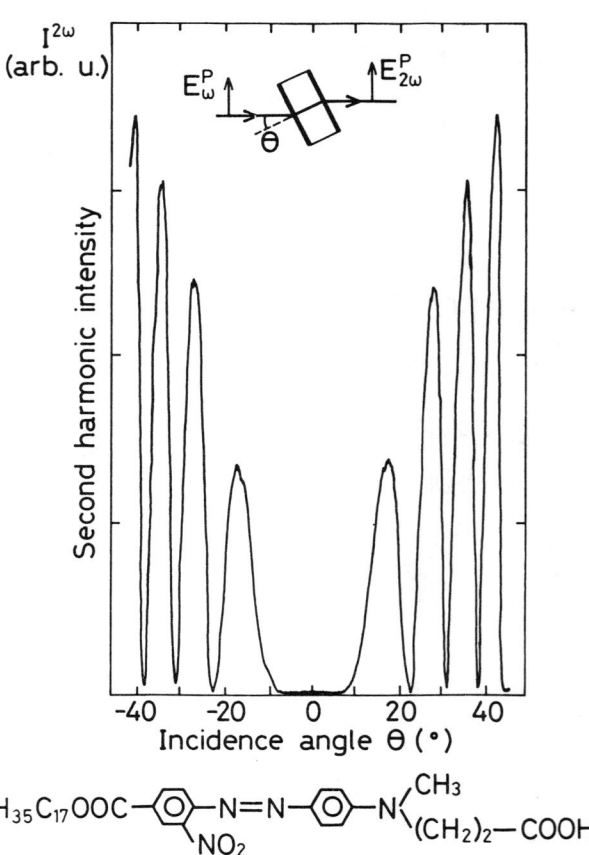

FIGURE 2: SHG fringe pattern for a 109054 monolayer.

From the envelope function of the SHG fringe pattern

we can infer both the β value of the L-B molecules and the average tilt angle φ between the charge-tranfer axis of the molecule and the normal to the substrate for a monolayer. β and φ are adjusted to give the best fit between the theoretical and the experimental envelope function of the fringes.

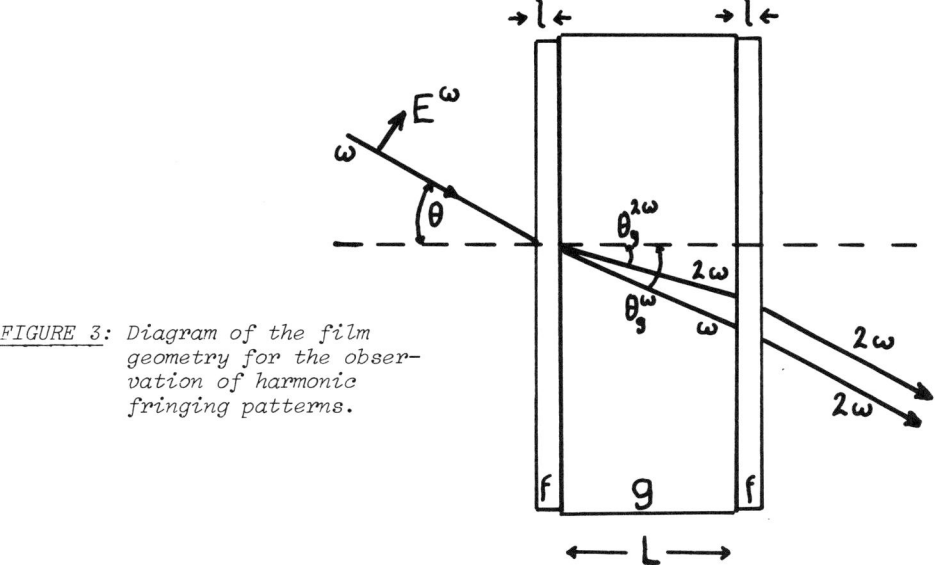

FIGURE 3: *Diagram of the film geometry for the observation of harmonic fringing patterns.*

The β values determined from EFISHG measurements at 1.06 μm are highly sensitive to resonance effects, and are not correlated directly to the "intrinsic" hyperpolarizability of the molecule. The dependence of β on the electronic transition energies of the compound is taken into account by using a two-level model (11) which is given by:

$$\beta \ (-2\omega ; \omega , \omega) = F(\lambda_o , \lambda) \beta_o$$

where $F(\lambda_o , \lambda) = \dfrac{1}{(1 -(2\lambda_o/\lambda)^2)(1 - (\lambda_o/\lambda)^2)}$

λ being the fundamental wavelength, λ_o the wavelength of the maximum of the absorption spectrum.

The quantity β_o is calculated by dividing the measured value of β by $F(\lambda_o , \lambda)$. Comparisons between β_o values of molecules with different transition energies are therefore not affected by resonant enhancements. Table I gives the λ_o , β_o and β values for three diazo dyes. As expected, the β_o value of 108782 is slightly smaller than those of 109054 and 107474 because of the lack of an electron attactor group in the para position. However, this value remains comparable to those of the other diazo dyes with an acid or ester function in the para position: the NO_2 group in the ortho position plays a crucial role in the charge transfer processes responsible for the quadratic nonlinearity of the molecules. The β_o values of 107474 and 109054 are quite similar, illustrating the weak influence of the aliphatic chains on the charge-transfer processes responsible for the quadratic non-linear response. These values are 4.5 times higher than that of paranitroaniline, to be accounted for by a greater number of τ-electrons involved in the charge-transfer process.

Code	108782	109054	107474
Formula	$H_{25}C_{12}$ ... NO_2 ... N=N ... N(CH$_2$)(CH$_3$), CH$_2$, COOH	$H_{35}C_{17}OOC$ ... NO_2 ... N=N ... N(CH$_2$)(CH$_3$), CH$_2$, COOH	HOOC ... NO_2 ... N=N ... N($C_{12}H_{25}$)(CH$_3$)
λ_{max} (nm)	460	440	450
β ($\times 10^{38}$ S.I.)	10±1.5	11±1.5	12±1.6
β_o ($\times 10^{38}$ S.I.)	2±0.3	2.9±0.4	2.8±0.4

TABLE I: Maximum absorption wavelength λ_{max}, hyperpolarizability at $1.06\,\mu m$ β and at 0-frequency β_o for 3 diazostilbene derivatives.

Table II gives the β values of polyenic derivatives. A significant increase of β is observed when going from the diazostilbene derivatives to the polyenic molecules. This phenomenon can be accounted for by a much better polarizability of the polyenic chains as compared to the stilbene structures, rather than by resonance effects, the maxima of the absorption bands being found at similar frequencies for all molecules studied here. In addition, the substituent

85

=C(CN)(COOH)   used in 109224, 111177 and 114355 has a   much
stronger acceptor   character   than   the acid   or   ester   substituents. However, the slightly smaller β_o value for 114355
as compared to that   of 109224 is   quite unexpected, for the
number of delocalized π -electrons   is larger   in the   former
molecule than in the latter one. It must be pointed out   that
L-B films made of polyenic molecules with a large number of
double bonds are not quite stable; either a partial   chemical
modification of the molecule or a poor stability of the layer
after deposition onto a solid substrate could be responsible
for this weakening of the harmonic signal, leading to a significant underestimation of the β values.

CODE	109224	111177	114355
FORMULA			
λ_{max}	430	470	450
β	27±4	35±5	11±1.5
β_o	7.8±1.3	6.2±0.9	2.6±0.15

TABLE II: λ_{max} (nm), β and β ($\times 10^{-38}$S.I.) values for 3 polyenic derivatives.

      In Table III are   reported the β values of two   other
families of nonlinear molecules, stilbazium salts and phenyl-

hydrazone derivatives (12). The nonlinear behaviour of these molecules is studied for various donor (R-O, R_2 N, R-COO) groups associated with the same acceptor ($\geq \overset{+}{N}$-R for stilbazium salts, $-NO_2$ for the phenylhydrazones). The results obtained on the two stilbazium salts confirm the strong electron-donor character of the disubstituted amino group $(H_{33} C_{16})_2 N$ when compared with the $H_{37} C_{18} O$ group of compounds 1. The electron-donor power of the $R_2 N$ group in 2 is slightly reinforced by the presence of two long aliphatic chains grafted onto the nitrogen atom. In addition, the presence of two hydrophobic tails on a L-B molecule instead of one should increase the stability and therefore the ordering of the monolayer. Compound 2 exhibits one of the highest quadratic nonlinearities ever measured, its β value coming just behind that of the merocyanine described in Ref 7 . In addition, the corresponding β_o is two times higher than the largest value reported for the polyenic derivatives. The coupling of the best donor and acceptor groups is largely responsible for these exceptional performances.

Molecular structure and designation	λ_{max}	β	β_o
$H_{37}C_{18}-O-\bigcirc-CH=CH-\overset{+}{\bigcirc}N-CH_3$ $I^{\ominus}$ **1**	360	6.25	3.0
$\begin{matrix} H_{33}C_{16} \\ \\ H_{33}C_{16} \end{matrix} N-\bigcirc-CH=CH-\overset{+}{\bigcirc}N-CH_3$ $I^{\ominus}$ **2**	475	83.4	13.5
$H_{35}C_{17}-\underset{O}{\overset{\parallel}{C}}-O-\bigcirc-CH=N-\underset{H}{N}-\bigcirc-NO_2$ **3**	412	4.6	1.5
$H_{37}C_{18}-O-\bigcirc-CH=N-\underset{H}{N}-\bigcirc-NO_2$ **4**	420	16.7	5.3
$H_{33}C_{16}-O-\bigcirc-CH=N-\underset{H}{N}-\bigcirc-NO_2$ **5**	420	13.7	4.4

TABLE III: λ_{max} (nm), β and β ($\times 10^{-38}$ S.I.) values for stilbazium salts and phenylhydrazone derivatives.

As far as phenylhydrazones are concerned, comparison
between 3 and 4 or 3 and 5 indicates that the ester substituent $H_{35}C_{17}COO$ is a very poor electron donor group, the
corresponding β value being much smaller than for the other
L-B molecules reported up to now for quasi-identical resonance effects. But the most important feature of these compounds is their large β values, comparable to those of diazostilbene derivatives, whereas the chain -CH=N-NH- is not
fully conjugated. The charge-transfer process is probably
achieved through the non-bonding electrons of the two nitrogen atoms. Therefore, phenylhydrazone derivatives appear to
be a new attractive class of efficient nonlinear molecules
for L-B films.

4 - STRUCTURAL AND ORDERING CONSIDERATIONS ON DIAZOSTILBENES

Some unexpectedly high SH emission from centrosymmetric Y-type layers of 107474 has stimulated the investigations
for structural and ordering properties of monolayers in various environments. We have studied the influence of the substrate by measuring the SH intensity from samples containing
one active layer of 107474 with a variable number N_0 of intermediate Y-type inactive layers placed as "buffers" between
the 107474 layer and the glass substrate. We have determined
the variation in β with N_0 for hydrophilic (Figure 4) and
hydrophobic (Figure 5) substrates. A strong decrease is observed from $N_0 = 0$ to $N_0 = 6$, the β value decreasing by a
factor of 3 in the hydrophilic case and of 2 in the hydrophobic one. For large N_0 values, the degree of disorder is approximately the same for both cases, giving rise to similar
values. The experimental value of the tilt angle ϕ between
the charge-transfer axis of the molecule and the normal to
the substrate is approximately the same (around 65°) for all
N_0's.

Electronic modifications to the charge-transfer process induced by the substrate cannot account for the behaviour of depicted in Figs. 4 and 5, the influence of N_0
on the absorption spectrum being negligible. A more relevant
interpretation is made possible by assuming that the upper
layers should be less ordered than the lower ones, giving
rise to a significant decrease of β. This hypothesis is
consistent with the results from other types of experiments,
such as thermodesorption (13), ellipsometric and plasmon resonance studies (14), in which a decrease in the binding energy of the upper layers has been observed. This enhancement of
the binding energy in the vicinity of the substrate extends
over 3 or 4 monolayers, corresponding to various interactions
such as the binding between the L-B film and the glass,
stronger (dipolar, hydrogen bonding) interactions between
different monolayers or van Der Waals forces extending over
long distances. This seems to account for the decrease in β
for an active 107474 monolayer. Interaction energy between

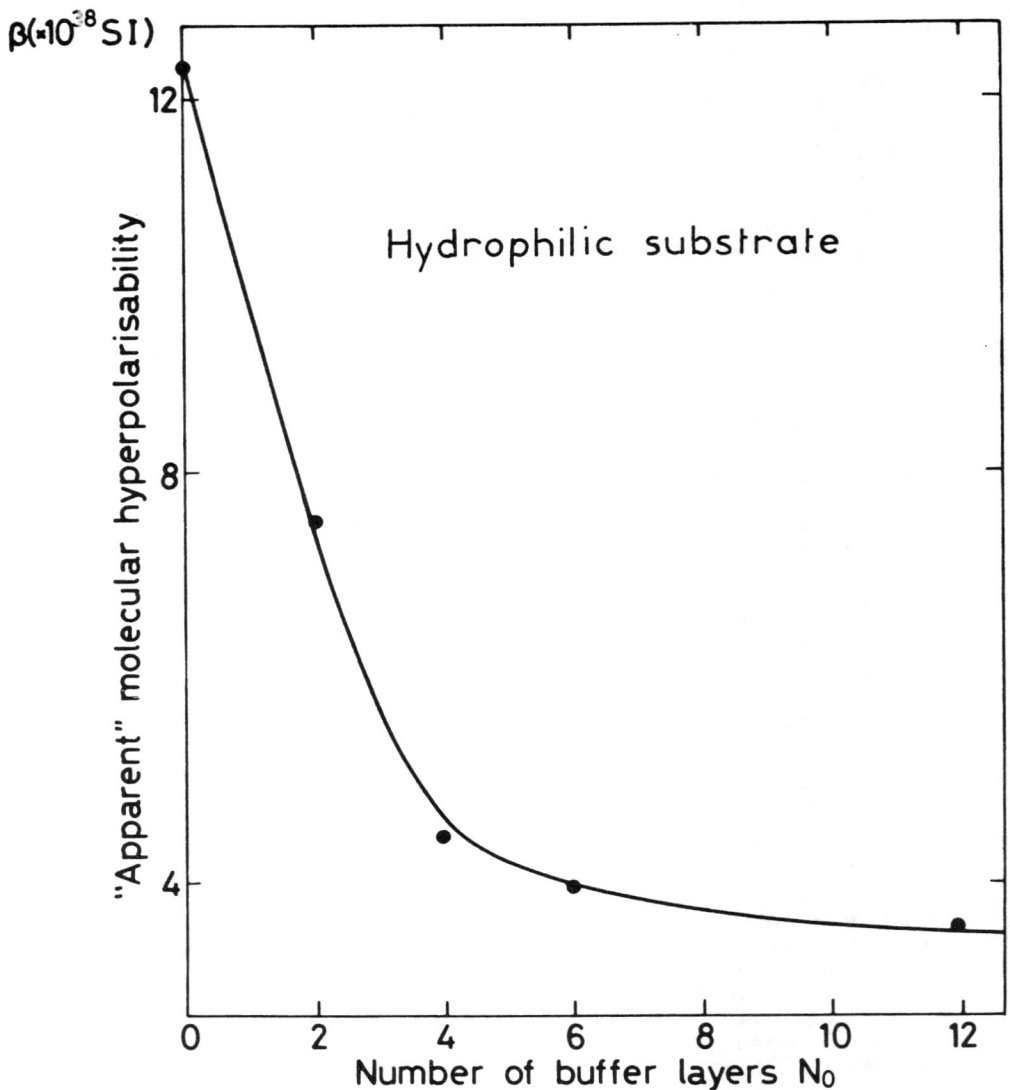

FIGURE 4: *Variation in the hyperpolarizability* β *of 107474 with the number* N_0 *of "buffer layers" of arachidic acid on an hydrophilic substrate.*

the layer and the substrate seems to be weaker in the hydrophobic case, the value for $N_0 = 0$ being significantly
smaller than for hydrophilic substrates.

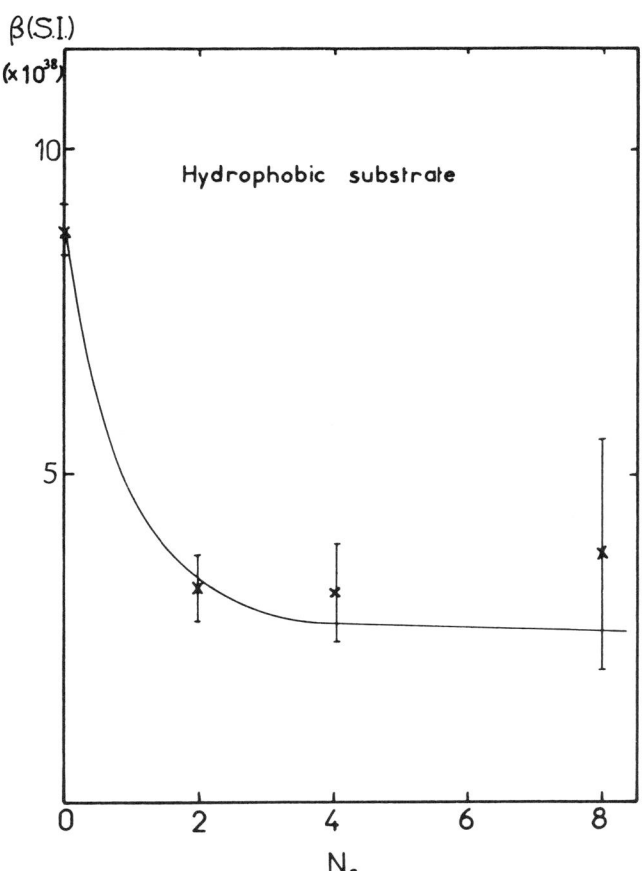

FIGURE 5: *Variation of β of 107474 with respect to N_0 in the case of an
hydrophobic substrate.*

5 - ADDITIONAL REMARKS

It must be pointed out that the abovementioned considerations on disorder and substrate effects have been developed from experiments carried-out on diazostilbene deriva-

tives. The behaviour could be significantly different for other families of molecules such as polyenic derivatives. In fact, these compounds give rise to strong SHG signals just after their deposition, but sometimes completely disappear within a few days, as confirmed by linear and nonlinear measurements. We have observed a tendency towards stabilization of a monolayer of 114355 when coating the film with an additional inactive layer (arachidic acid): $\chi^{(2)}$ coated = 2 $\chi^{(2)}$ uncoated. However, the corresponding β value remains small when compared to the results obtained on freshly prepared samples of 111177 carefully kept in the dark and under a nitrogen pressure.

6 – CONCLUSION

We have investigated by SHG measurements four families of molecules: diazostilbenes, polyenes, stilbazium salts and phenylhydrazones. High β values have been evidenced, especially in stilbazium salts. Structural and order effects in monolayers have been observed, evidencing the crucial role of the substrate/molecule interactions in the behaviour of the nonlinear response.
Availability of highly stable and nonlinear L-B films opens new perspectives for testing by SHG some L-B films in a waveguided configuration. The losses due to scattering phenomena occuring inside the layers are not yet compatible with a waveguided propagation inside the L-B film. However, it is possible to propagate inside a linear waveguide coated by a nonlinear L-B film, the SH light being generated by the evanescent wave in the active layer. Preliminary experiments evidenced a quite weak SH signal from a linear glass waveguide coated by a Z-type, highly diffusing L-B film of 107474. Additional improvements of the films quality have to be found in order to decrease the losses due to grain boundaries, so as to observe significant conversion rates in a waveguided configuration.

REFERENCES
1) K.B. Blodgett, J. Am.Chem.Soc. 57,1007 (1935)
2) G.G. Roberts, Advances in Physics, 34, n°4, 475 (1985)
3) L.M. Blinov, Russian Chemical Reviews, 52, n°8, 1263 (1983)
4) J. Zyss, J. Mol.Electron. 1, 25 (1985)
5) D.B. Neal, M.C. Petty, G.G. Roberts, M.M. Ahmad, W.J. Feast, I.R. Girling, N.N. Cade, P.V. Kolinsky and I.R. Peterson, Electron. Lett. 22, 460 (1986)
6) O.A. Aktsipetrov, N.N. Akhmediev, E.D. Mishina and V.R. Novak JETP Letters, 37, 207 (1983)
7) I.R. Girling, P.V. Kolinsky and C.M. Montgomery, Electron. Let. 21, 169 (1983)
8) I.R. Girling, N.N. Cade, P.V. Kolinsky, R.J. Jones, I.R. Peterson, M.M. Ahmad, D.B. Neal, M.C. Petty, G.G. Roberts and W.J. Feast, J. Opt. Soc. Am.B 4, 950 (1987)
9) I. Ledoux, D. Josse, P. Fremaux, J.P. Piel, G. Post, J. Zyss, T. McLean, R.A. Hann, P.F. Gordon and S. Allen, Thin Sol. Films, 160, 217 (1988)
10) F. Kajzar, J. Messier, J. Zyss and I. Ledoux, Opt. Comm.

45, 133 (1983)

11) J-L. Oudar and D.S. Chemla, J. Chem. Phys. 66, 2664 (1977)

12) D. Lupo; W. Prass, U. Scheunemann, A. Laschewsky, H. Ringsdorf and I. Ledoux, J. Opt. Soc. Am.B 5, 300 (1988)

13) L.A. Laxhuber, B. Rothenhaüsler, G. Schneider and H. Möhwald, Apll. Phys. A 39, 173 (1986)

14) V. Skita, W. Richardson, M. Filipowski, A. Garito and J.K. Blasie, J. Phys. (Paris) 47, 1849 (1986)

PREDICTION OF THIRD ORDER NONLINEAR OPTICAL ACTIVITY IN ORDERED POLYMERS

I. J. Goldfarb and J. Medrano

Materials Laboratory, Air Force Wright Aeronautical Laboratories,
Wright-Patterson AFB, OHIO 45433-6533 USA

1. INTRODUCTION

    The Air Force Ordered Polymer Research Program has produced a
family of rigid-rod and extended chain aromatic-heterocyclic polymers
with unique properties (1). In addition to outstanding environmental
resistance, these polymers, because of their rigid-rod nature, have
extremely high levels of orientability resulting in excellent mechanical
properties in fiber and film form. This orientability, combined with
the extended conjugation afforded by the aromatic heterocyclics, has
generated considerable interest in their nonlinear optical behavior. A
third order optical nonlinear susceptibility, $\chi(3)$, with a subpicosecond response, has been measured as high as 5 x 10^{-11} esu for
poly(p-phenylenebenzo[1,2-d:4,5-d']bisthiazole (PBT)(2).

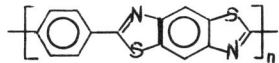

PBT

    The relatively high value of $\chi(3)$ for this polymer is noteworthy
considering that the measurements were made on film samples that were
prepared for use as structural materials with excellent thermal
stability, environmental and radiation resistance as well as significant
mechanical properties, i.e., tensile modulus and strength. No attempts
have been made to tailor the structure or the processing conditions to
optimize the optical behavior.

    As the first step in the development of polymers with improved
nonlinear optical activity, this laboratory has initiated research
efforts in the prediction of the nonlinear optical activity of ordered
polymers and their model compounds. This paper represents a preliminary
report of that effort.

2. THEORETICAL METHOD

    There are basically two approaches to the calculation of molecular
polarizabilities and hyperpolarizabilities of organic molecules. In the
Sum-Over-States (SOS) approach, the calculations are made for the
unperturbed molecule and perturbation theory is applied to these states
to calculate hyperpolarizabilities while in the Finite-Field (FF)
approach, determination of molecular wavefunctions are made on the

93

J. Messier et al. (eds.), Nonlinear Optical Effects in Organic Polymers, 93–99.

perturbed molecule including the field and the hyperpolarizabilities are obtained by numerical differentiation of the energy or the dipole moment with respect to the appropriate electric field components. In the SOS approach semi-empirical approximations have been utilized to obtain the various excited states which are then combined into the Sum-Over-States expression for hyperpolarizabilities (3). This approach has the advantage of computing frequency dependent hyperpolarizabilities, however it does require the utilization and proper definition of numerous excited states. The FF approach, on the other hand, has the advantage of only requiring ground states of the molecular wavefunctions and applies the perturbation directly on the Hamiltonian. This method has the disadvantage that it utilizes a homogeneous field which yields static hyperpolarizabilities which represent the extrapolation to low frequency and may be an adequate discription for the non-resonant case. This approach has recently been applied utilizing ab initio SCF theory (4).

In this paper, we utilize semi-empirical methods utilizing the MNDO approximation with the FF approach. It is our hope that, as a result, large molecules, even polymers, may be used in the calculations and predictions of at least the rank ordering of chemical structures be accomplished for molecules of interest to the Air Force Ordered Polymer Research Program. An outline of the steps in this approach are shown in Figure 1.

SEMI-EMPIRICAL FINITE FIELD
HYPERPOLARIZABILITY COMPUTATIONAL METHOD

1. Optimize Molecular Geometry (MNDO)

2. Obtain Molecular Hamiltonian for Optimized Geometry, including Field Dependent Perturbation

3. Obtain Self-consistent Field Dependent Wavefunction & Energy

4. Calculate Dipole Moment as Function of the Field

5. Change Field Strengths & Directions, recalculating Wavefunction, Energy and Dipole Moments in each Case

6. Obtain First, Second and Third Derivatives of Dipole Moments as Function of Field to obtain (Hyper)polarizabilities

Figure 1

The details of the mathematical deviation of the FF approach are given below:

According to the series expansion of the energy W of a charge distribution in an external potential $\Phi(\underline{x})$

$$W = q\ \Phi(0) - \underline{p}.\underline{E}(0) + \ .\ .\ .$$

where $\underline{E} = -\nabla\Phi$ and $\underline{p}$ is the dipole moment of the system in the external field $\underline{E}$. Of course we can always choose $\Phi(0) = 0$. Then, the complete Hamiltonian of the system, including the perturbation due to the external field is:

$$\hat{\underline{H}} = \hat{\underline{H}}^{(0)} + W(\underline{E}) = \hat{\underline{H}}^{(0)} - \underline{p}.\underline{E} = \hat{\underline{H}}^{(0)} + [\underset{n}{\Sigma}\ \underline{r}(n)].\underline{E}$$

The mean value of the energy will be

$$<H> \ = \ \varepsilon = \varepsilon^{(0)}_{el} + 2\underset{i}{\Sigma}\ E_i \underset{n}{\Sigma} < \psi_n(\underline{E})|x_i|\psi_n(\underline{E})>$$

where ψ_n (n = 1, occ.) are all the occupied molecular orbitals and x_i is i-th coordinate of the n-th electron pair. When expressed in terms of the basis set, the perturbation becomes

$$<W> \ = \ \underset{i}{\Sigma}E_i \underset{\mu}{\Sigma} \underset{\nu}{\Sigma}\ P_{\mu\nu}\ (\underline{E})<\mu|\ x_i\ |\nu>$$

where $P_{\nu\mu}$ are the elements of the density matrix obtained from the molecular orbitals. The full Hamiltonian has to be solved for self-consistency to find the molecular orbitals $\psi_a(\underline{E})$. Once this has been done the dipole moment $\underline{p}(\underline{E})$ can be calculated in the MNDO approximation.

Expanding the molecular energy in a series of the electric field:

$$\varepsilon(\underline{E}) = \varepsilon^{(0)} + \underset{i}{\Sigma}\ E_i(\partial\varepsilon/\partial E_i)_{E=0} + 1/2 \underset{i}{\Sigma} \underset{j}{\Sigma}\ E_iE_j \partial^2\varepsilon/\partial E_i\partial E_j)_{E=0} \tag{1}$$
$$+ 1/6 \underset{i}{\Sigma} \underset{j}{\Sigma} \underset{k}{\Sigma}\ E_iE_jE_k\ (\partial^3\varepsilon/\partial E_i\partial E_j\partial E_k)_{E=0} + \ .\ .\ .$$

Or:
$$\varepsilon(\underline{E}) = \varepsilon^{(0)} - \underset{i}{\Sigma}\ E_i\ \mu_i(0) - 1/2 \underset{i}{\Sigma} \underset{j}{\Sigma}\ E_i\ E_j\ \alpha_{ij}$$
$$- 1/6 \underset{i}{\Sigma} \underset{j}{\Sigma} \underset{k}{\Sigma}\ E_iE_jE_k\ \beta_{ijk} - \ .\ .\ .$$

Therefore
$$-\mu_i(\underline{E}) = \partial\varepsilon(\underline{E})/\partial E_i = -\mu_1(0) + \underset{j}{\Sigma}\ E_j\alpha_{ij} + 1/2 \underset{j}{\Sigma} \underset{k}{\Sigma}\ E_jE_k\ \beta_{ijk} + \ .\ .\ . \tag{2}$$

Comparing (1) and (2)
$$\mu_i(\underline{E}) = \mu_1(0) + \underset{j}{\Sigma}\ E_j\alpha_{ij} + 1/2 \underset{j}{\Sigma} \underset{k}{\Sigma}\ E_jE_k\beta_{ijk} + \ .\ .\ . \tag{3}$$

96

The dipole moment itself can be similarly expanded in a series of powers of the field resulting in the alternate definitions of the (hyper)polarizabilities:

$$\alpha_{ij} = (\partial \mu_1(\underline{E})/\partial E_j)_{E=0} = -(\partial^2 \epsilon(\underline{E})/\partial E_i \partial E_j)_{E=0}$$

$$\beta_{ijk} = (\partial^2 \mu_i(\underline{E})/\partial E_j \partial E_k)_{E=0} = -(\partial^3 \epsilon(\underline{E})/\partial E_i \partial E_j \partial E_k)_{E=0}$$

3. RESULTS AND DISCUSSION

At the outset it was important to determine how well the computational method agreed with experimental measurements. Results are shown in Table I for several polyenes.

TABLE I.

SECOND HYPERPOLARIZABILITIES COMPARED WITH EXPERIMENT

STRUCTURE	<GAMMA> THEOR.	<GAMMA> EXP.
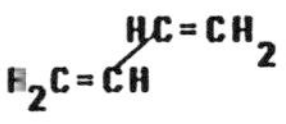 HC=CH$_2$ H$_2$C=CH	2.33 x 10^{36} e.s.u.	(5) 2.38 x 10^{36} e.s.u.
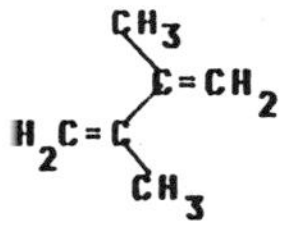 CH$_3$ C=CH$_2$ H$_2$C=C CH$_3$	2.2	(6) 4.9
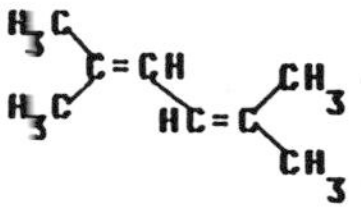 H$_3$C C=CH CH$_3$ HC=C CH$_3$	8.32	(6) 7.4
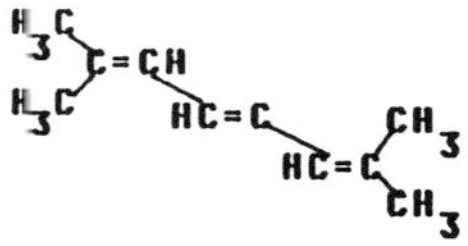 H$_3$C C=CH H$_3$C HC=C CH$_3$ HC=C CH$_3$	38.24	(6) 38

It should be noted that for this purpose, one looks for estimates of a space-fixed or scalar molecular second hyperpolarizability, usually measured in solution. The results for this series is quite promising considering that we are looking at the third order polarizabilities which is a very stringent test. These results encouraged us to tackle larger molecules as shown in Table II.

The structures shown in Table II represent model compounds to the rigid-rod ordered polymers, in particular related to PBT. Most of the compounds have been synthesized and experimental hyperpolarizabilities

TABLE II.

SECOND HYPERPOLARIZABILITIES
AS A FUNCTION OF STRUCTURE

STRUCTURE	CONFORMATION	GAMMA(XXXX)
	staggered	62×10^{-36}
	planar	241
	staggered	252
	planar	1248
	trans,staggered	219
	trans, planar	614
	trans, staggered	1026
	cis, staggered	1064
	trans, planar	3163
	cis, planar	3591
	trans, staggered	1527
	trans, planar	2930
	trans, planar (opt.)	3342
	cis, staggered	2011
	planar	1227×10^{-36}
	planar	342
		5.5
	n=0, twisted	62
	n=0, planar	122
	n=1, twisted	201
	n=1, planar	659

are being obtained for them and will be reported in the future. In each case we report the gamma (xxxx) which is the tensor component in the long molecular direction although the calculations produce all the components of the tensor. Also it should be noted that the molecular geometry has been optimized utilizing the MNDO approximation which tends to over-emphasize nonbonded interactions resulting in the optimized geometry having phenyl rings out of plane to the rest of the molecule. We have therefore calculated the hyperpolarizabilities for both the optimized or staggered conformation as well as the conformation which confines the phenyl rings to the same plane as the rest of the molecule.

It can be easily seen that the planar conformations have second
hyperpolarizabilities two to five times those of the staggered
conformations.

All the structures containing ethylenic linkages are calculated
with the substituents arranged trans- to the double bonds, the so-called
"cis" conformations involving the rotation of the single bond adjacent
to the heterocyclic ring by 180°. There seems to be a slight
improvement in the values in this "cis" conformation although this may
be too small a difference to be significant. A reasonable rule-of-thumb
description of the molecular length effect would be linear with number
of rings (double bond considered as a 1/2 ring). Similar remarks can be
made with the thiophene oligomers.

Finally, a series of structures were investigated of the type shown
below with various substituents at the x position:

These calculations are summarized in Table III.

TABLE III.

SECOND HYPERPOLARIZABILITIES
AS A FUNCTION OF SUBSTITUTION

SUBSTITUENT	GAMMA(XXXX)	HAMMETT SIGMA
H	218×10^{36} e.s.u.	0
CH_3	279	-0.170
NH_2	430	-0.660
NO_2	283	0.778
OCH_3	327	-0.268
CN	348	0.660

The values of second hyperpolarizability are compared with the substituent electron withdrawing effect as demonstrated by the substituent effect constant, sigma, from the Hammett equation. There seems to be reasonable correlation between electron releasing power (negative sigma) and gamma. The hyperpolarizabilities for substituents with positive sigmas also show an increase over the unsubstituted but since there are only two values it is unclear whether there is a relationship.

In conclusion, the results of computations of second hyper-polarizabilities on aromatic heterocyclic molecules show considerable promise in guiding the research toward the synthesis of new ordered polymers with improved nonlinear optical activity.

It is anticipated that experimental measurements for structures described in this paper will be available in the near future and should determine the accuracy of the computational method. Efforts to extend the computational method to consider polymers is currently in progress.

ACKNOWLEDGEMENTS

The authors wish to thank Lt. Scott Wierschke for his aid in the computations and his helpful discussions. We are indebted to Professor Paras Prasad for many valuable discussions and recommendations.

REFERENCES

1.   R. E. Evers, F. E. Arnold and T. E. Helminiak, Macromolecules, 14, 925 (1981).

2.   A. F. Garito, K. Y. Wong, Y. M. Cai, H. T. Man and O. Zamani-Khamiri, in Molecular and Polymeric Optoelectronic Materials: Fundamentals and Applications, edited by G. Khanarian, (Proceedings SPIE, 682, August 1986) pp. 2-11; D. N. Rao, J. Swiatkiecz, P. Chopra, S. K. Ghoshal, and P. N. Prasad, Appl. Phys. Lett. 48, 1187 (1986).

3.   A. F. Garito, J. R. Heflin, K. Y. Wong and O. Zamani-Khamiri, in Materials Research Society Symposium Proceedings Vol. 109 "Nonlinear Optical Properties of Polymers," edited by A. J. Heeger, J. Orenstein and D. R. Ulrich (1988), p. 91.

4.   P. Chopra, L. Carlacci, H. F. King and P. N. Prasad, submitted to J. Chem. Phys.

5.   J. F. Ward and D. S. Elliott, J. Chem. Phys., 69, 5438, (1978).

6.   S. H. Stevenson, D. S. Donald and G. R. Meredith in Materials Research Society Symposium Proceedings, Vol. 109, "Nonlinear Optical Properties of Polymers" edited by A. J. Heeger, J. Orenstein and D. R. Ulrich (1988) p. 103.

EFFECT OF π-ELECTRON DELOCALIZATION ON THE SECOND-ORDER POLARIZABILITY OF DISUBSTITUTED HYDROCARBONS

R.A. HUIJTS and G.L.J. HESSELINK

AKZO Corporate Research, Appl. Physics Dept., Velperweg 76, 6824 BM Arnhem, The Netherlands.

Insight in the molecular hyperpolarizabilities is of importance to the understanding and optimization of nonlinear optical organic polymers. Organic materials with extended delocalized π-electron systems exhibit large nonlinear optical responses (1). We present systematic measurements of the hyperpolarizability β of some disubstituted hydrocarbons with various extent of π-electron delocalization.

The length of the conjugated π-electron system is an important factor to produce (hyper)polarizability (2). In addition, charge transfer is another factor to produce hyperpolarizability (3). We investigated a series of organic molecules with a different number of conjugated π-electrons and the same functional groups as substituents: p-methoxy-nitrobenzene, 4-methoxy-β-nitrostyrene, α-p-methoxyphenyl-ω-p-nitrophenyl-polyene (4) with 1,2,3,4 π-bonds (ethene to octatetraene). The chemical structure of the molecules is given in Table 1. All the organic substances were synthesized, purified, and analyzed by standard chemical methods at our laboratories.

The molecular third-order nonlinearity μβ, the scalar product of the dipole moment μ and the vector part β of the quadratic polarizability tensor, was determined by electric field induced second-harmonic generation (dcSHG) measurements on liquid 1,4-dioxane solutions. The dcSHG experiments were performed by the wedge cell technique (5) with synchronous dc voltage pulses and a fundamental wavelength of 1064 nm . Accurate values (6) for μβ could be obtained by measuring the macroscopic third-order nonlinearity of the solutions over about nine different concentrations per molecule. We determined the dipole moments of the polyenes by dielectric measurements with a low-frequency capacitance bridge (7). The value of μ of the octatetraene compound was extrapolated from the results of the other polyenes. The solubility of this polyene was so low that the measured μ-value became very inaccurate. The values of β were obtained by dividing the values of μβ by

101

J. Messier et al. (eds.), Nonlinear Optical Effects in Organic Polymers, 101–104.
© *1989 by Kluwer Academic Publishers.*

TABLE 1. Number of π-bonds (n_π), dipole moments (μ), electronic transition energy (ω_0), and second-order polarizabilities (β_0) of donor-acceptor molecules with different conjugation length.
(a) μ for 1,2: A.L McClellan, Tables of experimental dipole moments (Freeman, San Fransisco, 1963); μ for 6: extrapolated from 3,4,5.

compound	n_π	$\mu^{(a)}$ (D)	ω_0 (eV)	β_0 (10^{-30} esu)
1. CH_3O—⟨⟩—NO_2	2	4.9	4.08	4±1
2. CH_3O—⟨⟩—⟍NO_2	3	5.5	3.58	6±1
3. CH_3O—⟨⟩—(⟍)$_1$—⟨⟩—NO_2	5	5.7	3.45	39±4
4. CH_3O—⟨⟩—(⟍)$_2$—⟨⟩—NO_2	6	6.0	3.16	53±6
5. CH_3O—⟨⟩—(⟍)$_3$—⟨⟩—NO_2	7	6.6	3.02	94±8
6. CH_3O—⟨⟩—(⟍)$_4$—⟨⟩—NO_2	8	6.9	2.88	89±26

the values of μ.

A comparison of the measured β-values among the different molecules is possibly affected by resonant enhancement (8). This effect is diminished in the quantity β_0,

$$\beta_0 = \beta(-2\omega;\omega,\omega)\,(\omega_0^2 - \omega^2)\,(\omega_0^2 - 4\omega^2) / \omega_0^4, \qquad (1)$$

in which the frequency dependence is factored out by assuming a dispersion relation for β given by the two-level model (9), and where ω is the fundamental laser frequency and ω_0 is the charge transfer transition frequency. The electronic charge transfer transition of the molecules was identified by UV-VIS absorption spectroscopy. Figure 1 shows β_0 versus the number of π-bonds between the two substituents CH_3O and NO_2 (i.e. the length of the conjugated π-electron system). A phenyl ring is counted as two π-bonds. Figure 1 clearly shows that β_0 varies as the third power of the number of π-electron bonds. This dependence possibly levels down at the longest chain length.

The third power dependence of β_0 on the conjugation length is in very good agreement with model predictions (10). The donor-acceptor substituents are thought to induce an electric field equivalent to $\delta\mu/\alpha$, where α is the linear polarizability of the molecule and $\delta\mu$ is the mesomeric moment. This electric field biases the third-order polarizability γ_0 of the π-electron system without substituents. Hence the second-order nonlinearity is given by (10)

$$\beta = 3 \gamma_0 \delta\mu / \alpha . \tag{2}$$

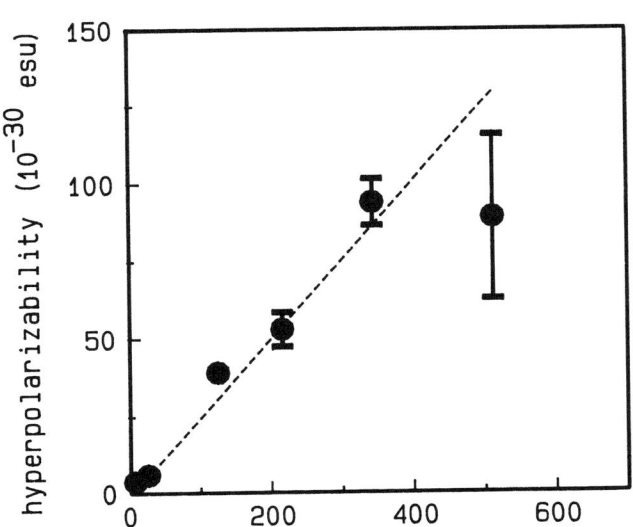

FIGURE 1. Effect of increasing length of conjugated π-electron system (n_π) on the second-order polarizability (β_0). The solid line is a fit to the experimental results $\beta_0(10^{-30}$ esu$)=1+n_\pi^3/4$.

104

Since α and γ_0 vary as the third and fifth power of the length of the conjugated π-electron system (11), and $\delta\mu$ is proportional to the conjugation length (12), β is expected to vary as the third power of the chain length.

The measured hyperpolarizability increases with increasing chain length over the range covered by our experiments. This is in qualitative agreement with calculated results (13) on similar molecules. These calculations predict a leveling down of the third-power dependence at long chain lengths. Our results possibly show the onset of this effect, but this needs to be verified more accurately.

Our experimental results make clear that the extent of π-electron delocalization can drastically influence the quadratic polarizability. This knowledge can be used, among others, to optimize nonlinear optical organic materials. Of course, one should keep in mind that the increase of hyperpolarizability by increasing the chain length is inevitably connected to some loss of transparancy.

ACKNOWLEDGEMENTS

We gratefully thank Dr. C.T.J. Wreesmann for synthesis of the molecules, Dr. L.W.Jenneskens for the UV/VIS data and Dr. J.N. Louwen for help with the data analysis.

REFERENCES:

1. Nonlinear Optical Properties of Organic Molecules and Crystals, Vols. 1 and 2, Eds. D.S. Chemla and J. Zyss (Academic, Orlando, 1987).
2. J.F. Nicoud and R.J. Twieg in Ref.1, p227.
3. J.L. Oudar and H. Le Person, Opt. Commun. 15(1975)258.
4. Y. Takeuchi, S. Akiyama and M. Nakagawa, Bull. Chem. Soc. Jap. 46(1972)2828.
5. B.F. Levine and C.G. Bethea, J. Chem. Phys. 63(1975)258.
6. K.D. Singer and A.F. Garito, J. Chem. Phys. 75(1981)3572.
7. C.J.F. Bottcher, Theory of electric polarization (Elsevier, Amsterdam, 1973).
8. C.C. Teng and A.F. Garito, Phys.Rev. A28(1983)6766.
9. J.L. Oudar, J. Chem. Phys. 67(1977)446.
10. J.L. Oudar and D.S. Chemla, Opt. Commun. 13(1975)164.
11. K.C. Rustagi and J. Ducuing, Opt. Commun. 10(1974)258.
12. K.B. Everard and L.E. Sutton, J. Chem. Soc. 625(1951)2818.
13. J.O. Morley, V.J. Docherty and D. Pugh, J. Chem. Soc. Perkin Trans. II 6(1987)1351.

STUDIES OF STRUCTURE-PROPERTY RELATIONS

G.R. Meredith and S.H. Stevenson

E.I. du Pont de Nemours & Co., Central Research and Development Dept.,
Experimental Station - 356, Wilmington, DE 19808 USA

1. INTRODUCTION

In the study of nonlinear optics in organic media structure-property relationships are of fundamental importance. In molecules, the relationships between perturbing fields and total dipolar polarization are often expressed as

$$\underline{p} = \mu + \alpha \cdot \underline{E} + \beta \cdot \cdot \underline{EE} + \gamma \cdot \cdot \cdot \underline{EEE} + \ldots \qquad (1.1)$$

Limiting attention to the electronic portion, in the Born-Oppenheimer approximation, there is probably nothing involved that couldn't be calculated with a powerful enough computer. The Hamiltonians are known as are methods to approach the many body solutions. However, the reliability of such calculations is not unquestionable and it is unrealistic to suggest they might now be used to scout trends without experimental checks. This must be true particularly with semiempirical methods for which the specific parameterizations which work well in some cases perhaps partly due to compensations of the methods' approximations may not be so good for other classes of molecules or other calculated quantities. On the other hand, good experimentation is fraught with difficulties (1, 2). Certainly techniques have not evolved to the point of ease and reliability that "measurements" of β and γ are common. There is undeniably a need for a better empirical database against which to test our knowldege of nonlinear mechanisms. However, there are fundamental limitations to information that can be extracted from solution measurements, which are most often used to investigate larger molecules. No matter how well the solution susceptibilities are determined, there is uncertainty in the application of models of the liquid dielectric behavior. Even if the trivial models were correct, only a minimal part of the tensors' properties can be determined. Thus theoretical anaylsis is essential to complete the picture of molecular polarization mechanisms and to predict the distribution of magnitudes within polarizability tensors.

In this paper we present specific theoretical and experimental examples of structure-property relationships. We also review a formal solution of the microscopic-macroscopic dipole polarization relationships. This is important in that it serves to illustrate the complexity of the relationships and the level of approximations required to extract molecular hyperpolarizabilities from measured bulk susceptibilities. Also it reminds us that the pursuit of structure-property relationships is not limited to molecules. The ability to describe macroscopic properties in terms of the molecular constituents is a key, but not yet seriously challenged, aspect of this field. This review is placed as appendix to allow a more coherent flow of molecular structure-property concepts.

2. MOLECULAR PROPERTIES: THEORY

It has become commonplace to utilize quantum-chemical programs to calculate molecular hyperpolarizabilities. While this activity is very important, we think that two related activities are also very important. First is the empirical

J. Messier et al. (eds.), Nonlinear Optical Effects in Organic Polymers, 105–122.

testing of such predictions and the generation of a larger database of empirical hyperpolarizabilities, there now being many more reported calculations than corresponding molecular polarizability data. Second is the use of other analytic tools and models to render more physical the effects illustrated by calculations. Below we will first illustrate the later point with some examples from our own computations, then illustrate some simple empirical relationships arising from systematic studies made possible by our recent developments in solution characterization techniques.

2.1. Topologically Induced β

Using a CNDO/Finite-Field method (3), it has been predicted that second-order hyperpolarizabilities as large as those of p-nitroaniline can be induced in pure hydrocarbons of comparable size (4). Such a phenomenon comes about if the π-electron system of a molecule can be made to have spatially displaced centers of high electron affinity and ionization potential. Examples being,

, etc.

Insight can be gained into the reasons for such behavior from considerations of simple properties of π-orbitals (5, 6). According to Huckel theory, and other approaches, the mth eigen-orbital is described as a sum of admixtures of the atomic π-symmetry orbitals,

$$\Psi_m = \Sigma_i c^m_i \varphi_i \ . \tag{2.1}$$

π-electron systems may be classified as alternant or nonalternant depending on whether the atomic sites may be separated into two groups wherein members are bonded only to members of the other group. This connectivity is important since it will be impressed onto the form of the secular determinant. There is a theorem which states that the π-electron density in a single-determinantal ground state is exactly unity on every site of an alternant compound,

$$\Sigma_m 2 |c^m_i|^2 = 1 \tag{2.2}$$

(summation over filled orbitals only, the factor of two taking into account spin degeneracy). Thus charge is evenly distributed. Also due to the conservation of the Hilbert space, a similar theorem holds for the unoccupied orbitals. Thus there is little chance that an asymmetrical polarization will be induced for oppositely polarized electric fields, that is, even-order nonlinear polarizabilities will vanish. On the other hand, for nonalternant systems these rules do not hold and it is likely that nonvanishing and unequal dipole moments will be observed in the ground and excited states, thus even-order hyperpolarizabilities such as β will have finite magnitudes.   In the examples cited, it is clear that ring stresses will lead to differences in electronegativity and thus, as our calculations indicate, consequential asymmetric polarizability β.

Another picture of this phenomenon can be considered.   Recall that there is "orbital pairing" in alternant π-systems.   This gives rise to a pseudosymmetry label: pseudoparity.   Since the two-electron interaction connects only similarly labelled orbitals, even when configuration interaction is included the states of an alternant π-system can be classified even or odd under pseudoinversion. Electronic operators are odd or even depending on the relative pseudoparities of functions they connect with nonvanishing matrix elements.   The eletric dipole operator,   as   with   systems   described   by   real   physical   inversion   sysmmetry, connects states only of different (pseudo)parity.   Thus, since β as   calculated   by the sum-over-states method is a sum of terms each proportional to a product of three matrix elements of the dipole operator taken sequentially between three states, $<0|\mu|i>$ $<i|\mu|j>$ $<j|\mu|0>$, it must be zero-valued by parity selection rules in physically centrosymmetric molecules and zero-valued by pseudoparity selection rules in alternant π-electron systems (to the extent these are valid descriptions of the electronic states).   Thus we expect that for alternant hydrocarbons, even though inversion symmetry does not hold, β should be minimal as is seen, for instance, in phenanthrene.   (See also the case of quinoline below.)

2.2. Bystander Effect

Another prediction from our CNDO/Finite Field calculations is that the addition or repositioning of a substituent can have major effect on the magnitude of β (4). An example is the prediction:

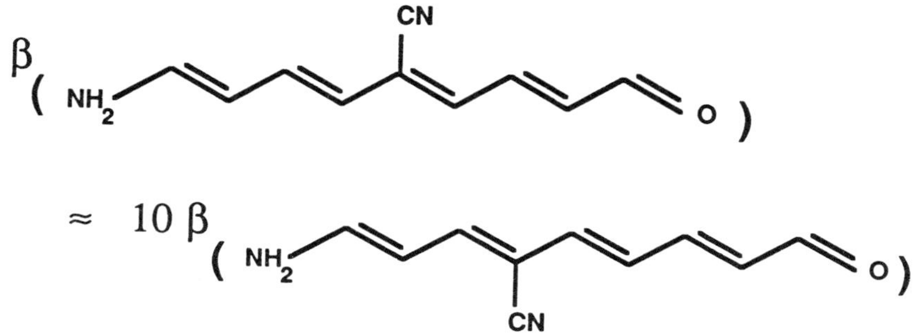

Other examples of the bystander effect are contained in reports of calculations by others.   For example, results of PPP calculations of Dirk, Twieg and Wagniere (7) can be summarized as follows.   Aza substitution at positions 2 or (2 and 6) diminish β while at positions 3 or (3 and 5) enhance β in   1-amino-4-nitrobenzene. Similarly substitution at the α position of the ethylene group (nearest the amino) enhances β while β-substitution or simultaneous (3, 5, 2' and 6') substitution diminish β.   In contrast to our example, these changes are all less than a factor of two from the parent compound's β.

There is a simple rationalization of enhancements, which is appealing in these cases, but not universally applicable to all derivatization-induced hyperpolarizability changes.   For molecules which exhibit very intense, low-lying charge-transfer bands in their absorption spectra (such as seen in p-nitroaniline, DANS and others), a large contribution to β is contained in the two-level, or "diagonal", contributions:

$$\beta(\omega_1,\omega_2)_{\text{diagonal}} = \Sigma_e \, \beta_e^{TL}(\omega_1,\omega_2) \qquad (2.3)$$

$$\beta_e^{TL}(\omega_1,\omega_2) = D_e \, \mu_{e0}^2 \, [\mu_e - \mu_0] \qquad (2.4)$$

$$D_3 = \frac{4\pi^2 \, [3\omega_e^2 - (\omega_1 + \omega_2)^2 + \omega_1\omega_2]}{(h/2\pi)^2 \, [\omega_e^2 - \omega_1^2] \, [\omega_e^2 - \omega_2^2] \, [\omega_e^2 - (\omega_1 + \omega_2)^2]} \qquad (2.5)$$

As has been discussed extensively this contribution is maximized by a compromise between increasing $\Delta\mu = [\mu_e - \mu_0]$ and decreasing $|\mu_{e0}|^2$, plus variations of the dispersion function, in simple donor-conjugation-acceptor systems as the strengths of the donor and/or acceptor are increased. However, the extreme zwitterionic and nonpolarized resonance structures used to illustrate the importance of these mechanisms, have long ago been shown to be inaccurate. The dipole moment changes are the result of systematic charge displacement including the conjugation system (8). When one looks at the density and differential-density contours of, say, the orbitals in 1-amino-4-nitrobenzene (Figure 1) which make up the charge-tranfer configuration which dominates the charge-transfer state (after configuration interaction) which dominates in determining β, one is struck by the large number of zero- or low-values systematically placed throughout the molecule. This appears to be a consequence of the nodes in the orbitals. As is well known to chemists, one can predict much

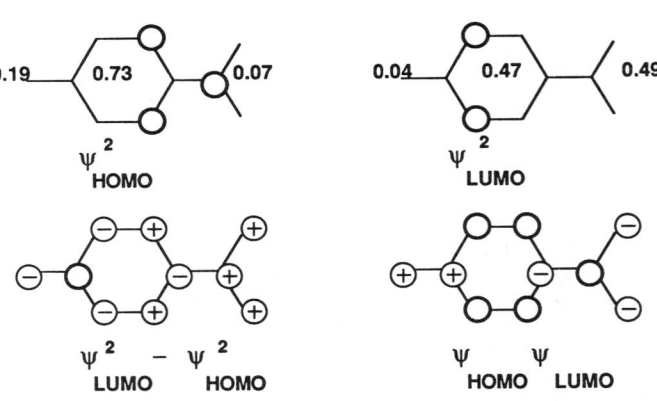

FIGURE 1. Electron Densities Associated with Two States of p-Nitroaniline (after Ref. 3). The odd-alternant hydrocarbon structure from which it may be thought to derive is shown. Electron densities within and between the HOMO and LUMO are depicted. Densities calculated within the amino, nitro and phenyl groups are listed. Zero- (or very small-) valued contributions at atomic sites are shown by empty circles. The differential densities show signs of significant site-contributions.

from orbital nodal patterns. Most trivial, but important spectroscopically and for the determination of hyperpolarizability, is the fact that very large oscillator strength is observed in "linear systems" due to the large transition dipole moments which connect orbitals differing by one node count (see below). Here it is suggested that an important mechanism of β enhancement is disruption of these nodal patterns which increases the oscillator strength of charge-transfer transitions.

As has been done extensively in the early years of color chemistry, consider the π-electron systems to be described in the MO picture with the effects of substituents to be perturbations of the corresponding pure hydrocarbon case (5, 6). Normally this is done to describe energies (e.g. Dewar's rules). These energy variations were suggested to be the dominant mechanism, through the dispersion factor D_e, for the variation of β with aza-substitutions in p-nitroaniline and its stilbene analog (7, 9). However, when tested quantitatively under the approximation of the single two-level model of Eqs. 2.4 and 2.5 appropriate to SHG of Nd/YAG laser radiation, the deviations between predictions and the PPP values are, on average, ± 22 percent. Wavefunction perturbations should then also be considered. This particularly since the spectra of the cyano-push-pull-polyenes are not so variable.

The above examples of "bystander effect" are related to odd alternant hydrocarbons. Odd alternants have unequal numbers in the two connecting groups as defined above and illustrated for p-nitroaniline in Figure 1. By convention the members of the larger group are labelled "starred" and the smaller "unstarred". Due to orbital pairing, there is a nonbonding orbital (an orbital which is not stabilized by electron delocalization below the one-site energy) which can be shown to have vanishing admixture coefficients at the unstarred sites, $c^{nb}_{i=unstarred} = 0$. On doubly filling all orbitals to obtain the single determinantal ground state, this orbital turns out to be the highest-lying occupied molecular orbital (HOMO). There are two effects which limit the magnitude of the dipole moment $\mu_{L \leftarrow H}$ between this electron configuration and that where one electron has been excited from the HOMO to the lowest-lying unoccupied molecular orbital (LUMO). Disregarding site polarization, ($\underline{R}_i$ is the position vector of the ith atomic site)

$$\mu_{L \leftarrow H} = \langle L|\mu|H \rangle \approx \Sigma_i e \, \underline{R}_i \, c^{L*}_i c^H_i \qquad (2.6)$$

There are two factors which effect this quantity. First, note that the admixture coefficients generally appear as spaced sampling points on sine-like waves of differing periods along the molecule. This is the n to (n+1) node problem. The majority of the result comes from the ends of the molecule where the wavefunctions "rephase" and overlap coherently, but with changed sign. Second, the occurrence of exact zeroes on atomic positions further reduces the possibility of additions to this sum from the unstarred locations when H is a "nonbonding" HOMO. If a perturbation can disrupt these effects, there will be an increase of . $\mu_{L \leftarrow H}$. An examination of the transition electron density contours for p-nitroaniline (see Figure 1) shows the dominant contributions to be at the molecular ends, but with deviation of the zero rules at perturbed carbon positions. One expects the strongest effect to occur when a substitution occurs at one of the unstarred locations, as observed in all calculated "bystander" examples above. A significant role of the substituent in this "bystander effect" is to enforce a coherence of the wavefunctions in the middle region of molecules and thus to modify oscillator strength. Both variations in dispersion factor and wavefunction can be seen to be important with this simple picture. Also the qualitative rule it creates is important, unless one really wants to do calculations before every molecular modification consideration.

The concept of "bystander interference" is important in development of second-order nonlinear polymers. First it provides a fairly volume-efficient means to enhance nonlinearity. Second, there is an implicit warning that "grafting" of nonlinear chromophores to polymers, even with alkyl groups, may not be an innocuous process. One should consider whether the choice of grafting methods will help (or hinder). A phenomenon related to the "bystander effect" is the modification of "anharmonicity" by substitution illustrated below.

3. MOLECULAR PROPERTIES: EMPIRICAL STRUCTURE-FACTOR TRENDS

As stated above, collection of a larger database of molecular hyperpolarizabilities is important. Over the past few years we have developed the techniques of dc electric-field-induced second-harmonic generation (EFISH) and third-harmonic generation (THG) to allow rapid and precise determinations of the third-order susceptibilities of liquids (1, 2, 9). Most recently we have devised a method for simultaneous EFISH and THG (SET) characterizations (2). We have modified the procedure to remove the need for fitting algorithms, though they do improve the precision, by separating the amplitude and coherence-length aspects of the characterization into separate experiments. This eases the requirements on signal-to-noise for individual data points since it is not neccesary to fit a sinusoidal function when the amplitude is determined. That is important since our EFISH signals arise at the technically ackward spectral region of 954 nm. Since the liquid's dispersion can be seen whether or not it is an SHG source, the data for coherence length determinations may be improved by using crystalline quartz windows, which produce much larger and more stable signals. Application of the simplistic liquid models allows the hyperpolarizabilities of interesting molecules to be estimated. It is important to remember that the techniques are still evolving and that prior works in this area suffered from various errors of methodology or analysis. Hopefully these results will stand the test of time.

It is important to remark that our measurements are done with a 1.908 μm wave length fundamental. Dispersion is a dominant factor in the nonlinear response of molecular systems (11). To uncover global structure-property relationships, it is vital to work in a regime where dispersion enhancement due to specifics of the low-lying electronic excitations are unimportant. Similarly, it is important to work at frequencies above the vibrational resonance enhancements. 1.908 μm is virtually the best choice under these criteria. As illustrated in other parts of this workshop, dispersion is so important in highly conjugated polymers that it may not make sense to speak of a nonresonant susceptibility. In normal molecules, on the other hand, the concept of the low-frequency hyperpolarizability is a meaningful quantity to investigate and from which to see the global effects of structural variations.

3.1. Bond Additivity

Though not of much significance technologically, the behavior of saturated compounds is interesting. In analogy to the treatment of linear polarizability many decades ago, it is interesting to pursue the idea that the nonlinear polarizabilities of such compounds might be described by local bond, or group, polarizabilities (10, 11). In this model one assumes that the molecular hyperpolarizability can be expressed as the sum of the hyperpolarizability of these units. For isolated molecules there is a very serious problem in that the induced linear polarization of neighboring groups by the true nonlinear polarization will be as large as the latter. (Linear polarizabilities are approximately equal to the volume of a system, and the dipole fields scale as (p/r^-), that is, the proportional magnitude of the inducing fields are approximately equal to the inverse of the volume. Thus the linearly induced polarization is comparable to the "original" bond or group hyperpolarization.) How one might take account of this, considering the very anisotropic distribution of polarizability around bonds of a molecule, is a subject to debate. For molecules in solution, a mean local field is already present and used in the simple liquid models to determine molecular hyperpolarizabilities. Whether one assumes it is established by the mutual polarization of neighboring molecules or whether it is created by the mutual polarization of neighboring groups or bonds is simply a matter of choosing the identity of the microscopic polarizable units. Therefore the problem is much less significant than for isolated molecules.

The third-order polarizabilties of some alkane systems were studied by THG in liquids (10). The corresponding study by EFISH (or one of the nonlinear refractivity methods) would be interesting, but difficult to interpret since, due to conformational flexibility, the molecular polarization response would include a part due to the redirection of noncentric bonds which would be difficult to separate. Within each of the homologous series, a very good linear dependence of the orientationally averaged hyperpolarizability on the number of carbon atoms in the compound was observed. This is the prediction of the additivity model. That is, when a mean local-field model is adopted, the orientationally-averaged hyperpolarizabilities of the molecule and of the groups or bonds are simply related,

$$\gamma = \ <\gamma>_{1111} \ = \ <\Sigma_i\, \gamma_i>_{1111} = \Sigma_i <\gamma_i>_{1111} \ . \qquad (3.1)$$

The latter equation holds because the orientational average is taken over an isotropic distribution. Though not generally true, this case can be proved by noting that only the L=0 scalar irreducible-spherical-tensor components embedded in the γ tensor survive the averaging (12). These scalar components are independent of coordinate system and, therefore, of geometric details of molecular bonding and conformation.

If additivity of nonlinear polarizability does indeed hold, then the precision of the technique would be reflected in the goodness of the linear behavior of the molecular hyperpolarizabilities as function of m in saturated hydrocarbon series containing m ($-CH_2-$) units. Linear behavior with very little scatter of data was observed. The parameters for the best lines according to the least squares criteria are listed in Table 1 for several series. The slopes of these lines represent the change on addition of one ($-CH_2-$) unit. The measured differences in the slopes are significant within the limits of this technique. As can be seen from the subtle variations between series, a high redundancy of parameters is essential if significant bond-polarizability values are to be derived.

TABLE 1. Least Squares Best Linear Fit Parameters To $<\gamma>_{1111}$ In Homologous Hydrocarbon Series. $<\gamma>_{1111}$ was determined in liquids using 1.908 micron wavelength fundamental and employing Lorentz-Lorenz local field models.

Hydrocarbon Series	(Size Range)	Slope $(10^{-36}\,esu)$	Intercept $(10^{-36}\,esu)$
n-alkanes	(m=6-16)	0.495 ± .006	0.49 ± .07
cycloalkanes	(m=6-8,10)	0.464 ± .006	0.23 ± .05
n-alkylcyclohexanes	(m=7-10)	0.502 ± .021	0.40 ± .19
1-alkanols	(m=1-5,7-12,15)	0.506 ± .003	0.41 ± .03
1-chloroalkanes	(m=3,5-10,14,16)	0.513 ± .003	0.49 ± .03

We do not wish to dwell on the implications of the slight, but experimentally valid differences in slopes, which may be speculatively attributed to progressively diminishing perturbations down the alkyl chains by the substituents or to microscopic breakdown of the mean local field model around the substituents. What is most significant to us was the observation of severe dependence on isomeric form. Isomers of simple alkanes have the same number of carbons, hydrogens, carbon-carbon bonds and carbon-hydrogen bonds. Despite this, as shown in Table 2, in marked contradiction to the additivity model, the apparent hyperpolarizabilities vary substantially.

TABLE 2.   Values Of $< \gamma >_{1111}$ In Some Branched Alkanes Compared To Bond Additivity Predictions.   The BAA predictions come from the n-alkane series.   Units are 10^{-36} esu.

Alkane	(Size)	Observed	Predicted	Deviation (percent)
2-methylpentane	(m = 6)	3.53	3.46	+ 2
3-methylpentane	(m = 6)	3.40	3.46	- 2
2 3,4-trimethylpentane	(m = 8)	4.55	4.45	+ 2
2 2,4-trimethylpentane	(m = 8)	5.17	4.45	+ 1 6
2 2,3-trimethylbutane	(m = 7)	4.31	3.96	+ 9
hexamethylethane	(m = 8)	6.46	4.45	+ 4 5

One may speculate on the cause of this breakdown of additivity. First, of course, one may challenge the liquid model suggesting that the treatment of local fields is too naive. Second, one might consider that primary, secondary and tertiary carbons are not equivalent and that the additivity model should acknowledge these differences by adopting more parameters. This view is appealing as a means to begin rationalizing both this behavior and the nature of the intercept values in the various homologous series. Finally, one might recognize that the additivity model would break down if marked differences between a valence-bond and molecular-orbital picture of the electronic system existed. For example, the molecular-orbital nature of delocalization through sigma-bonds has recently been identified to be an important aspect of silanes. This is an appealing manner to view the dependence of averaged hyperpolarizability in the cyclic alkane series, an aspect which cannot be attributed to variation of carbon type (though one might attempt to save this picture by including strain dependence of bond properties). We do not attempt to provide the answer, but suggest that this is an interesting problem for more rigorous theoretical investigation.

3.2. Structurally Induced Anharmonicity
Organic materials have been of substantial interest for second-order nonlinear optics due to the very large microscopic polarizability that can be created and due to the great flexibility of composition, assembly, structure and fabrication which they provide. There are currently serious efforts at technological implementation. On the other hand the potential utility of third-order nonlinear optical effects in organic materials is still an open question. A major reason for this is the limited magnitude of total nonlinear

response that can be usefully created with organic structures. In the absence of clear guidelines as in second-order, a major research area is the search for factors which contribute to third-order nonlinearity in π-electron systems.

Enthusiasm for nonresonant third-order response in organic materials derives from work on linear conjugated molecular and polymeric systems. Over a decade ago, it was shown that the "nonresonant" electronic response of these systems is quite large owing to delocalization, holding out the prospect of low loss, high speed optical processing with devices built from related materials. (13) Unfortunately, since that time, the magnitudes of known usable nonresonant responses (other than through dispersion enhancement) have not been substantially improved, though many interesting studies of resonance effects, guided-wave effects and processing of materials have been reported. Logically, several groups are pursuing aspects of the nonlinearity in conjugated systems. We would like to summarize the results of an empirical test of an overlooked viewpoint, due to Rustagi (14), which merits attention in this regard (15).

Often, the delocalization of electrons along linear conjugated systems, is cited as a major cause of large nonresonant third-order susceptibilty. Delocalization, of course, is effectively limited by electron correlation, vibronic coupling (electron-phonon coupling) and by structural modifications such as alternating and superalternating bond lengths. Rustagi developed a different viewpoint through a series of simple calculations. He calculated zero frequency, electronic, linear and third-order polarizabilities for polarization along the chains of some model one-dimensional systems. Of course the expected increase of polarizability with length, that is, with delocalization, was observed. However, an important insight came from plotting the calculated nonlinear polarizabilities against the calculated linear polarizabilities. A power-law relationship was observed,

$$\gamma_{zzzz} \approx K_{series}\, \alpha_{zz}{}^{n} \quad ; \quad n \sim 7/3 \quad . \tag{3.2}$$

The constant K_{series} is taken to be a constant for a particular chemical series, but varies between classes of molecules. Large nonlinearity then is the result of two factors: a large linear polarizability and a large "structural anharmonicity factor", K_{series}. In the calculations $K_{polyene}$ was seen to be much greater than $K_{cyanine}$ (simple symmetric cyanines). However, the much lower calculated polarizability of the polyene pi-system, was seen to severely limit the magnitude of nonlinear polarizability. On the other hand, the linear polarizability of simple cyanines were seen to be much larger, due to the near equivalence of the bonds (caused by the two resonance structures of cyanines) and thus large delocalization. It was observed, though, that the small magnitude of $K_{cyanine}$ indicated that the electron cloud acted as if it were bound by a nearly perfectly harmonic potential against polarization deformation. Larger nonlinearity would be achieved if large polarizability could be preserved while K_{series} was somehow enhanced. Rustagi predicted that replacement of a CH group in the center of the cyanine ion by a nitrogen atom would accomplish this. Considering the value of such an approach, we have empiricaly tested these predictions.

Table 3 contains some results. It is necessary before viewing these data, though, to recall that these results were determined by THG with the harmonic appearing at 636 nm wavelength. Dispersion is a limitation for both the objective interpretation of these trends and as an effective limitation on the length of molecule which can be studied. For cyanine dyes, unlike polyenes, the optical edge continues to close as lengths increase, thus preventing examination of a long-chain member of the series. Secondly, these results are the idealized orientational averages using a mean local-field approximation, and are the total molecular response including the sigma electrons in addition to molecular fragments besides the simple linear-chain chromophores. These extra molecular

TABLE 3. Linear and Nonlinear Polarizabilities of Conjugated Molecules. Optical THG was performed with 1.908 μm wavelength fundamental. The wavelength at the maximum of the lowest-lying electronic absorption bands are also indicated. N is the number of double bonds in the linear chain system. Abbreviations used for polyenes are: TMP2 = 2,5-dimethyl-2,4-hexadiene; TMP3 = 2,7-dimethyl-2,4,6-octatriene; TMP4 = 2,9-dimethyl-2,4,6,8-deca-tetraene; DMP2 = 2,3-dimethyl-1,3-butadiene.

Compound	N	λ_{max} (nm)	α (10^{-23} esu)	γ (10^{-36} esu)
simple cyanines (I)	2	312	2.05	2 ± 1.5
	3	416	2.95	52 ± 1
	4	519	3.95	-510 ± 50
simple azacyanine (II)	2	268	1.64	$5.7 \pm .3$
thiazoles (III)	2	336	3.15	16 ± 5
	3	448	4.45	-100 ± 50
azathiazole (IV)	2	279	2.86	110 ± 80
benzthiazoles (V)	2	423	5.3	-30 ± 30
	3	561	10.8	-470 ± 300
azabenzthiazole (VI)	2	369	4.92	490 ± 40
DMP2	2	226	1.13	$4.9 \pm .1$
TMP2	2	242	1.57	$7.4 \pm .1$
1,3,5-hexatriene	3	258	1.18	$9.1 \pm .4$
TMP3	3	262	2.55	$38 \pm .5$
TMP4	4	309	~6	400 ± 50
beta-carotene	11	452	127	$11,000 \pm 2500$

pieces were required for chemical reasons and to allow the testing of enough accessible cyanines to prove the general validity of this prediction.

Clearly, despite the inclusion of sigma contributions, which would serve to enhance the short oligomer regime, very strong length-dependence of the magnitude of hyperpolarizability is observed in polyenes, simple cyanines, and thiazole and benthiazole cyanine derivatives. The occurrence of negative-valued quantities is also significant, but not discussed here. The important observation here is the substantial effects of symmetrical aza-substitution. Despite the fact that for all three compounds listed the longest wavelength transition moved to higher energy and the polarizability, as determined from refractive indices, decreased from the parent compound, the hyperpolarizability increased by an order of magnitude. This verifies the essence, if not the details, of Rustagi's predictions.

It is important to realize that the "structural anharmonicity factors" are not universal for specific substitutions across different classes of conjugation systems. For example, the simple cyanine dyes are differentiated from polyenes not only by their charge but more significantly by the occurrence of nearly equal bond lengths. In cyclic compounds benzene has that trait in common; the two resonance structures of alternating single and double bonds are a crude picture indicating that the aromatic bonds are all equivalent in benzene. Aza substitution in benzene produces pyridine, but it is known that this substitution reduces the orientationally averaged γ from 3.85×10^{-36} esu to 3.36×10^{-36} esu, in direct contrast to the cyanine case just illustrated.

Implications of this viewpoint pertaining to the development of nonlinear polymerics are significant. Larger nonresonant nonlinearity can be achieved by pursuit of the structure-property relationships of K_{series}. It is therefore important to investigate derivates and doped modifications of more highly polarizable polymers and their shorter oligomeric relatives. These derivatives must be such that they cause significant variants within the conjugation system, rather than being distant pendant group changes which simply give rise to variations in conformation of the same basic conjugated backbone. It is variation of the microscopic potential in the delocalized electronic system which gives rise to the nonlinearity enhancement in Rustagi's mechanism.

3.3. Ionic Character

Besides the need to uncover traits which will substantially enhance γ and $\chi^{(3)}$, there is a need to learn more details about the nonlinear polarization response of smaller molecular enties. It's fair to say that in this area, the level of knowledge is very rudimentary. From an earlier work, in which the power of a new method of THG characterization was demonstrated, a rudimentary observation was made

TABLE 4. Group Additivity Values of Third-Order Hyperpolarizability. Values are calculated from 1.908 μm wavelength THG in neat liquids. For benzene substituents values are calculated as $\{\gamma_{X-\phi} = \gamma_{X\phi} - \gamma_\phi + \gamma_{H-C}\}$. For nonaromatics γ_{X-C} were calculated from γ's among various simple compounds.

Substituent, X	$\gamma_{X-\phi}$ (10^{-36} esu)	γ_{X-C} (10^{-36} esu)
fluorine	-0.69	
chlorine	0.46	0.77
bromine	1.52	
iodine	4.31	
methyl	0.67	0.55
amino	1.83	
nitro	1.49	
cyano	0.23	0.33
(pyridine)	-0.52	

about the effects of simple substitutions of the benzene ring (11). Applying the concept of bond additivity to a small set of saturated compounds, a set of bond polarizabilities was determined. As noted above these are suspect, coming from such a small set. However, when these X-C hyperpolarizabilities are contrasted to the correspondingly derived X-benzene hyperpolarizabilities a clear trend is observed as shown in Table 4. Nonlinearity is enhanced over that expected based on observations of saturated compounds for π-donating groups and is diminished in π-accepting groups. (There is also an indication of reduction due to inductive or attractive perturbation, e.g., in pyridine and the lower halobenzenes.) It is well known that linear and nonlinear polarizability of atomic systems increase in the order cation < neutral < anion. This trend has also been predicted in calculations for organic molecules (16) though the simple interpretation that the electron cloud is more loosely or tightly bound depending on the ionic charge is less complete for large molecules. Since the dominant contribution to hyperpolarizability is expected from π systems, attributing this trend in mono-substituted benzenes largely to the effect of the variation of hyperpolarizability with ionic nature seems warranted.

3.4. Asymmetric Dielectric Properties: Naphthalenes

Using the SET methodology the variation of second-order hyperpolarizability in structurally related series of molecules have also been investigated. For example, one might ask whether since rather crude resonance-structure pictures of charge-transfer lead to choice of the best isomeric form of, e.g. nitroaniline, do they apply to related polyacenes? Or, within a specific isomeric form, e.g. 4,4'-substituted stilbenes, what are the consequences of variation of the donor or acceptor? Is the area of molecular second-order nonlinearity "solved" or will unkown aspects be observed?

In Figure 2 some aspects of substitution and derivatization of asymmetric dielectric properties of naphthalene are summarized. In interpreting such data it is important to remember that the β_{EFISH} is the resultant effective molecular nonlinearity after some complicated approximation of the behavior of a solution. It is, ideally, equal to the projection onto the permanent dipole moment of the L=1 (vector) irreducible spherical-tensor portions of the full SHG β tensor. Consequently, simple empirical comparisons of β_{EFISH} magnitudes, even though measured at low optical frequencies to allow dispersion-insensitive global comparisons between compounds, would not be accurate but must be moderated by a sense of the tensorial nature of the polarization and projection processes. The trends seen in these cases are simple enough that the effects of interest are clear.

Naphthalene has zero-valued μ and β. Replacement of a CH unit by a nitrogen atom to obtain quinoline creates substantial ground-state electronic polarization as indicated by the resultant approximately 2 D dipole moment. This is, however, not greater than the dipole moment of pyridine, indicating, as will also be seen below for 1-amino-4-nitronaphthalene, that the second ring is not of much significance for polarization. Curiously, the large dipole moment is not accompanied by large β. This is perhaps a consequence of a relation of the excited-state wavefunctions of quinoline to those of naphthalene. The pseudoparity selection rules in alternant hydrocarbons which diminish β (see above) would apply in that event. α-nitro substitution on naphthalene is seen to be a much stronger perturbation producing larger μ and β. 5-nitro substitution of quinoline continues this series. It is seen that competition between aza and nitro electron affinities results in a smaller μ. Interestingly, however, these competing electron withdrawing groups set up a charge asymmetry that results in a large value of β (half of p-nitroaniline). Since undoubtedly the lowest lying unoccupied molecular orbitals have a large admixture of the nitro-group's π^* orbital character, the excess π-electron density which is created around the nitrogen might be considered to act as an electron "donating" region. Though

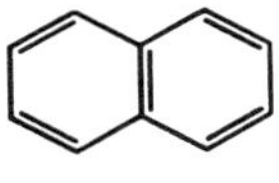

μ = 0
β = 0

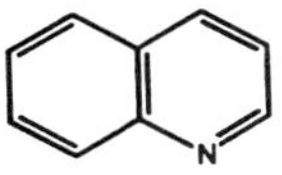

μ = 1.9
β = .25

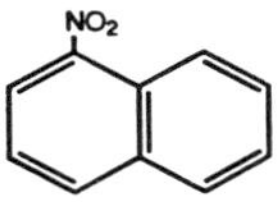

μ = 3.7
β = 2.7

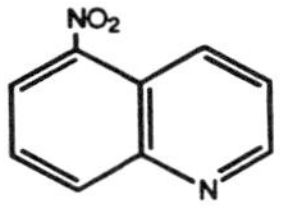

μ = 2.2
β = 6

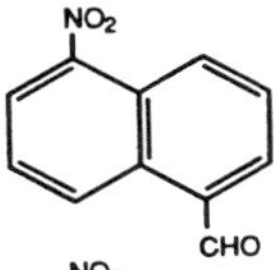

μ = 2.2
β = 13

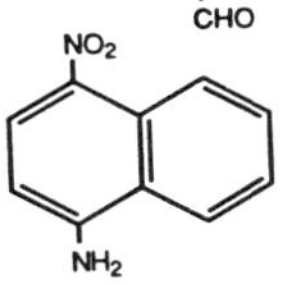

μ = 5.6
β = 10

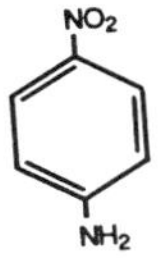

μ = 6.5
β = 11

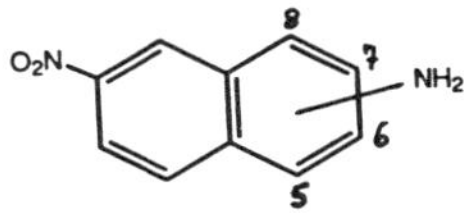

Position	μ	β
5	(4.9)	(8)
6	6.4	13
7	5.2	8.1
8	4.2	17

FIGURE 2. Asymmetric Dielectric Properties of Naphthalene-Related Compounds. SET optical measurements were performed with a 1.908 μm wavelength fundamental. Sovent was p-dioxane. μ and β are in units of Debye and 10^{-30} esu, respectively.

this line of argument forms an analogy to the prototypical p-nitroaniline, the aza substitution is by no means a π-electron donor. Rather it is preferable to note that the centroid of electron-affinity density is displaced from the centroid of ionization-potential density in this class of compound, a recipe for enhancement of β, as indicated earlier. Taking this picture one step farther, the use of a recognized weak electron acceptor as substituent on naphthalene in place of the heterocyclic nitrogen, (1-formyl-5-nitronaphthalene) results in an increase of β by a factor of 2. Thus double acceptor-substitution is seen to yield appreciable second-order nonlinearity. This effect will again be illustrated below with stilbene compounds.

Continuing the probe of the naphthalene class, one naturally would like to know if the resonance-capable isomers enhance donor-acceptor susbstituted naphthalenes as they do in p-nitroaniline (17). For simultaneous amino- and nitro-substitution, it is seen that the 1,4-isomer of naphthalene possesses μ and β not significantly different than in the same isomer of benzene. The second ring and all of the additional electronic states which accompany the change from the single ring system, seem to be inconsequential for polarization along the intermediate molecular axis. The question of the isomeric dependence when both rings are substituted is more interesting.

Based on resonance structures one expects the (1,3-) or (2,8-), (1,5-), and (2,6-) isomers to be β-enhancing. However, our Finite-Field calculations indicate this may not be the case. The author's have also been privately informed that pK_a's of the related hydroxynitronaphthalene's do not follow the implications of these resonance structures. Therefore, isomers of the aminonitronaphthalenes have

118

been obtained and characterized. As seen in Figure 2 the (2,6-) and (1,3-) or (2,8-) isomers are the most nonlinear. Considering the changes of dipole moment directions, as mentioned above, this is not an entirely rational comparison. The effect of reduced symmetry is to distribute hyperpolarizability magnitude onto additional tensor elements. Given this, though, (and disregarding the prospects of negative signed interfering contributions) the EFISH comparison biases observation in favor of the strongest expected resonance-structure effect, the (2,6-) isomer. That this is not an overwhelming result indicates that the additional states or, more physically, the extra ring polarizability available when both rings are actively involved must be considered in a description of the origins of β of these isomers.

3.5. Asymmetric Dielectric Properties: Nitrostilbenes

Another set of systematically varied compounds which has been investigated consists of 4-substituted-4'-nitrostilbenes. It allows to see variations of dielectric properties without concern for geometric variations. Crystals of these compounds have also been characterized regarding crystal growth, crystal structures and SHG powder activity (18). Several are promising and fuller investigations are underway. Asymmetric dielectric properties determined by the SET methodology are listed in Table 5. Supplemental considerations indicate that these are trans isomers in solution.

TABLE 5. Asymmetric Dielectric Properties of 4-Substituted-4'-Nitrostilbenes. SET optical measurements utilize a 1.908 μm fundamental wavelength. Sovent was p-dioxane.

Substituent	μ (Debye)	β_{EFISH} (10^{-30} esu)
formyl	2.63	27
chloro	2.45	31
methoxy	4.62	37
hydroxy	4.77	41
bromo	2.42	43
dimethylamino	7.55	78

Largest μ and β occur with the strongest electron donor, dimethylamino. This is the well-known DANS compond identified to be highly active about a decade ago (19). In the other members of this series, as would be expected in molecules capable of charge-transfer along the molecular long axes and which are only weakly polarized in their ground states, decreasing π-donating strength leads generally to smaller values. While this trend is probably intuitively obvious, some important material aspects are present. For instance, the methoxy compound retains substantial hyperpolarizability, but has decreased color compared to DANS. Additionally, though not directly of interest to polymers, the smaller dipole moment was thought to be a good reason for investigating crystals (18). Other aspects are also important.

First, the magnitudes of μ and β are not necessarily correlated. This is a fact previously emphasized in N-oxides of heterocyclics (20). Analogously, it is seen here that the use of Br as both a π-donor and a σ-inductor allows large β with much smaller μ than the comparably nonlinear methoxy compound. Second, as with the naphthalenes, there are examples of reasonably large β when strong donors are absent On a volume-normalized basis the chloro and formyl compounds display ~3/2 the hyperpolarizability density of p-nitroaniline. Though nonlinearity dispersion has not been studied, it is significant that this occurs without a loss of transparency in the blue spectral region. It is important to find such molecular approaches which do not fit the sometimes stated generalization that increased nonlinarity is accompanied by increased absorption-edge wavelength (the efficiency-transparency trade-off).

An interesting derivative of MONS is CMONS, the β-cyano derivative. At first glance it appears to be a candidate for demonstration of "bystander effects". The measured β_{EFISH} is, however, 10% less than the hyperpolarizability of MONS. The cyano substitution occurs at a starred location, so according to the discussion above, a diminution rather than enhancement is observed. CMONS, however, demonstrates very important and interesting crystal behavior (18).

A number of other structural dependences of second-order nonlinearity in other derivatives of stilbenes, styrenes, polyenes and diverse molecular systems have been investigated. Polymeric and surface-ordered second-order nonlinear materials are very appealing in part due to the ability to choose and modify in fine detail molecular compositions and structures and then obtain correspondingly planned variations of the macroscopic properties (21, 22, 23). This can be expected without suffering unpredictable gross changes in structure as occurs in crystals. The rational design of the molecular components of these systems requires more and finer detail about molecular polarizability behavior, both as "isolated" molecules and subject to intermolecular interactions. The power of the SET methodology, as illustrated above, allows empirical investigations of many finer aspects and variations from the lmited set of well-know and documented molecular systems.

4. CONCLUSION

In this paper a range of theoretical and empirical molecular hyperpolarizability structure-factor relationships have been presented and discussed. As a serious study of the literature will show, our knowledge of electronic hyperpolarization accross the broad spectrum of organic compounds is really quite rudimentary at this time. This is not a major shortcoming in our ability to currently find, or design and assemble, highly nonlinear second-order materials. But, as illustrated herein by the "bystander effect", there are aspects of these activities which are more subtle than might be expected from straightforward application of "rules" for large β. It is our belief that the verified details of electronic third-order nonlinear polarizability in general molecular systems are really quite scarce. Given the current shortage of organic materials with exceptionally large susceptibilities, it is a useful correlary activity to the search for such materials among conjugated polymers to investigate the more basic molecular properties as we have illustrated.

APPENDIX: MICROSCOPIC-MACROSCOPIC RELATIONSHIPS

In this appendix a straightforward derivation of the nonlinear dielectric constants of a molecular material is summarized (24). With the intention to use these expressions where no zero-frequency perturbations of the ensemble result, a frozen-gas model is adopted. Should this not be the case, the final expressions could be assumed to be the zero-perturbation expressions for the ensemble and additional "ensemble distribution perturbations" could be added by a straightforward Taylor expansion.

The starting point is the recognition that the microscopic field which polarizes molecule κ is the sum of the Maxwellian electric field plus contributions from the dipole moments on all of the other molecules λ in the ensemble,

$$E_m{}^\kappa = E + \Sigma_\lambda \; L^{\kappa\lambda} \cdot p^\lambda \quad . \tag{A.1}$$

If one assumes point dipoles, one can adopt the relationship

$$L^{\kappa\lambda} = (3 r^{\kappa\lambda} \, r^{\kappa\lambda} - |r^{\kappa\lambda}|^2) / |r^{\kappa\lambda}|^5 \quad . \tag{A.2}$$

However, it is well known that the point dipole approximation is not good for larger, anisotropic molecules which are densely arranged. Nevertheless, in the spirit of single polarizability tensors and a uniform field acting on each molecule, it is assumed there are some $\mathbf{L}^{\kappa\lambda}$ which describe this relationship between $\mathbf{p}^{\lambda}$ and its contribution of field at molecule κ. If one defines the tensors $\mathbf{M}^{\kappa\lambda}$ and their "inverses" $\mathbf{R}^{\lambda\mu}$,

$$\mathbf{M}^{\kappa\lambda} = \delta_{\kappa\lambda} \mathbf{U} - \alpha^{\kappa} \cdot \mathbf{L}^{\kappa\lambda} , \tag{A.3a}$$

$$\Sigma_{\lambda} \mathbf{M}^{\kappa\lambda} \cdot \mathbf{R}^{\lambda\mu} = \delta_{\kappa\mu} \mathbf{U} , \tag{A.3b}$$

one may obtain a tensor,

$$\mathbf{N}^{\kappa} = \Sigma_{\lambda} (\mathbf{R}^{\lambda\kappa})^{\mathsf{T}} , \tag{A.4}$$

which will be identified below as the "local field tensor".

The dipole moment on each molecule is a sum of the permanent, linearly induced and the nonlinearly induced moments. Generalizing Eq. 1.1,

$$\mathbf{p}^{\kappa} = \mu^{\kappa} + \alpha^{\kappa} \cdot \mathbf{E}_m{}^{\kappa} + \mathbf{p}^{(nl)\kappa} \tag{A.5}$$

$$\mathbf{p}^{(nl)\kappa} = \beta^{\kappa} \cdots \mathbf{E}_m{}^{\kappa} \mathbf{E}_m{}^{\kappa} + \gamma^{\kappa} \cdots \mathbf{E}_m{}^{\kappa} \mathbf{E}_m{}^{\kappa} \mathbf{E}_m{}^{\kappa} \cdots . \tag{A.6}$$

Simple substitution and rearrangement yields

$$\Sigma_{\kappa} \mathbf{p}^{\kappa} = \Sigma_{\kappa} \{ (\mathbf{N}^{\kappa})^{\mathsf{T}} \cdot \mu^{\kappa} + \alpha^{\kappa} \cdot \mathbf{N}^{\kappa} \cdot \mathbf{E}$$
$$+ (\mathbf{N}^{\kappa})^{\mathsf{T}} \cdot [\beta^{\kappa} \cdots \mathbf{E}_m{}^{\kappa} \mathbf{E}_m{}^{\kappa} + \ldots] \} . \tag{A.7}$$

One sees how with only linear polarizability $\mathbf{N}^{\kappa}$ appears formally to convert the Maxwellian to the microscopic field. It is defined only in terms of the linear polarizability, being not a bad approximation in general since nonlinear polarizability is small in comparison except in very intense fields. It is convenient to collect some terms into a new tensor

$$\mathbf{Q}^{\kappa\mu} = \Sigma_{\lambda} (\mathbf{R}^{\lambda\kappa})^{\mathsf{T}} \cdot \mathbf{L}^{\lambda\mu} , \tag{A.8}$$

which functions as a "dressed dipole transfer tensor". Two subsequent relationships, which demonstrate this functionality,

$$\mathbf{E}_m{}^{\kappa} = \mathbf{N}^{\kappa} \cdot \mathbf{E} + \Sigma_{\mu} \mathbf{Q}^{\kappa\mu} \cdot (\mu^{\mu} + \mathbf{p}^{(nl)\mu}) , \tag{A.9}$$

$$(\mathbf{N}^{\kappa})^{\mathsf{T}} \cdot \mu^{\kappa} = \Sigma_{\lambda} (\delta_{\kappa\lambda} \mathbf{U} - \alpha^{\lambda} \cdot \mathbf{Q}^{\lambda\kappa}) \cdot \mu^{\kappa} , \tag{A.10}$$

also suggest the definition of another tensor,

$$\mathbf{D}^{\kappa} = \Sigma_{\lambda} \mathbf{Q}^{\kappa\lambda} \cdot \mu^{\lambda} \tag{A.11}$$

which looks like the induced electric field at molecule κ due to all of the permanent dipoles - and their distributions - in the ensemble.

With the above expressions and definitions one can collect terms by powers in the Maxwellian field and identify them as the bulk constants.

$$\mathbf{P}_0 = v^{-1} \Sigma_{\kappa} (\mathbf{N}^{\kappa})^{\mathsf{T}} \cdot \{ \mu^{\kappa} + \beta^{\kappa} \cdots \mathbf{D}^{\kappa} \mathbf{D}^{\kappa} + \gamma^{\kappa} \cdots \mathbf{D}^{\kappa} \mathbf{D}^{\kappa} \mathbf{D}^{\kappa}$$
$$+ 2 \beta^{\kappa} \cdots \mathbf{D}^{\kappa} (\Sigma_{\lambda} \mathbf{Q}^{\kappa\lambda} \cdot \beta^{\lambda} \cdots \mathbf{D}^{\lambda} \mathbf{D}^{\lambda}) + \ldots \} \tag{A.12}$$

$$\chi^{(1)} = V^{-1} \Sigma_\kappa \{ (\alpha^\kappa)^T \cdot \mathbf{N}^\kappa$$

$$+ (\mathbf{N}^\kappa)^T \cdot [2 \ \beta^{\kappa \cdot \cdot} \ \mathbf{D}^\kappa \ \mathbf{N}^\kappa + 3 \ \gamma^{\kappa \cdot \cdot \cdot} \mathbf{D}^\kappa \ \mathbf{D}^\kappa \ \mathbf{N}^\kappa$$

$$+ 2 \ \beta^{\kappa \cdot \cdot} \ (\Sigma_\lambda \ \mathbf{Q}^{\kappa \lambda} \cdot \beta^{\lambda \cdot \cdot} \ \mathbf{D}^\lambda \ \mathbf{D}^\lambda) \ \mathbf{N}^\kappa + ...]\} \qquad \text{(A.13)}$$

$$\chi^{(2)} = V^{-1} \Sigma_\kappa \ (\mathbf{N}^\kappa)^T \cdot \{ \beta^{\kappa \cdot \cdot} \ \mathbf{N}^\kappa \ \mathbf{N}^\kappa + 3 \ \gamma^{\kappa \cdot \cdot \cdot} \ \mathbf{D}^\kappa \ \mathbf{N}^\kappa \ \mathbf{N}^\kappa$$

$$+ 2 \ \beta^{\kappa \cdot \cdot} \ \mathbf{D}^\kappa \ (\Sigma_\lambda \ \mathbf{Q}^{\kappa \lambda} \cdot \beta^{\lambda \cdot \cdot} \ \mathbf{N}^\lambda \ \mathbf{N}^\lambda)$$

$$+ 4 \ \beta^{\kappa \cdot \cdot} \ \mathbf{N}^\kappa \ (\Sigma_\lambda \ \mathbf{Q}^{\kappa \lambda} \cdot \beta^{\lambda \cdot \cdot} \ \mathbf{D}^\lambda \ \mathbf{N}^\lambda)$$

$$+ 6 \ \delta^{\kappa \cdot \cdot \cdot} \mathbf{D}^\kappa \ \mathbf{D}^\kappa \ \mathbf{N}^\kappa \ \mathbf{N}^\kappa + ... \qquad \text{(A.14)}$$

$$\chi^{(3)} = V^{-1} \Sigma_\kappa \ (\mathbf{N}^\kappa)^T \cdot \{ \gamma^{\kappa \cdot \cdot \cdot} \ \mathbf{N}^\kappa \ \mathbf{N}^\kappa \ \mathbf{N}^\kappa$$

$$+ 2 \ \beta^{\kappa \cdot \cdot} \ \mathbf{N}^\kappa \ (\Sigma_\lambda \ \mathbf{Q}^{\kappa \lambda} \cdot \beta^{\lambda \cdot \cdot} \ \mathbf{N}^\lambda \ \mathbf{N}^\lambda)$$

$$+ 4 \ \delta^{\kappa \cdot \cdot \cdot} \mathbf{D}^\kappa \ \mathbf{N}^\kappa \ \mathbf{N}^\kappa \ \mathbf{N}^\kappa + ... \} \qquad \text{(A.15)}$$

etc.

The large number of terms arises because the microscopic fields occur 1) due to $\mathbf{E}$ and $\mathbf{N}^\kappa$, 2) due to μ^μ, and 3) due to $p^{(nl)\mu}$ (see Eq. A.9). If only the first mechanism were active, there would be a one-to-one relationship between orders of microscopic nonlinear polarizability and macroscopic nonlinear susceptibilities. However, the other two sources of microscopic fields respectively "pull down" higher-order molecular nonlinearities into lower-order susceptibilities and "cascade" lower-order molecular nonlinearities into higher-order susceptibilities. These two mechanisms are almost never included in descriptions of molecular behavior, only the first terms of Eqs. A.12-A.15 usually being assumed. Considering the discovery that highly nonlinear molecular species exist, one can see the importance of their inclusion, since mutual polarization phenomena are so important in condensed-phase media.

In Eqs. A.12-A.15 multiplicative numeric factors occur because terms are generated wherein the ordering of factors are interchanged. Since the presentation above does not specifically label frequency, the tensors contain all frequencies. If one wishes to move to frequency labelling, such as $\beta(-2\omega, \omega, \omega)$, in the expressions, one must further count the additional identical terms which arise as frequency components of the fields are specifically listed, as occurs in the susceptibility descriptions of nonlinear polarization density.

These expressions are not very useful in their infinite summation form. Given the complexity of their calculation, it is realistic to acknowledge the form and adopt ensemble averaging processes, perhaps using them in a semiempirical manner.

Several points need to be made. 1) The enhancement of the ensemble's nonlinear polarization due to the linear polarizability (i.e. the local fields and mutual polarizations) is as large as or larger than the sum of molecular nonlinear polarizations to the Maxwellian fields. Thus uncertainty in this effect severely limits our ability to accurately and realistically a) predict macroscopic properties and b) extract molecular properties from macroscopic susceptibilities. 2) In crystals since only one or a few crystallographically unique molecular sites occur, the averaging is trivial. However, the formalism requires that the local field tensor be determined at the site, not at the unit cell level. It cannot be factored from inside the summation over sublattices and thus the formalism used by many workers to discuss effects of molecular orientation is not correct, though probably supportable. 3) In fluids there is a distribution of fluctuating environments. The process of averaging over environments and orientations requires giant leaps to be able to generate the often used expressions of e.g. $\chi^{(3)}$ in which all five factors of the first term in Eq. A.15 appear as independently averaged quantities. Are correlations and correlated fluctuations insignificant in these processes?

REFERENCES

1.   Meredith GR: in Nonlinear Optical Properties of Organic and Polymeric Materials, ed. by Williams DJ. Washington, D.C.: American Chemical Society, 1983, p. 27.
2.   Meredith GR, Hsiung H, Stevenson SH, Vanherzeele H, Zumsteg H: in press
3.   Van Catledge FA: unpublished.
4.   Bernstein M: unpublished.
5.   Griffiths J: Colour and Constitution of Organic Molecules. New York: Academic, 1976.
6.   Murrell JN: The Theory of the Electronic Spectra of Organic Molecules. London: Chapman and Hall, 1963.
7.   Nicoud JF and Twieg RJ: in Nonlinear Optical Properties of Organic Molecules and Crystals, Vol. 1, ed. by Chemla DS and Zyss J. New York: Academic, 1987, p. 254.
8.   Lalama SJ and Garito AF: Physical Review A20, 1179, 1979.
9.   Meredith GR: Class notes, unpublished, Nonlinear Optics: Materials and Devices, International School of Materials Science and Technology, Erice, Siciliy, 1985
10.   Stevenson SH and Meredith GR: SPIE Proceedings 682, 147, 1986.
11.   Meredith GR, Buchalter B and Hanzlik C: Journal of Chemical Physics 78, 1533 and 1543, 1983.
12.   Jerphagnon J, Chemla D and Bonneville R: Advances in Physics 27, 609, 1978.
13.   Ducuing J: in Nonlinear Optics, ed. by Harper PG and Wherrett BS. New York: Academic, 1977, p. 11.
14.   Menendale SC and Rustagi KC: Optics Communications 28, 359, 1979.
15.   Stevenson SH, Donald DS and Meredith GR: Materials Research Society Proceedings Vol. 109, 103, 1988.
16.   Waite J and Papadopoulos MG: Journal of Physical Chemistry 89, 2291, 1985.
17.   Levine BF: Dielectric and Related Molecular Processes 3, 73, 1977.
18.   Wang Y, Tam W, Stevenson SH, Clement RA and Calabrese J: Chemical Physics Letters 48, 136, 1988.
19.   Oudar JL: Journal of Chemical Physics 69, 446, 1977.
20.   Zyss J, Chemla DS and Nicoud JF: Journal of Chemical Physics 74, 4800, 1981.
21.   Meredith GR: in Nonlinear Optical Properties of Organic and Polymeric Materials, ed. by Williams DJ. Washington, D.C.: American Chemical Society, 1983, p. 109.
22.   Popovitz-Biro R, Hill K, Landau EM, Lahav M, Leiserowitz L, Sagiv J, Hsiung H, Meredith GR and Vanherzeele H: Journal of the American Chemical Society 110, 2672, 1988.
23.   Meredith GR, Hsiung H, Stevenson SH and Vanherzeele H: Materials Research Society, in press.
24.   Meredith GR: SPIE Proceedings 824, 126, 1987.

SYNTHESIS OF NEW NONLINEAR OPTICAL LADDER POLYMERS

L. R. DALTON

Department of Chemistry, University of Southern California, Los Angeles, California 90089-0482, USA

1. INTRODUCTION

It is well-known that electron delocalization associated with π-orbital overlap contributes to third order susceptibility (1-6). It is thus reasonable to attempt to optimize nonlinear optical activity in centrosymmetric polymers by control of conformation to maximize π-orbital overlap between adjacent sp^2 centers and to minimize interruptions of delocalization by minimizing sp^3 defects. Steric interactions frequently force nonplanar conformations in linear-chain polymers and many systems such as polyacetylene have yet to be prepared in a form free of sp^3 defects such as cross bridges. Ladder polymers, such as shown in Fig. 1, represent structures which should lead to extensive electron delocalization and interesting nonlinear optical activity. Some theoretical evidence exists (6,7) to support this contention. Clearly, orbital overlap is optimized in ladder polymers with a zero dihedral angle between adjacent phenyl rings and for systems where nitrogen atoms exist in the deprotonated (imine) form.

Although ladder polymers have been known for some time (8-10), poor solubility and low molecular weights have discouraged consideration of these materials for structural or electronic applications. Poor solubility can likely be associated with strong Van der Waals interactions between the delocalized π electron clouds on adjacent polymer chains. Evidence of this interaction is also provided by the anisotropy of the electrical conductivity measured for doped ladder polymer systems. The maximum conductivity is observed normal to the plane of the ladder polymer consistent with strong interchain π electron interaction (11). It is thus reasonable that solubility of ladder polymers can be enhanced by introducing steric interactions to destabilize polymer-polymer interactions. Groups which influence polymer self-association are readily introduced into the final polymer by using derivatized monomers as shown in Fig. 2. This figure illustrates that a wide variety structures, hence physical properties, can be achieved. Although the present research focuses upon polymers with phenyl rings in the backbone, the polycondensation scheme is amendable to producing a variety of structures as is shown in Fig. 3. Clearly an advantage of polycondensation is that branching and cross linking is avoided. Because the condensation events in the polymers studied in this work are asymmetric, either open-chain or fully fused rings can be produced. Also, a variety of oligomeric materials can be produced by using end-capped monomers or by using of tosylate blocking and

J. Messier et al. (eds.), Nonlinear Optical Effects in Organic Polymers, 123–141.

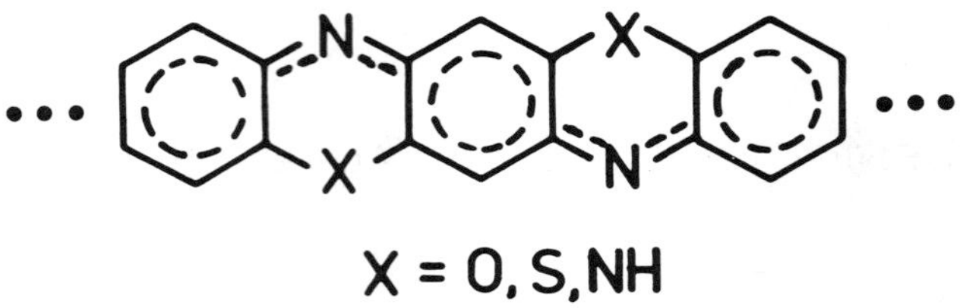
N
X
X
N
X = O, S, NH

Fig. 1. Electron delocalization in a representative ladder polymer.

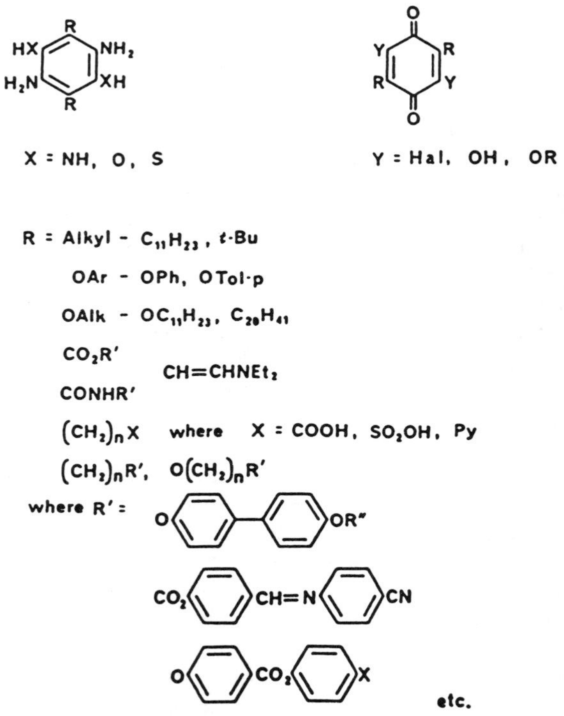
R
HX
NH2
H2N
XH
R
X = NH, O, S
O
Y
R
R
Y
O
Y = Hal, OH, OR
R = Alkyl - C11H23, t-Bu
OAr - OPh, OTol-p
OAlk - OC11H23, C20H41
CO2R'
CH=CHNEt2
CONHR'
(CH2)nX where X = COOH, SO2OH, Py
(CH2)nR', O(CH2)nR'
where R' =
O
OR"
CO2
CH=N
CN
O
CO2
X
etc.

Fig. 2. Examples of derivatized monomers used to synthesize derivatized polymers.

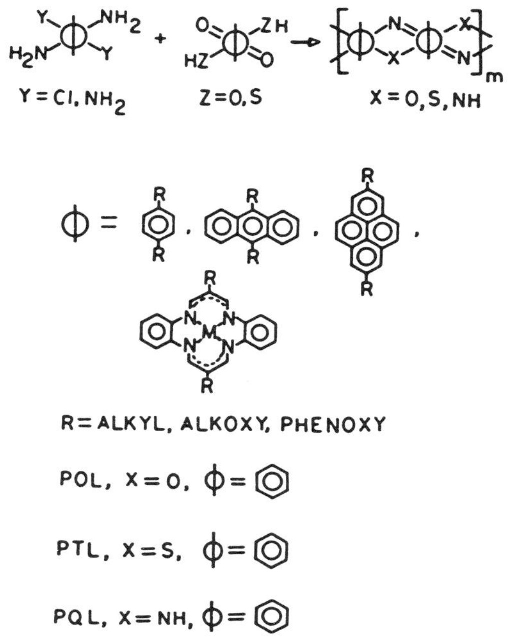

125

Fig. 3. General polycondenation scheme with representative examples.

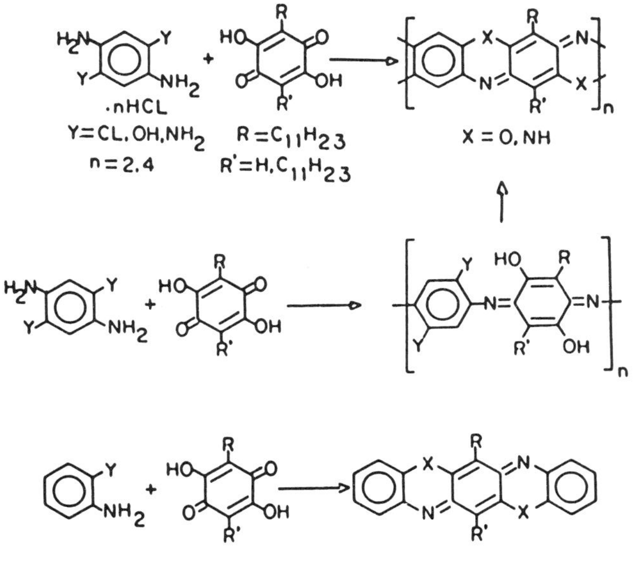

Fig. 4. Synthesis of fully fused ring ladder polymers, prepolymers, and oligomers with alkyl subtituents.

and subsequent deblocking. An example of this versatility is provided in Fig. 4 which focuses on our early work on alkyl-derivatized polymers (12-14). Conversion of the open-chain precursor polymer to the fully-fused ring polymer can be accomplished either in solution or in the solid state. The solid state conversion is conveniently followed by thermal gravimetric analysis (TGA) as is shown in Fig. 5.

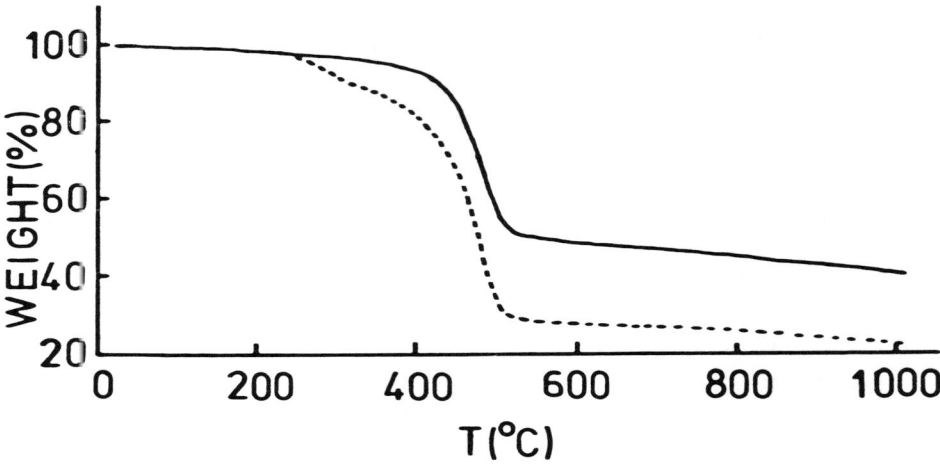

Fig. 5. TGA of alkyl-derivatized ladder polymer (solid line) and open-chain prepolymer (dashed line).

Fortunately, the final condensation step to convert the open-chain polymer to the corresponding ladder polymer occurs in the temperature range 250-325 C while dealkylation does not occur until 400-525 C. Alkyl derivatization improves solubility with little or no effect upon the electronic properties of the polymer. This observation appears to correspond to the effect of such derivatization on the properties of polythiophene (15-17). Of greater interest is the effect of vinylamine substituents which not only improve solubility but also enhance polymerization kinetics and produce an enhancement in nonlinear optical activity. Thus, we shall focus the present discussion upon the preparation and characterization of materials with vinylamine substituents.

The general scheme for preparing such materials is shown in Fig. 6. A Mannich reaction is used to prepare dichloroquinones derivatized with a variety of vinylamine substituents. Although we have prepared all of the materials indicated in Fig. 6, we shall focus our attention upon several representative examples. One does note that in all cases vinylamine derivatized monomers can be reacted to yield polymers under much milder

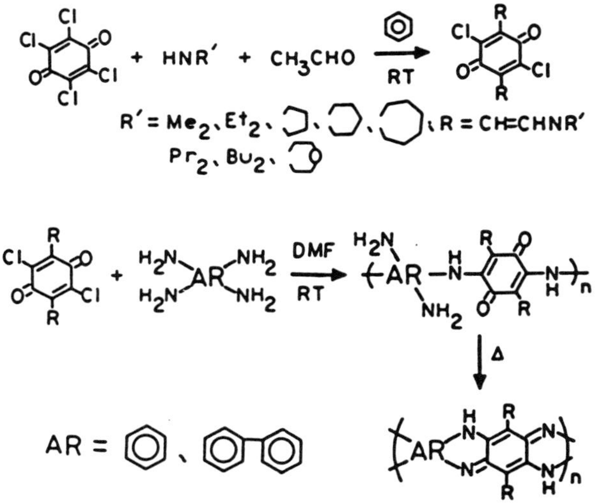

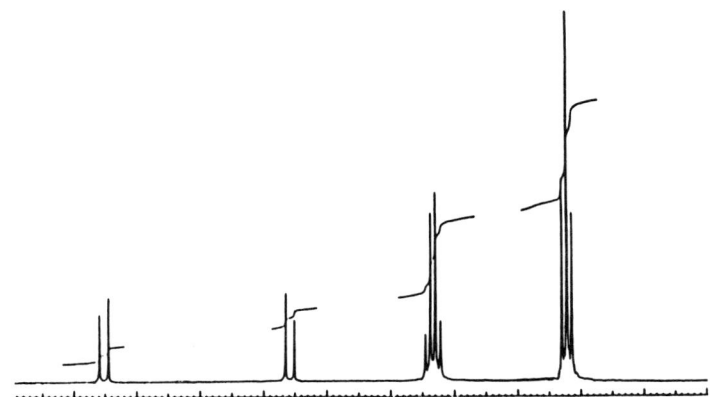

Fig. 6. General scheme for synthesis of vinylamine-derivatized ladder polymers with representative examples.

Fig. 7. Proton NMR of monomer I.

conditions (organic solvents and near ambient temperatures) than required for the polymerization of the underivatized monomers.

2. PROCEDURE

2.1. Synthesis and structural characterization

2.1.1. Monomers

2.1.1.1. 2,5-Dichloro-3,6-bis(2-diethylaminovinyl) benzoquinone (I). This monomer (hereafter referred to as I) was made by an improved literature procedure (18). Diethylamine (2.9 g) was added to a stirred solution of chloranil (4.92 g) and freshly distilled acetaldehyde (1.80 g) in 1500 mL toluene at 30 C. A blue solution was formed within 10 min. and another 2.9 g of diethylamine was added. After 5 hrs, 0.5 N sulfuric acid was used to extract the amine salt and the toluene layer was dried and concentrated by evaporation. The product was purified by chromatographic separation using deactivated alumina. The melting point of the product was observed to be 129-131 C. Elemental analysis: Calc.: C, 58.22; H, 6.47; N, 7.55; Cl, 19.14; Found: C, 58.30, H, 6.53; N, 7.52; Cl, 19.16%. The ^{1}H (see Fig. 7) and ^{13}C NMR, FTIR, and mass spectral analysis are consistent with the expected structure.

2.1.1.2. 2,5-Dichloro-3,6-bis(2-piperidinovinyl) benzoquinone (II). 4.29 g of chloranil and 1.8 g of acetaldehyde were added to 1400 mL of toluene and stirred for 5 min. To this solution was added dropwise 3.44 g of piperidine. The solution became blue after stirring for 10 min. Another 3.44 g of piperidine was added and the color gradually changed to purple. The reaction was carried out for 7 hrs at 30 C and was terminated by adding 0.5 N sulfuric acid to extract the amine salt. The toluene layer was washed with water and dried over 3 Å molecular sieves. The solution was concentrated by evaporation to about 60 mL. Needle-like crystals were collected by suction and washed with cold toluene. Recrystallization from toluene gave 1.5 g of product with a mp of 180-182 C. NMR, FTIR, etc. confirmed the expected structure. Elemental analysis: Calc.: C, 60.76; H, 6.08; N, 7.09; Cl, 17.97; Found: C, 60.83; H, 6.18; N, 7.06; Cl, 17.89%.

2.1.1.3. 2,5-Dichloro-3,6-bis(hexamethyleneiminylvinyl)-1,4-benzoquinone (III). Chloranil (1.24 g, 5.05 mmol) was dissolved in 800 mL of toluene. Acetaldehyde (0.56 mL, 10 mmol) and hexamethyleneimine (2.27 mL, 20.2 mmol) were added. The purple solution was stirred for 22 hrs, filtered, reduced to 5 mL on a rotary evaporator and chromatographed on activity II neutral alumina with toluene/methylene chloride. The product was obtained as brown-green needles upon stripping the solvent: yield 449 mg (21.0%); mp 172-175 C; ^{1}H NMR (CDCl$_3$, TMS, see Fig. 8 for designation of particular nuclei) δ 1.5-1.9 (broad m, 16H, H$_a$), 3.47 (m, 8H, H$_b$), 5.57 (d, J$_{HH}$ = 12.5 Hz, 2H, H$_{c\ or\ d}$), 8.61 (d, J$_{HH}$ = 12.5 Hz, 2H, H$_{c\ or\ d}$); ^{13}C NMR (CDCl$_3$) 25.56 (C$_a$), 26.73 (C$_a$), 27.77 (C$_a$), 30.24 (C$_a$), 48.00 (C$_b$), 56.84 (C$_b$), 93.97 (C$_{c\ or\ d}$), 120.50 (C$_{e,f,\ or\ g}$), 141.51 (C$_{e,f,\ or\ g}$), 154.19 (C$_{c\ or\ d}$), 177.67 (C$_{e,f,\ or\ g}$) ppm; IR (KBr) 961 (out-of-plane C-H bend of a trans H-C=C-H), 1149, 1187, 1412, 1510, 1582, 1630 (carbonyl stretch), 2852, 2911, 2928, 3060, 3087 cm^{-1}. The sp^2 nature of the nitrogens is indicated by the presence of 6 methylene resonances in the ^{13}C

NMR, i.e., roatation about the N-vinyl carbon is slow on the NMR time scale. The ^{13}C NMR also provides conclusive evidence that the substitution pattern is para as in Fig. 8A and not meta as in 8B; the former has 3 ring signals whereas the latter gives 4 resonances. Elemental analysis: Calc.: C, 62.41; H, 6.67; N, 6.62; Cl, 16.57; Found: C, 62.04; H, 6.66; N, 6.48; Cl, 16.86%.

2.1.1.4. 2,5-Dichloro-3,6-bis(2-dipropylaminovinyl) benzoquinone (IV). This monomer was prepared and characterized in a manner analogous to I. Elemental analysis: Calc.: C, 61.83; H, 7.49; N, 6.56; Cl, 16.63; Found: C, 61.93; H, 7.54, N, 6.91, Cl, 16.68%.

2.1.1.5. 1,2,4,5-Tetraaminobenzene (TAB). 1,2,4,5-tetraaminobenzene tetrahydrochloride was prepared by reduction of 2,4-dinitro-1,5-diaminoben.zene using stanneous chloride dihydrate in concentrated hydrochloric acid. To remove trace amount of tin ions in the final product, H_2S was bubbled through the TAB.4HCL solution. The free amine was prepared using Vogel's procedure yielding while flat crystals with mp of 266-267 C. Elemental analysis: Calc.: C, 52.17; H, 7.24; N, 40.58; Found: C, 52.11; H, 7.09; N, 40.31%.

2.1.1.6. 3,3-Diaminobenzidine (DABD). This monomer was obtained from Aldrich Chemical Company and was purified by recrystallizing twice from acetonitrile/water mixture.

2.1.2. Prepolymers

2.1.2.1. Prepolymer IA. This prepolymer is prepared from the reaction of I with TAB. To 100 mL of DMF was added 5.000 g of I and 1.855 g of TAB; the solution was then stirred at room temperature overnight. The brown solution obtained can immediately be used to cast films on glass slides. To remove trace amounts of unreacted materials which may exist, the solution was poured into 500 mL of chloroform. The precipitate was collected by filtration and washed with chloroform and ethanol. The dark brown prepolymer was dried at 50 C under vacuum (yield 95%). An intrinsic viscosity of 1.4 dl/g was measured in DMF indicating a relatively low molecular weight polymer; however, quite nice free standing films can be cast from this polymer solution. Elemental analysis: Calc.: C, 56.58; H, 6.68; N, 16.50; Cl, 13.95; Found: C, 56.89; H, 6.45; N, 16.69; Cl, 13.71%.

2.1.2.2. Prepolymer IIA. This prepolymer was prepared by reaction of II with TAB analogous to IA except that polymerization was carried out at 60 C under helium. Elemental analysis: Calc.: C, 58.53; H, 6.38; N, 15.75; Cl, 13.32; O, 6.00; Found: C, 55.49; H, 5.96; N, 16.38; Cl, 12.53; O, 9.64%. Clearly, partial air oxidation of this polymer occurs. A subsequent preparation carefully avoiding air exposure to the time of elemental analysis yielded improved agreement with the expected structure. Samples were also prepared maximizing air oxidation.

2.1.2.3. Prepolymer IIB. This prepolymer was prepared by reaction of II with DABD analogous to IA except that polymerization was carried out at 60 under helium. Elemental analysis: Calc.: C, 63.47; H, 6.28; N, 13.88; Cl, 11.74; Found: C, 62.59; H, 5.86; N, 13.61; Cl, 11.51%.

2.1.3. Ladder Polymers

2.1.3.1. Polymer IA. Ladder polymers can be prepared in either of two ways. The first was is to reflux the prepolymer in DMF solution overnight. A black powder precipitates out and was collected directly from DMF using a siphon extractor and was dried at 100 C under vacuum (yield 90%). The second method involves heating a prepolymer film or powder at 140 C for 6 hrs and at 260 C for 10 hrs. FTIR spectra showed that these two procedures give the same polymer. Elemental analysis: Calc.: C, 64.36; H, 4.92; N, 17.21; Cl, 13.49; Found: C, 65.13; H, 4.49; N, 16.70; Cl, 13.97%.

2.1.3.2. Polymer IIA. This ladder polymer was prepared analogous to IA. Elemental analysis: Calc.: C, 65.09; H, 6.26; N, 17.52; Cl, 11.11; Found: C, 65.45; H, 4.80; N, 16.50; Cl, 10.90%.

2.1.3.3. Polymer IIB. This ladder polymer was prepared analogous to IA. Elemental analysis: Calc.: C, 72.38; H, 5.08; N, 15.83; Cl, 6.69; Found: C, 73.87; H, 4.90; N, 14.86; Cl, 6.32%.

2.1.3.4. Polymer IIIA. This polymer was prepared from the reaction of monomer III with TAB to form the corresponding prepolymer followed by thermal treatment to product ladder polymer IIIA.

2.1.3.5. Polymer IVA. This polymer was prepared from the reaction of monomer IV with TAB, followed by appropriate thermal treatment.

The elemental analyses given above were performed by Galbraith Laboratories. Materials were also sent to Atlantic and to Wright Patterson (AFWAL/MLBP) for analysis. The polymerization process could be followed conventiently either by UV-Vis spectroscopy (recorded using either a Shimadzu UV-260 spectrophotometer or a Perkin-Elmer Lambda 4C spectrophotometer) or by NMR spectroscopy). The optical spectra of monomer II and prepolymer IIB are given in Fig. 9. The NMR linewidths increase upon polymerization. A differential rate of broadening for the vinylamine nuclei for prepolymer IA indicates that the ethyl groups attached to the amine nitrogen undergo significant rotation. The conversion of the prepolymer to the polymer is conventiently followed by FTIR (see Table I) and by TGA. Note in particular that before heating a prepolymer, a carbonyl group IR absoption band at 1657-1658 cm^{-1} can be observed which disappears on heating and is replaced by a C=N absorption at 1617-1626 cm^{-1}. As shown in Table II, EPR linewidth changes also reflect polymer conformation; these trends arise from increased electron delocalization and greater polaron concentrations in the fully-cyclized ladder polymers.

UV/Vis spectra of both prepolymer and polymer in methanesulfonic acid show similar structure but the spectra of the polymers exhibit a lower energy band edge consistent with greater electron delocalization existing in the polymer.

2.2. Nonlinear optical measurements

2.2.1. Ancillary measurements. Refractive indices were measured by measuring back reflection from the polymer-air interface at the operating wavelength (532 nm). This reflection could be measured with an accuracy better than 3%, giving the refractive indices listed in Table III. The film

131

Table I. Tentative Assignments of FTIR Spectra of Prepolymer IIA and Polymer IIA (from Piperidinovinyl Dichloroquinone and TAB).

Absorption peaks of Prepolymer cm^{-1}	Remarks	Absorption peaks of Polymer cm^{-1}
3370-3320 (s)	N-H, H_2O associated with KBR	3400 (s)
2931-2850 (m)	>CH_2, stretch	2930-2850 (w)
1657 (s)	>C=O, stretch	disappears
	>C=N-, stretch	1626 (s)
1605 (w)	>C=C<, stretch	1605 (w)
1511 (s)	>NH_2, δ	disappears
1471 (m)	-C-H, >CH_2, δ	1471 (w)
1435 (m)	H-C-H, bending	1435 (w)
1246 (s)	C-N, primary amine, δ	very weak
1184 (s)	aromatic C=C-H (1,2,4,5-substituted benzene in-plane deformation)	1184 (m)
1098 (s)	?	disappears
1029 (s)	C-C (skeletal vibrations in piperidine ring)	1029 (w)
947 (s)	C=C , δ	947 (w)
853 (s)	Benzene Ring	974 (s)
661 (m)	?	very weak
555 (m)	?	disappears

Table II: g Values and Linewidths of Prepolymers and Polymers

	Prepolymer		Polymer	
	IIA	IIB	IIA	IIB
g	2.00510	2.00428	2.00312	2.00312
ΔH(G)	10.05	9.50	7.50	6.00

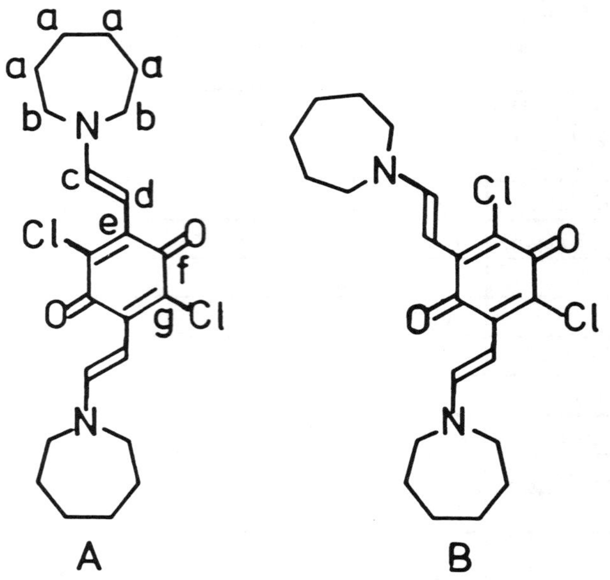

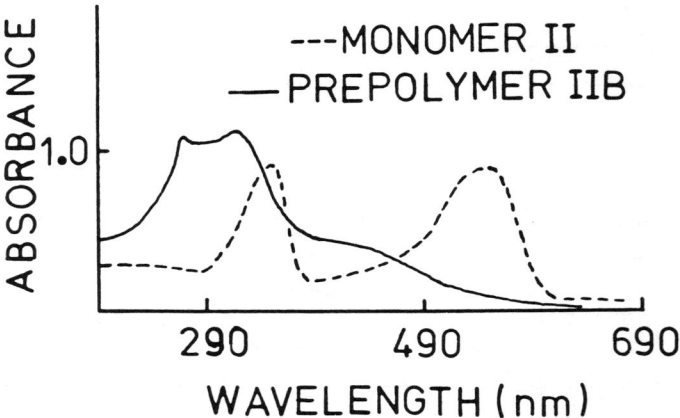

Fig. 8. Indication of nuclei for NMR assignments for monomer III. Two isomers shown corresponding to para (A) and meta (B) substitution.

Fig. 9.. Optical spectra of monomer II and prepolymer IIB.

thicknesses were measured with a step profiler (Tencor Instruments' Alpha Step Profiler Model 1000) and were verified by electron microscopy. Representative thicknesses are also given in Table III.

Table III: Refractive Index, Film Thickness, and Absorption Data for Representative Vinylamine Polymers.

Material	n	L(μ)	a(μ⁻¹)
Prepolymer IA	1.75 ±0.05	0.4±0.1	1.7±0.1
Polymer IA	1.75 ±0.05	2.4±0.2	1.1 ±0.1
Polymer IIA	1.80 ±0.05	4.0 ±0.3	1.4 ±0.1

2.2.2. Measurement of Third Order Susceptibilities. Although some measurements of nonlinear optical activity were made by third harmonic generation, the bulk of the analysis relied upon degenerate four wave mixing (DFWM) measurements. Hellwarth and coworkers (19) have introduced a new variation of the method for measuring optical nonlinearity which permits measurement of all tensor components, c_{ijkl}, of the fast component of the nonlinear susceptibility $\chi^{(3)}$, as well as of a slow component which gives, among other things, the acoustic velocity and damping coefficients. A "phase-conjugation" geometry for DFWM is employed in which all four beam polarizations can be varied to give independent data. Three input beams are formed from 532 nm laser pulses (derived from 5 Hz rep. rate mode-locked pulses from a Quantel Model YG471-C Nd:YAG laser frequency-doubled to 532 nm) having energy approximately 2 mJ and duration approximately 25 psec. In this geometry delaying each of the three input-beam pulses separately also gives three different and independent sets of data for the strength and polarization of the fourth "signal" beam that is generated. The three input beams (see Fig. 10) are denoted "F" for the forward beam, "B" for the backward pump beam and "P" for the probe beam (which impinged on the polymer sample at a variable angle of approximately 10 to 20 degrees to beam F). We have performed experiments with essentially all combinations of beam polarization and delay. The detailed discussion of this effort will be presented elsewhere (19). As expected, results are consistent with the polymer films exhibiting the characteristics of amorphous, isotropic media. The signal pulse energy varies as the cube of the energy as expected for pure $\chi^{(3)}$ nonlinearity. Typical plots of signal pulse energy versus the delay time (plus and minus) of the forward pump beam F are shown in Fig. 10. Each data point was an average over ten to twenty pulses; laser pulse energy fluctuations were of order ±20%. In Fig. 10, all polarizations are horizontal and both fast and slow components are clearly visible. In Fig. 11, the backward pump beam B is delayed instead and its polarization is made orthogonal to that of the forward pump F and probe P beams. About 25 psec after the fast component there appears a much larger peak which is verified to be the first of a series of peaks in an acoustical density variation having a

134

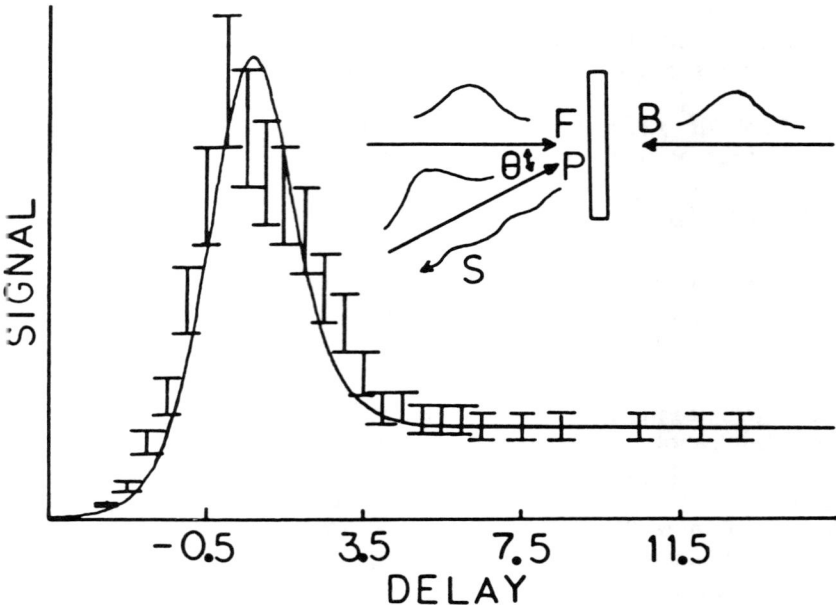

Fig. 10. Phase conjugate signal (arbitrary units) versus forward pump
delay is shown. All polarizations are horizontal. Solid line is
the theoretical fit discussed in (19).

(small) wavevector equal to the difference of the wavevectors of the two
simultaneous parallel-polarized F and P beams.

The polymers studies in this work have ususually high optical damage
thresholds (approximately 1 GW/cm^2 for 20-25 psec pulses) and show no
saturation of the nonlinear response before damage.

Representative third order susceptibilities are summarized in Table IV.

Table IV. Optical Nonlinearity Data for Representative
Vinylamine Polymers.

Material	C_{1111} x10^{-10}(esu)	C_{1221} x10^{-10}(esu)
Polymer IA (air-oxidized)	0.6	0.33
Prepolymer IA	1.0	0.3
Polymer IA	3.0	1.34
Polymer IVA	2.8	1.4
Polymer IIA	12.5	4.37
Polymer IIB	12.8	4.5
Polymer IIIA	11.2	3.7

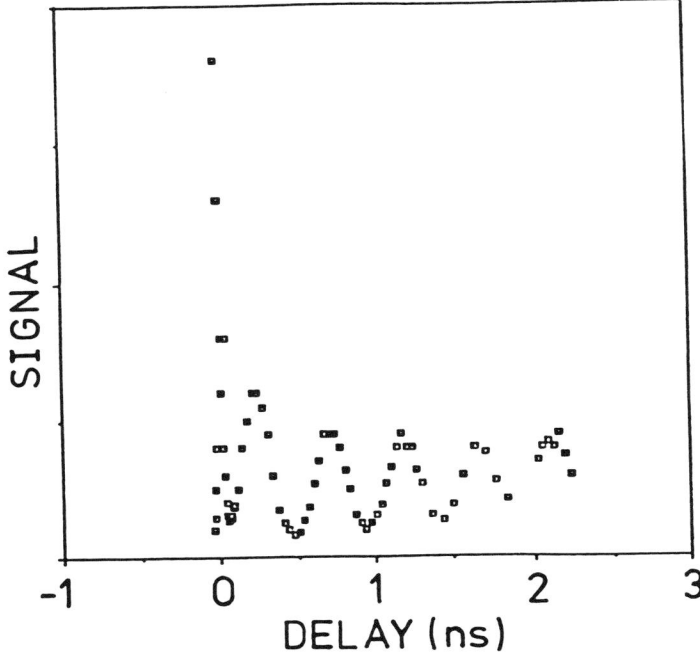

Fig. 11. Phase conjugate signal versus backward pump delay is shown.

3. DISCUSSION OF RESULTS
    The elemental analysis clearly demonstrates the difficulty in obtaining perfect polymer and prepolymer structures. The major problem appears to be oxidation at nitrogen although some loss of side groups and incomplete condensation may also contributed to lack of complete agreement between calculated and observed elemental compositions. Because of the difficulty in obtaining perfect structures caution must be applied to the interpretation of NLO results. Nevertheless, several trends appear to exist. Third order susceptibilities do appear to depend on polymer conformation and increase as a planar conformation (zero degree dihedral angle) is approached. This is evident by comparing the data for prepolymer IA and polymer IA given in Table IV. Ideally, the comparison of data for polymers IIA and IIB would also provide insight into this dependence. However, partial air oxidation in polymer IIA probably sufficiently restricts the electron delocalization that this comparison is not warranted. The dependence of third order susceptibility upon the extent of air oxidation reflects the dependence of $\chi^{(3)}$ upon electron delocalization; air oxidation introduces sp^3 defects (at nitrogen) which, of course, interrupt π-conjugation. Certainly, air-sensitivity represents a significant problem for certain of the ladder polymers and the exact nature of this problem requires further investation. Perhaps encapsulation or the

development of copolymer or composite systems may ultimately provide a resolution of the dilemma encountered with the more air-sensitive materials. On the positive side, it should be realized that the largest third order susceptibilities reported in Table IV likely represent lower limit values.

From our extensive studies of the dependence of third order susceptibility upon the nature of the vinylamine substituent (see Fig. 6 and Table IV), it appears that third order susceptibility increases as the conjugation of the vinylamine π-orbitals with those of the polymer backbone is enhanced. Torsional motion (as independently detected by NMR) of the side groups appears to the the major contributor to decorrelating the orbital interaction. Side group dynamics may even be reflected in the substituted dichloroquinone monomers. When vinylamine substituents are terminated by alicyclic groups the monomers can be crystallized while they cannot if the vinylamine substituents are terminated by dialkyl groups. However, the general observation that larger third order susceptibilities are observed for alicyclic vinylamine-derivatized polymers than for dialkyl vinylamine-derivatized polymers must be somewhat tempered by an unknown extent of oxidation in some of these polymers. Rather interestingly, the magnitude of the thermal component appears to correlate with the extent of side chain motion suggesting that laser heating of local modes as well as general phonon modes may be important. Again, this observation requires further investigation.

Obviously, there exists a significant resonance contribution to the observed third order susceptibility. This is clear from preliminary measurements made at longer wavelength. Clearly, the dispersion of $\chi^{(3)}$ requires further investigation.

4. FUTURE STUDIES

A logical first objective of future research should be to identify polymers less sensitive to air oxidation. This involves investigating a variety of aromatic amines. We have already prepared 1,4-diamino-2,5-dithiolbenzene following the procedure of Wolfe (20) and have reacted this with various substituted dichloroquinones to prepare the corresponding prepolymers and polymers. Films have been fabricated and preliminary DFWM measurements have been performed. This data, together with the fundamental characterization data, is being analyzed at the time of this writing. Diaminodihydroxybenzene should also be employed as a reaction monomer as well as derivatized aromatic tetraamines such as shown in Fig. 12 (21).

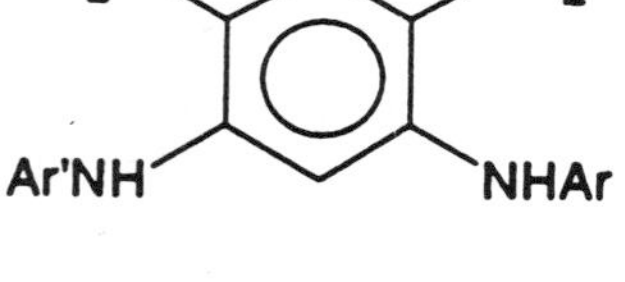

Fig. 12. A derivatized aromatic tetraamine.

Work should also be pursued on other aromatic amines, i.e., based upon anthracene and pyrene units. When the problem of oxidation is more defined it will be useful to measure third order susceptibilities as a function of side group including studies of both alkyl and vinylamine substituents.

We have carried out preliminary studies of the effect of chemical doping with electron accepting or donating dopants upon nonlinear optical activity. We have generally observed enhancements in nonlinear optical activity. However, these materials must be viewed as even more undefined than the oxidized ladder polymers and more controlled doping must be effected. In this regard, it would be useful to carry out electrochemical doping. We have already performed cyclic voltammetry measurements on polymer films deposited upon platinum electrodes. It would seem reasonable to extend these measurements to correlated electrochemical doping/nonlinear optical measurements. The observation of third order susceptibilities on the order of 10^{-7} to 10^{-8} esu for organic donor-acceptor complexes (22) clearly provides motivation for such studies as does the observation of doping-induced visual transparency for some ladder polymers (11). In this regard, the synthesis of new organometallic polymers should be of considerable interest and some promising starts have been made in our laboratory in this regard. In addition to doping involving transfer of electrons, it is interest to investigate the effect of protonic doping for polymers with imine nitrogens (6) as such doping has already been observed to have a dramatic effect upon electronic properties.

Attention needs to be paid to the development of oriented polymer films by either employing Langmuir-Blodgett thin film methods or by processing from liquid crystalline phases. In both cases, this is a matter of choice and synthesis of the right side groups. We have already demonstrated Langmuir-Blodgett processing for prepolymers asymmetrically derivatized with alkyl groups. Moreover, we have recently demonstrated the preparation of water soluble prepolymers. Clearly, the investigation of substituent variation for advanced processing methodology needs to be enlarged upon. An example of a synthetic scheme of interest is shown in Fig. 13.

Finally, we note that more work must be done on oligomeric compounds. Single crystals can be obtained for some of the smaller oligomers and the structure can be determined exactly. To date, it has not been possible to obtain crystals suitable for DFWM measurements but clearly the study of these materials should be pursued.

The preparation of copolymer and composite materials is of interest and examples of possible schemes are given in Fig. 14 and 15. It may prove interesting to prepare two component systems (involving materials with large and small nonlinearities $n_2 = \chi^{(3)}/c\varepsilon\pi^2$) where the linear indices of refraction, n_0, are matched. The material should be transparent at low powers and opaque at higher powers.

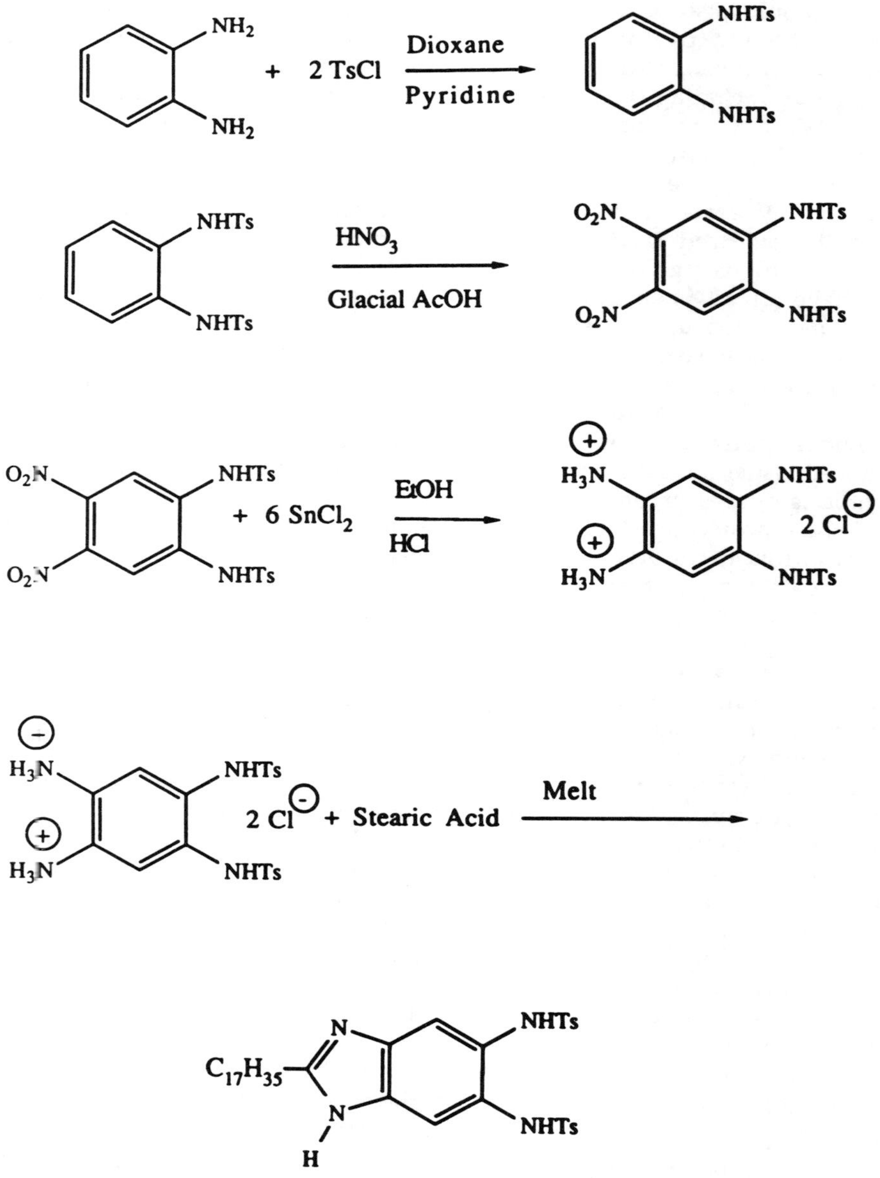

138

Fig. 13. Reaction scheme for producing an alkyl terminated ladder polymer.

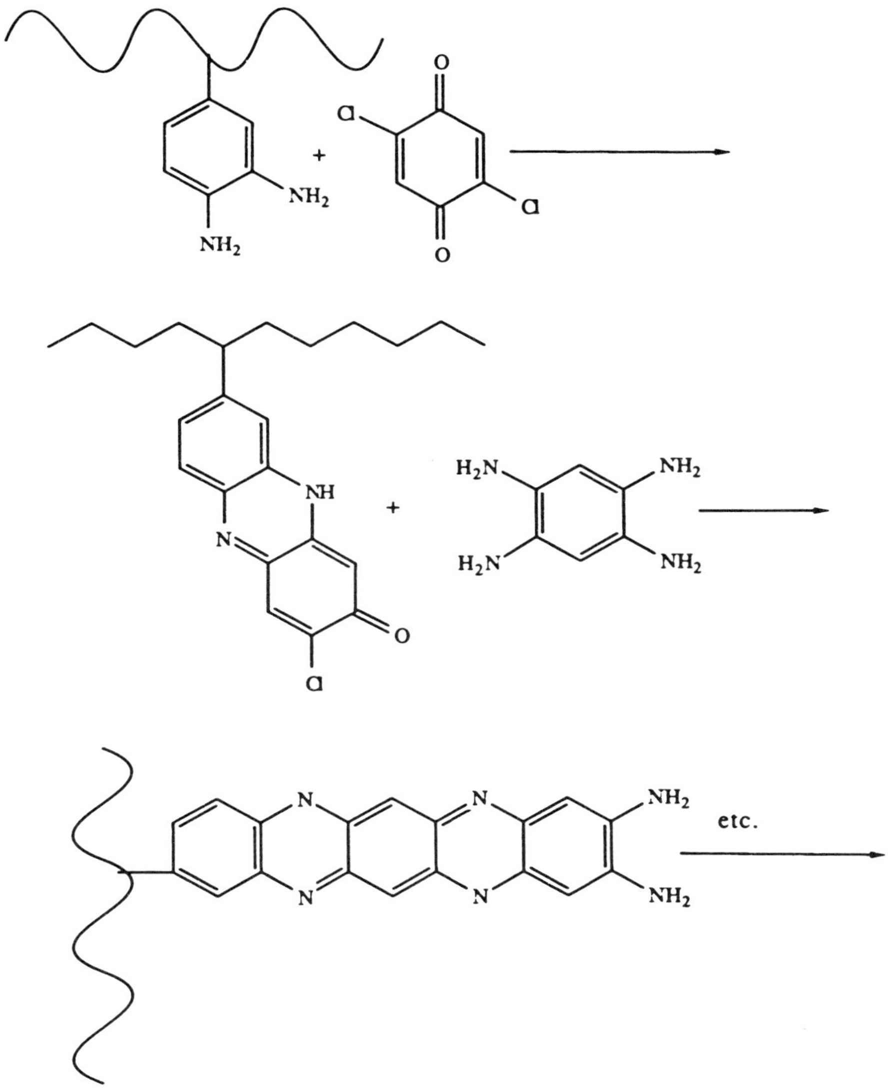

Fig. 14. Solid state synthesis of a ladder/fexible chain copolymer.

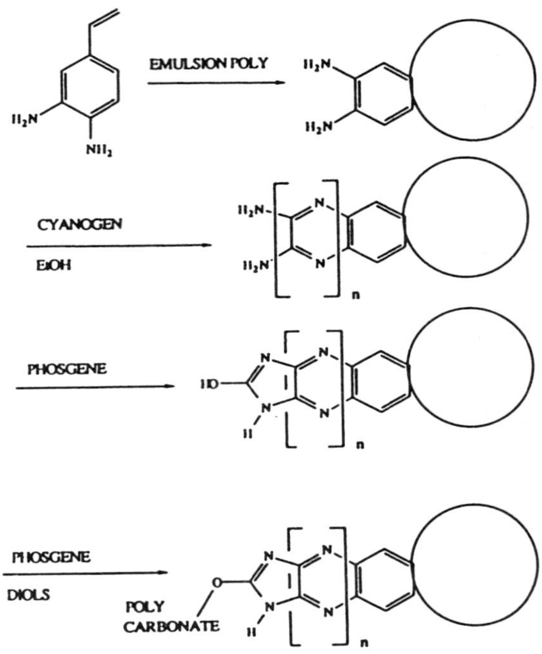

140

Fig. 15. Synthesis of ladder polymer coated polystyrene beads and the coupling of these to polycarbonate.

ACKNOWLEDGEMENTS

This work was supported by Air Force Office of Scientific Research contracts F49620-87-C-0100, F49620-85-C-0096, and F49620-88-C-0071. Special thanks are due Dr. Donald Ulrich for stimulating this research. This is a preliminary report of a substantial amount of work carried out by my graduate students and collaborators. The synthetic work is largely that of Mr. L. -P. Yu and a more detailed treatment of this effort will hopefully be published shortly. The efforts of Dr. R. W. Hellwarth and his graduate students are gratefully acknowledged. We wish to thank M. R Unroe and her colleagues at AFWAL/MLBP for carrying out a variety of characterization measurements on our materials.

REFERENCES

1. Williams, D. J. (ed): Nonlinear Optical Properties of Organic and Polymeric Materials. ACS Symposium Series 233. Washington, DC: Amer. Chem. Soc., 1983.
2. Chemla, D. S. and Zyss, J. (ed): Nonlinear Optical Properties of Organic Molecules and Crystals, Vol. 1 and 2, New York: Academic Press, 1987.

3. Khanarian, G. (ed): Proc. SPIE, <u>682</u>, 1986; Proc. SPIE, <u>567</u>, 1985; Proc. SPIE, <u>878</u>, 1988; Sandman, D. J. (ed): Solid State Polymerization: Structures and Properties of Polymers Produced by Lattice Controlled Processes. ACS Sym. Series. Washington, DC: Amer. Chem. Soc., 1986.

4. Prasad, P. N. and Ulrich, D. R. (ed): Nonlinear Optical and Electroactive Polymers. New York: Plenum Press, 1988.

5. Heeger, A. J., Orenstein, J. and Ulrich, D. R. (ed): Nonlinear Optical Properties of Polymers. Pittsburgh, Materials Research Society, 1988.

6. DeMelo, C. P. and Silbey, R., Chem. Phys. Lett., <u>140</u>, 537, 1987; Boudreaux, D. S., Chance, R. R., Wolf, J. F., Shacklette, L. W., Bredas, J. L., Themans, B., Andre, J. M. and Silbey, R., J. Chem. Phys., <u>85</u>, 4584, 1986; Chance, R. R., Boudreaux, D. S., Wolf, J. F., Shacklette, L. W., Silbey, R., Themans, B., Andre, J. M., and Bredas, J. L., Syn. Met., <u>15</u>, 105, 1986.

7. Medrano, J. and Wierschke, private communication of results of molecular orbital calculations.

8. Stille, J. K. and Mainen, E., Polym. Lett., <u>4</u>, 39, 1966; Stille, J. K. and E. Mainen, Macromolecules, <u>1</u>, 36, 1968; Stille, J. K. and Freeburger, M. E., J. Polym. Sci. A-1, <u>6</u>, 161, 1968; Szita, J. and Marvel, C. S., J. Pol. Sci. A-1, <u>7</u>, 3203, 1969; Patai, S. (ed): The Chemistry of the Quinonoid Compounds, Vol. 1 and 2, New York: Interscience, 1974; Patai, S. (ed): The Chemistry of the Amino Group, New York: Interscience, 1968; Mital, R. L. and Jain, S. K., J. Chem. Soc. (C), <u>1971</u>, 1875; Manecke, G. and Rotter, U., Makromol. Chem., <u>175</u>, 1695, 1974.

9. Kim, O. K., J. Pol. Sci. Pol. Lett. Ed., <u>23</u>, 137, 1985.

10. Van Deusen, R. L., Polym. Lett., <u>4</u>, 211, 1966; Arnold, F. E. and Van Deusen, R. L., Macromolecules, <u>2</u>, 497, 1969; Arnold, F. E. and Van Deusen, R. L., J. Appl. Polym. Sci., <u>15</u>, 2035, 1971.

11. Davidov, D., private communication.

12. Dalton, L. R., Proc. SPIE, <u>682</u>, 77, 1986.

13. Dalton, L. R., Thomson, J. and Nalwa, H. S., Polymer, <u>28</u>, 543, 1987.

14. Dalton, L. R. in: Nonlinear Optical and Electroactive Polymers, Prasad, P. N. and Ulrich, D. R. (ed), New York: Plenum Press, 1988, p. 301.

15. Patil, A. O., Shenoue, Y., Wudl, F., and Heeger, A. J., J. Am. Chem. Soc., <u>109</u>, 1858, 1987.

16. Elsenbaumer, R. L., Jen, K. Y., Miller, G. G. and Shacklette, L. W., Syn. Met., <u>18</u>, 277, 1987.

17. Sato, M., Tanaka, S. and Kaeriyama K., Syn. Met., <u>18</u>, 229, 1987; Kaeriyama, K., Sato, M. and Tanaka, S., Syn. Met., <u>18</u>, 233, 1987.

18. Buckley, D., Henbest, H. B. and Slade, P., J. Chem. Soc., <u>1987</u>, 4891.

19. Hellwarth, R. W., Cao, X. F., Jiang, J. P., Bloch, D. F., Yu, L. P. and Dalton, L. R., to be published.

20. Wolfe, J. F. and Loo, B. H., Macromolecules, <u>14</u>, 915, 1981.

21. Kane, J. J., private communication.

22. Blau, W., Appl. Phys. Lett., <u>51</u>, 2183, 1987; Blau, W., Phys. Technol., <u>18</u>, 250, 1987.

NONLINEAR OPTICAL PROPERTIES OF POLY(P-PHENYLENE VINYLENE) THIN FILMS

C.Bubeck, A.Kaltbeitzel, R.W.Lenz[#], D.Neher, J.D.Stenger-Smith[##], G.Wegner

Max-Planck-Institut für Polymerforschung, Postfach 3148, D-6500 Mainz, FRG

[#] University of Massachusetts, Department of Polymer Science and Engineering, Amherst, MA 01003 USA

[##] University of Massachusetts, Department of Chemical Engineering, Amherst, 01003 USA

1. Introduction

Nonlinear optical investigations on conjugated polymeric systems have shown strong nonlinear effects and fast response times (1). For application of these phenomena it is necessary to fabricate thin films with high uniformity of thickness, low optical scattering, photostability and low absorption in the near IR. Recently the conjugated polymer poly(p-phenylene vinylene) (PPV) was described (2-7). Third harmonic generation (THG) measurements at 1.85 μm (8) and femtosecond degenerate four wave mixing (DFWM) experiments at 580 and 602 nm (9) on PPV thin films demonstrated high optical nonlinearity and subpicosecond response time. Therefore PPV is an interesting system for further nonlinear optical investigations at other wavelength regions.

2. Preparation of PPV thin films

Thin films of PPV were obtained by spin coating or casting from aqueous solution of the precursor polymer followed by thermal treatment in vacuum (24h). For our experiments we used the new tetramethylene sulfonium chloride (THTCl) precursor polymer (6,7). The conversion reaction follows the scheme:

J. Messier et al. (eds.), Nonlinear Optical Effects in Organic Polymers, 143–147.

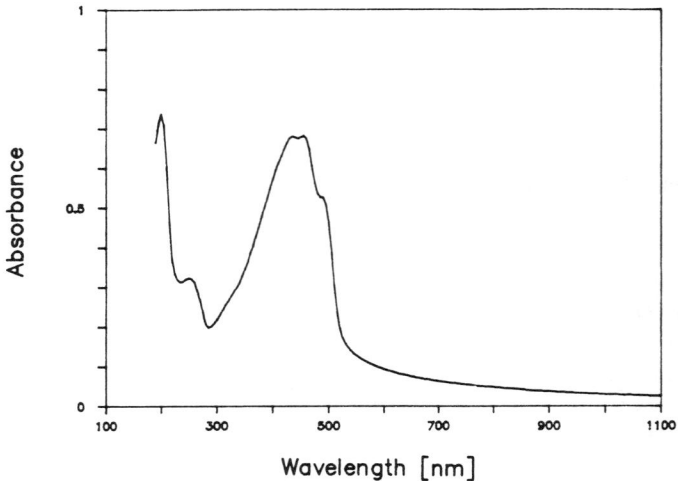

Fig.1 : UV/VIS Spectrum of a PPV thin film, thickness ≅ 42nm

The film thickness was determined with a step profiler. The optical absorption spectrum is shown in Fig.1 . Vibronic side-bands can be clearly seen in the range of the main absorption band around 500 nm. This is a significant difference to the rather broad absorption spectra of PPV films obtained, if the dimethyl sulfonium chloride precursor polymer is used. (3-5).

3.Third harmonic generation measurements
The experimental setup was based on the Maker fringe method. Infrared light pulses of 0.4 mJ and 20 ps length generated by an active/passive modelocked Nd:YAG laser were focused on the sample with the focal regime in air. The sample was mounted on a rotation stage. The generated harmonic light was filtered by a monochromator and detected by a photomultiplier. We adjusted the energy of the incident infrared beam with calibrated attenuation filters giving a THG signal of approximate 100 photons per pulse. The measured harmonic intensity as function of the rotation angle was fitted using the evaluation procedure described recently (10,11) including the air contribution. The reference was a 1 mm thick fused silica substrate with $\chi^{(3)}(-3\omega;\omega,\omega,\omega) = 3.11 \cdot 10^{-14}$ esu (10). Fig.2a shows the Maker fringe pattern of the fused silica reference together with the theoretical fit. The peak power density of the infrared beam was ≅ 1 GW/cm^2 at the focal point. The measured THG-intensity of a 37 nm thick PPV film on the back side of a glass substrate as function of the rotation angle is shown in Fig.2b. Because the peak power density of the IR beam was reduced by a factor 12 compared to the reference measurement, no Maker fringes can be seen in this measurement.

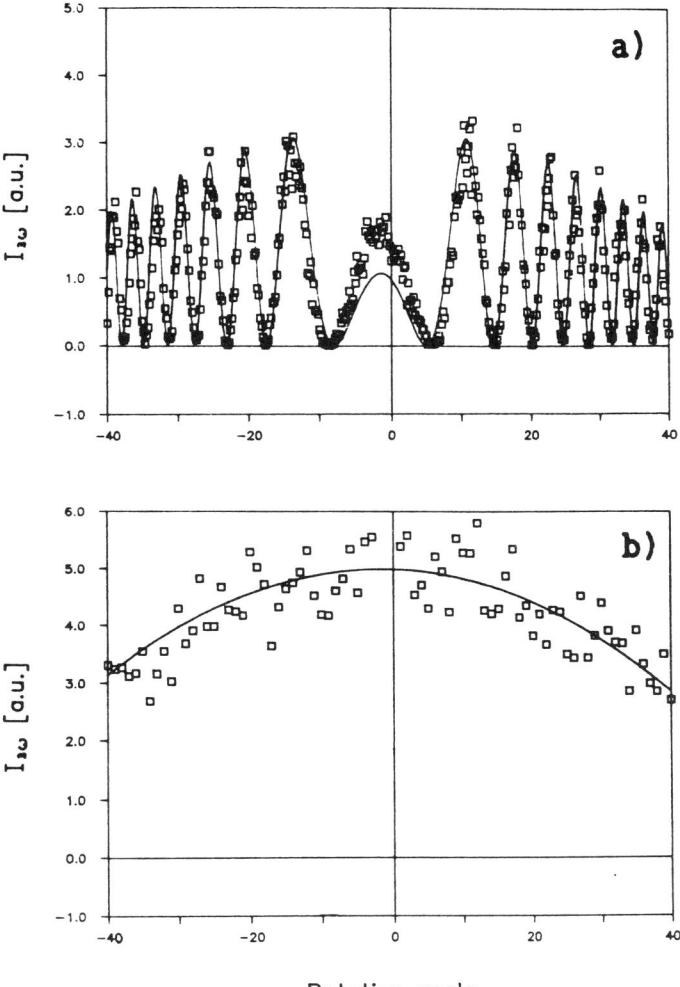

Fig.2 : Measured third harmonic intensity as function of the incidence angle (squares) together with the theoretical fit (full line) for:
(a) 1 mm thick fused silica reference
(b) 37 nm thick PPV film on the back side of a glass substrate. Here the intensity of the incident IR beam is reduced by a factor 12 compared to (a).

For the evaluation of $\chi^{(3)}$ it is necessary to know the real part n and the imaginary part κ of the refractive index at the fundamental and harmonic frequencies (11). These optical constants were obtained by a Lorentz-Lorenz analysis (12) of the PPV absorption spectrum combined with ellipsometric measurements at λ = 632.8 nm giving n(1064nm) = 1.7 $\pm$ 0.2, κ(1064nm) = 0.05 $\pm$ 0.01, n(353.8nm) = 0.9 $\pm$ 0.2 and κ(353.8nm) = 0.57 $\pm$ 0.05. The calculated curve in Fig.2b was obtained with $\chi^{(3)}(-3\omega;\omega,\omega,\omega)$ = (1.5 $\pm$ 0.6)$\cdot 10^{-10}$ esu for PPV. The $\chi^{(3)}$ values of different samples were nearly independent from film thickness in the range from 35 to 125 nm. This indicates a small amount of light scattering in PPV thin films.

4.Degenerate four wave mixing experiments (DFWM)
DFWM experiments were performed using a folded boxcar configuration. The output of a synchronously pumped and cavity dumped dye laser system was split into three beams with variable delay and focused on the sample. Discrimination between the scattering light and the nonlinear signal was achieved by a lock-in technique. A minishaker causes an oscillation of the time delay between the two beams that create the transient grating. This leads to a modulation of the nonlinear signal without affecting the stray light.

The DFWM-intensity versus time delay from a $\cong$ 1 μm thick PPV cast film is shown in Fig.3 (wavelength: 647nm, intensity 2GW/cm^2, pulse duration: 0.4ps).

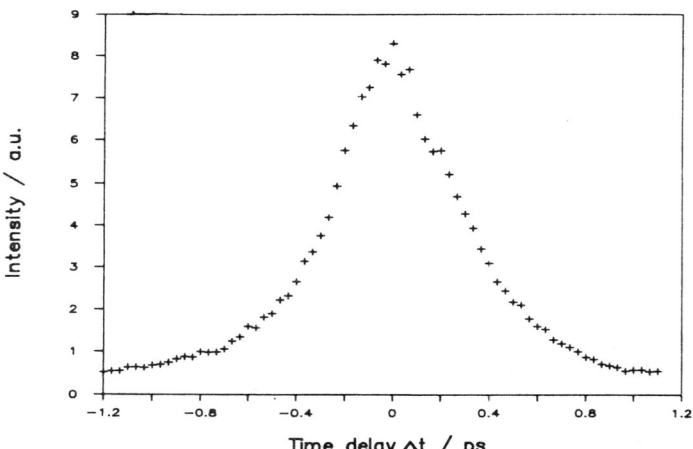

Fig.3 : DFWM intensity as function of time delay from a $\cong$ 1 μm thick PPV cast film at 647 nm. The peak power density was 2 GW/cm^2 with a pulse duration of 0.4 ps.

The experimental curve has a nearly symmetrical shape indicating a subpicosecond response time of PPV that could not be resolved in this experiment. Further measurements showed that a small contribution with a longer time constant exists. The comparision with CS_2 ($\chi^{(3)}(-\omega;\omega,-\omega,\omega) = 5 \cdot 10^{-13}$ esu (13)) gives a $\chi^{(3)}$ value of approximately 10^{-10} esu for PPV similar to the results of the THG measurements. During the laser experiments a decomposition or degradation of the PPV thin films was not observed. The samples showed only minor stray light in the DFWM experiments.

5. Conclusion
We found that PPV is a promising material for nonlinear optical applications that combines the possibilities to prepare thin stable films of good optical quality with nonresonant $\chi^{(3)}$ values in the range of 10^{-10} esu and subpicosecond response time.

Acknowledgements
This work is supported by the BMFT under the project 03M4008E9. The authors thank H.J.Menges for considerable technical support and J.D.Swalen for help and advice with the evaluation of the refractive indices.

References
1. D.S.Chemla, J.Zyss, Editors,"Nonlinear Optical Properties of Organic Molecules and Crystals", Academic Press, New York (1987)
2. R.A.Wessling, J. Polym. Sci (Polym. Symp.) 72 (1986) 55
3. J.Obrzut, F.E.Karasz, J.Chem. Phys 87,4 (1987) 2349
4. D.D.C.Bradley, J. Phys. D : Appl. Phys. 20 (1987) 1389
5. D.R.Gagnon, J.D.Capistran, F.E.Karasz, R.W.Lenz, S.Antoun, Polymer 28 (1987) 567
6. R.W.Lenz, C.C.Han, J.D.Stenger-Smith, F.E.Karasz submitted to J. Polym. Sci.
7. J.D.Stenger-Smith, R.W.Lenz, G.Wegner submitted to Polymer
8. T.Kaino, K.Kubodera, S.Tomaru, T.Kurihara, S.Saito, T.Tsutsui, S.Tokito, Electron. Lett. 23 (1987) 1095
9. B.P.Singh, P.N.Prasad, submitted to Polymer
10. F.Kajzar, J.Messier, Rev. Phys. A 32,4 (1985) 2352
11. P.A.Chollet, F.Kajzar, J.Messier, Solid State Commun. 61,4 (1985) 11
12. J.D.Swalen, J. Mol. Electron. 2 (1986) 155
13. N.P.Xuan, J.-L. Ferrier, J.Gazengel, G. Rivoire, Opt. Commun. 51 (1984) 433

THE SYNTHESIS AND PROPERTIES OF SOME NOVEL DIACETYLENE MONOMERS AND POLYMERS.

G H W MILBURN

Napier Polytechnic of Edinburgh, Department of Applied Chemical Sciences, Colinton Road, EDINBURGH EH10 5DT

1. INTRODUCTION

In recent years there has been great interest in the synthesis of novel monomers and polymers with the intention of producing materials having large non-linear optical properties [1],[2],[3].

In particular, attention has focussed on the design of molecules to maximize $\chi^{(2)}$ and $\chi^{(3)}$ effects, [4] and in the case of the former certain well identified molecular characteristics [5] can be related to the desired property, namely, non-centrosymmetric space groups, conjugated systems and electronic push-pull groups leading to delocalized π electronic systems having charge asymmetry.

Diacetylenes and polydiacetylenes have great theoretical potential as optically non-linear materials and are currently undergoing intensive investigation [6],[7] by many research groups. As well as fulfilling the molecular geometry requirements, these materials may also exhibit topochemical polymerisation [8] and are capable of forming liquid crystals, films and single crystals.

The synthesis of suitable molecules is simply the first step in developing material which may be used in devices for the optical processing of data with applications in laser systems [9].

It is also important to be able to tailor the material to a suitable form such as thin films or large single crystals [10],[11]. The chemical flexibility of diacetylenes allows them to be incorporated into a variety of other molecules [12],[13] while retaining their own intrinsic properties leading to a versatility of application and design. Unfortunately, it is not possible to design the space-groups of materials and the production of any $\chi^{(2)}$-active material is consequently still one of hit-and-miss techniques with the main exception of chiral molecules which must inevitably crystallize in non-centrosymmetric space groups when the handedness of the molecule is retained [14].

In order to try to establish correlations between structure and properties a programme of synthesis of symmetrically and unsymmetrically disubstituted diacetylenes [15], liquid crystalline diacetylenes [16], chiral diacetylenes and diacetylene-metal complexes [17] was undertaken. The products of this programme were

149

J. Messier et al. (eds.), Nonlinear Optical Effects in Organic Polymers, 149–158.
© *1989 by Kluwer Academic Publishers.*

then examined by single crystal X-ray diffraction methods, thermal analysis, hot-stage polarising microscopy, and evaluation of second harmonic generation using a pressed KCl disc and a Q-switched laser technique[18] which enables a rapid screening of materials to be carried out prior to a more detailed analysis of those samples showing promise.

Some of the results of this study are described below.

2.   SYNTHESIS - R_1-C≡C-C≡C-R_2

Standard methods of synthesis were used with some variations introduced in order to optimize the produce yields.

2.1 Symmetrical Diacetylenes

Glaser[19] coupling of propargyl alcohol followed by treatment with sulphonyl or acid chlorides was used to introduce the R_1=R_2 substituents.

Hay's[20] method was also used as a variation of the above.

2.2 Unsymmetrical Diacetylenes

The Cadiot-Chodkiewicz[21] method was used to produce a variety of products where R_1 = NO_2⟨◯⟩ ≠ R_2. By this means electronic push-pull groups were introduced into the molecule.

2.3 Chiral Diacetylenes

4-toluene sulphonic acid monohydrate was used to protect the $-NH_2$ group of the amino acids followed by reaction with propargyl alcohol and the use of Hay's method to give chiral amino acid diacetylenes.

The following examples represent some of the novel materials which were synthesised:

2.4 Symmetrical Diacetylenes

2.4.1 Methyl substituted benzene sulphonates[22-25]:

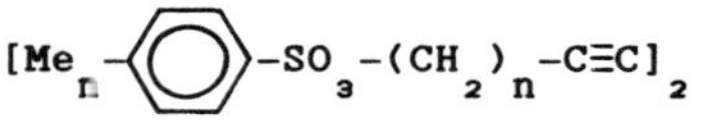

$[Me_n$⟨◯⟩$-SO_3-(CH_2)_n-C≡C]_2$

2.4.2 Methyl substituted benzoates[26]:

$[Me_n$⟨◯⟩$-CO_2-(CH_2)_n-C≡C]_2$

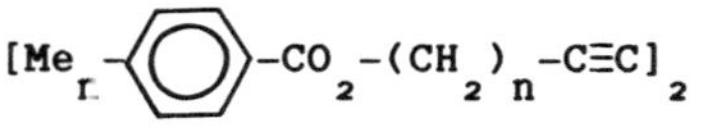

2.4.3 Methoxy substituted benzoates[26]:

$[MeO-$⟨◯⟩$-CO_2-(CH_2)_n-C≡C-]_n$

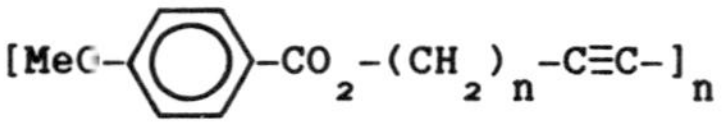

2.4.4 Various substituents R_1 and R_2 for use as metal ligands[27]:

2,4-hexadiynylene bis thiophenate
2,7-diamino 2,2,dimethyl 3,5-octadiyne
1,4-bis(4-pyridyl)butadiyne
1,4-bis(4-aminocyclohexyl)butadiyne
2,4-hexadiynylene-bis(furoate)

2.4.5 <u>Various substituents R_1 and R_2 to enhance liquid crystallinity</u>[16,28]:

<u>Type 1</u> $[ChOOCO(CH_2)_m-COO-CH_2-C\equiv C-]_2$

where Ch = the cholesteryl group.
The formate, succinate and glutarate have been synthesised.

<u>Type 2</u> $[CH_3(CH_2)_m-OOC-(CH_2)_n-COO-CH_2-COOCH_2-C\equiv C-]_2$

here m and n may be varied.

<u>Type 3</u> $[(CH_3(CH_2)_n-O-\bigcirc-COO-\bigcirc-COO-CH_2-C\equiv C-]_2$
Akoxybenzoate diester.

<u>Type 4</u> $[CH_3(CH_2)_n-O-\bigcirc-COO-CH_2-C\equiv C-]_2$

2.5 <u>Unsymmetrical Diacetylenes</u>[29,30]

The general form of the molecule that was investigated is:

$NO_2-\bigcirc-C\equiv C-C\equiv C-\bigcirc-R_1$

Other work in this area has also been published recently by a group working in Paris[31]. In the present studies the following groups were used as substituents:

where   1.   R' = NH_2, OMe, Br, OH

         2.   R' = $-N=CH-\bigcirc-OMe$

              $-N=CH-\bigcirc$
                         OH

              $-N=CH-\bigcirc-OH$
                       OMe

              $-N=CH-\bigcirc-O$
                          O–CH_2

              $-N=CH-\langle\!\!\!\!\bigcirc_O$

              $-N=CH-\bigcirc-OMe$

In addition to the above materials a variety of 1. Schiff's Bases and 2. amides were prepared:

1. $\bigcirc-C\equiv C-C\equiv C-\bigcirc-N=CH-\bigcirc$
    NO_2 R

2. $O_2N-\bigcirc-C\equiv C-C\equiv C-\bigcirc-N-H$
                                        $C-OR_1$

2.6 Chiral Diacetylenes[(32,33)] [R'CONH.CHR.CO$_2$CH$_2$C≡C−]$_2$

Chiral diacetylenes were prepared from the N‑acylated amino acids.
Type 1 from L‑valine
Type 2 from L‑alanine
Type 3 from L‑phenylalanine
where R = −CH(CH$_3$)$_2$ and R' = CH$_3$,

−CH$_3$ Ph,

−CH$_2$Ph   4−NO$_2$C$_6$H$_4$.

3. PHYSICAL ANALYSIS

3.1 Optical non‑linearity − χ$^{(2)}$ values

3.1.1 **Chiral Materials** χ$^{(2)}$ evaluation of the chiral polymers is being correlated at the present time and this programme is currently under development.

3.1.2 Butadiyne Monomers	SHG Signal

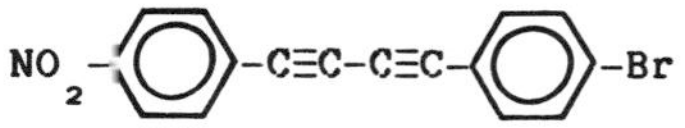

NO$_2$−⟨○⟩−C≡C−C≡C−⟨○⟩−Br   0.01

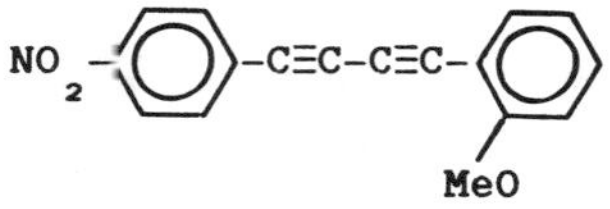

NO$_2$−⟨○⟩−C≡C−C≡C−⟨○⟩   0.02
            MeO

NO$_2$−⟨○⟩−C≡C−C≡C−⟨○⟩−OMe   0.03

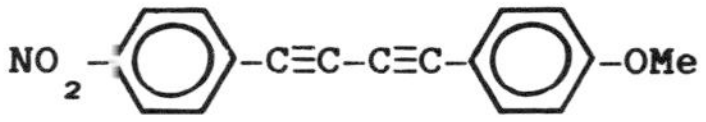

3.1.3 **Liquid Crystals**

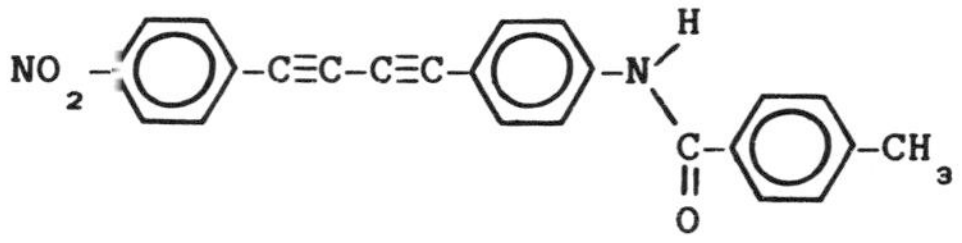

NO$_2$−⟨○⟩−C≡C−C≡C−⟨○⟩−N(H)−C(=O)−⟨○⟩−CH$_3$ 0.01

NO$_2$−⟨○⟩−C≡C−C≡C−⟨○⟩−N(H)−C(=O)−(CH$_2$)$_6$CH$_3$ 0.07

NO$_2$−⟨○⟩−C≡C−C≡C−⟨○⟩−N=CH−(furan)   0.10

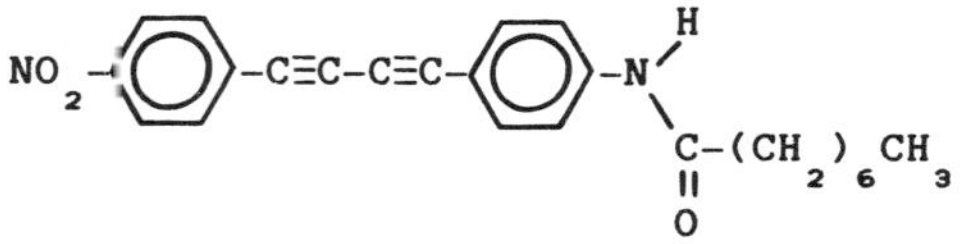

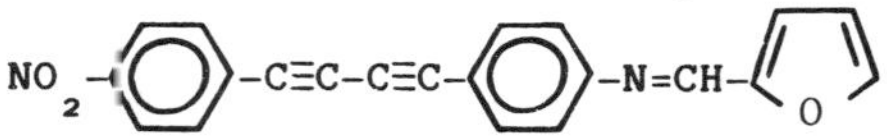

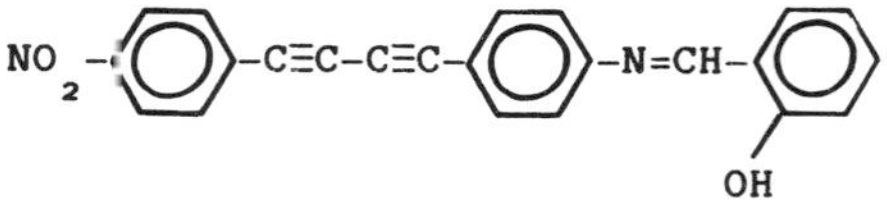

NO$_2$−⟨○⟩−C≡C−C≡C−⟨○⟩−N=CH−⟨○⟩   0.33
                          OH

3.1.4 **Standard**
2‑methyl‑4‑nitroaniline   4.32
3‑nitroaniline   0.62
2‑chloro‑4‑nitroaniline   0.02

3.2 Polymerization

(a) The nine chiral diacetylenes polymerized under a variety of procedures; pressure, UV irradiation, heat, and γ-irradiation to give characteristic highly coloured products in yields varying between 2 and 97%.

(b) ⟨◯⟩ –C≡C–C≡C– ⟨◯⟩

R_1 R_2

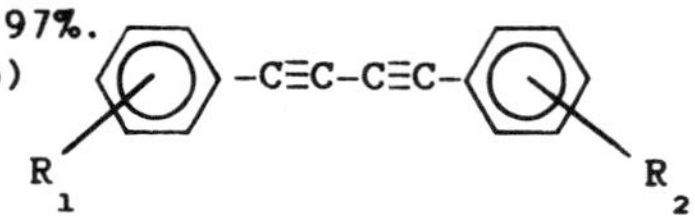

Unsymmetrically substituted diacetylenes with the following substituents polymerised in the solid state:

R_1 = 2-NO$_2$; R_2 = 2'-OMe

2'-NH$_2$ UV /visible light

4'-OMe

R_1 = 4-NO$_2$; R_2 = 4'-NH$_2$ Thermally.

All the unsymmetrically substituted diacetylenes polymerised in the liquid state (Figure 5 illustrates the liquid crystalline case).

3.3 Crystal Structures

Single crystal structure analysis of a variety of diacetylene molecules have been carried out in our laboratory using X-ray diffraction methods at both room temperature and liquid oxygen temperature, in an attempt to correlate property structure relationships of these materials [34]. Figures 1, 2, 3, and 4 give examples of this.

Criteria [7] for the topochemical polymerization of diacetylenes have been described in terms of d (the distance between monomer centres),

γ (the angle d makes with the carbon chain) and D (the C_1-C_4' distance between monomers). The methoxy benzene structure (Figures 1 & 2) is interesting in that it lies precisely in the D = 4Å curve (the proposed outer limit for polymerisation) and during the collection of X-ray data a red polymeric film formed on the surface of the crystal undergoing irradiation.

Figure 4, a low temperature structure illustrates the interesting feature of a slightly non-linear diacetylene group arising from the crystal geometry of the overall molecule [28].

Summary

Knowledge on the property structure relationships of solid state diacetylenes is continuing to be accumulated in the area of symmetrically and unsymmetrically disubstituted diacetylenes; non-centrosymmetric space groups and chiral materials, and polymer liquid crystalline diacetylenes. In particular the preparation of materials as films and single crystals is continuing on the basis of the information obtained from X-ray diffraction, thermal analysis and polarising microscopy.

154

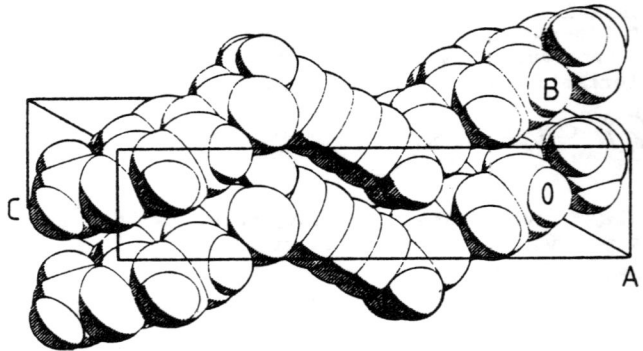

FIGURE 1. A space filling drawing based on Van der Waals radii showing the close approach of the diacetylene chains in the crystal structure of Hexa-2,4-diyne-1,6-diyl bis(4-methoxybenzoate)

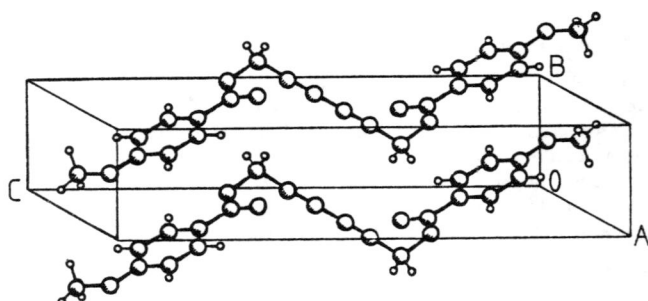

FIGURE 2. A ball and stick representation of Hexa-2,4-diyne-1,6-diyl bis(4-methoxybenzoate)

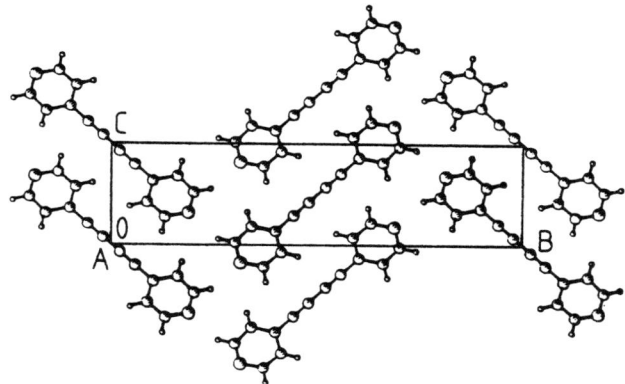

FIGURE 3. The crystal structure of Hexa-2,4-diyne-1,6-diyl bis(4-pyridine).

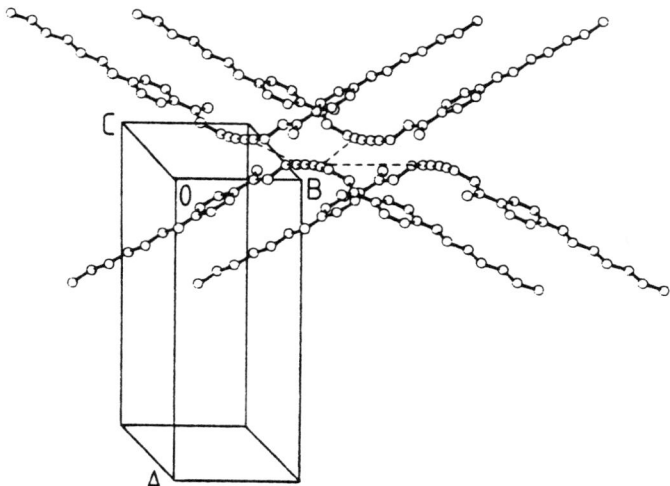

FIGURE 4. The crystal structure of Hexa-2,4-diyne-1,6-diyl bis(4-hexoxybenzoate) at 185K

156

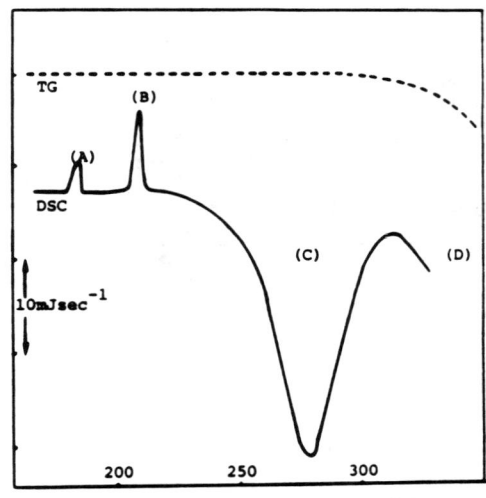

FIGURE 5. A typical DSC trace to illustrate the behaviour of the unsymmetrically di-substituted diacetylenes displaying liquid crystalline behaviour.

Peak A – crystal/crystal transition
Peak B – crystal/smectic liquid crystal transition
Peak C – polymerization of the nematic phase

REFERENCES

1. Bassler H, Sixl H, Enkelmann V: Advances in Polymer Science, 63, Cantow H-J (ed), Springer Verlag, 1984.
2. Takashi Kurihara, Hisao Tabei, Toshinumi Kaino: J.Chem.Soc.Chem. Comm, 959, 1987.
3. Desai KN, McGhie AR, Panackal AA, Garito AF: Mol.Cryst.Liq.Cryst. Letters. Vol.1(3-4), 89, 1985.
4. Garito AF, Teng CC, Wang KY, Zammani Khamiri O: Mol.Cryst.Liq. Cryst. Vol.106, 219, 1984.
5. Bloor D, Chance RR (ed): NATO ASI No. 102, Martinus Nijhoff Publ. 1985.
6. Cheula DS, Zyss J: Non Linear Optical Properties of Organic Molecules and Crystals: Academic Press, N.Y., Vols. 1 & 2, 1987.
7. Bloor D: Developments in Crystalline Polymers I. p.151. Bassett DC(ed) London: Applied Science Publishers 1982.
8. Wegner G: Faraday Discussions of the Chemical Society, 68, 495, 1979.
9. Smith SD: Nature, Vol.307, No.5949, p.315, 26 January 1984.
10. Baughman RH: Journal of Polymer Science: Polymer Physics Edition, Vol. 12, 1511, 1974.
11. Williams DJ: Angew Chem. Int. Ed. Engl. 23, 690, 1984.
12. Ziegler Jr CB, Harris SJ, Baldwin JE: J. Org. Chem. 52, 443, 1987.
13. Ozcayir Y, Blumstein A: Mol. Cryst. Liq. Cryst. Vol. 135, 237, 1986.
14. Zyss J: Journal of Non-Crystalline Solids, 47, 2, 211, 1982.
15. Bloor D, Ando DJ, Norman AP, Motevalli M, Hursthouse MB, Milburn HW, Werninck AR, Blair E: Acta Cryst. C42, 1051, 1986.
16. Hardy Gy, Milburn GHW, Nyitrai K, Horvath J, Balazs G, Varga J, Shand AJ: New Polymeric Materials, in press 1988.
17. Milburn GHW, Allan JR, Beaumont PC, Macindoe LA, Werninck AR, Barrow MJ: Inorganica Chemica Acta, Vol. 148, No. 1, p.85, 1988.
18. Bailey RT, Blaney S, Cruickshank FR, Guthrie SMG, Pugh D, Sherwood JN: J.App.Phys.B, in press 1988.
19. Glaser C: Ber 2, 422, 1869; Ann 137, 154, 1870.
20. Hay AS: J.Org.Chem. 27, 3320, 1962.
21. Chodkievicz S, Cadiot P: Compt. Rend. 241, 1055, 1955.
22. Werninck AR, Blair E, Milburn HW, Ando DJ, Bloor D, Motevalli M, Hursthouse MB: Acta.Cryst. C41, 227, 1985.
23. Day RJ, Ando DJ, Bloor D, Norman PA, Blair E, Werninck AR, Milburn HW, Motevalli M, Hursthouse MB: Acta.Cryst. C41, 1456, 1985.
24. Day RJ, Bloor D, Ando DJ, Norman PA, Hursthouse MB, Motevalli M, Milburn HW, Blair E, Werninck AR: Acta.Cryst. C42, 336, 1986.
25. Motevalli M, Norman PA, Hursthouse MB, Werninck AR, Milburn HW, Blair E, Bloor D, Ando DJ: Acta.Cryst. C42, 1049, 1986.
26. Lough A: PhD Thesis, Napier Polytechnic, Edinburgh 1988.
27. Macindoe L: PhD Thesis, Napier Polytechnic, Edinburgh 1988.
28. Milburn GHW, Barrow MJ, Lough AJ, Hardy G, Nyitrai K, Horvath J: Acta.Cryst. in press 1988.
29. Tsibouklis J: PhD Thesis, Napier Polytechnic, Edinburgh 1988.
30. Milburn GHW, Tsibouklis J, Werninck AR, Shand AJ: Chemtronics, in press 1988.

158

31. Fouquey C, Lehn J-M, Malthete J: J.Chem.Soc., Chem. Commun. 1424-, 1987.
32. Bolton E: PhD Thesis, Napier Polytechnic, Edinburgh 1988.
33. Milburn GHW, E Bolton: Chemtronics, in press 1988.
34. Milburn GHW, Lough AJ, Werninck A, Barrow MJ, Tsibouklis J, Hardy Gy, Horvath J, Nyitrai K: British Crystallographic Association Annual Meeting, Heriot Watt University, Edinburgh 1987.

DESIGN AND OPTICAL PROPERTIES OF A LOW ENERGY GAP
CONJUGATED POLYMER: POLYDITHIENO(3,4-b:3',4'-d)THIOPHENE

C. TALIANI, R. ZAMBONI, G. RUANI

Istituto di Spettroscopia Molecolare, CNR, Castagnoli 1, 40126 Bologna, Italy

P. OSTOIA

Istituto LAMEL, CNR, Castagnoli 1, 40126 Bologna, Italy

A. BOLOGNESI, M. CATELLANI, S. DESTRI and W. PORZIO

Istituto di Chimica delle Macromolecole, CNR, Bassini 15/a, 20133 Milano, Italy

1 INTRODUCTION

Although the scientific field of conducting conjugated polymers is only ten years old, the knowledge to which a growing number of scientists has contributed is such that it may be regarded as an already mature field. A consolidated theory of the charge transport mechanism has become available both for polymers with degenerate and non-degenerate ground states based on the notion of solitons and polaron-bipolaron excitations respectively [1-4]. At the same time a relatively large number of new conducting polymers [5] has been discovered whose respective optical properties may all be accounted for under the same general theoretical framework.

A further step in the development of conjugated polymers as useful new materials both in organic charge transport and non-linear optic fields, depends on the capability to control the physical properties of the resulting polymer by designing new polymeric structures.

In view of this, a certain number of theoretical and experimental papers [6-9] have examined the problem of the relationship between the molecular architecture and the transport properties.

A remarkable aspect of the conducting conjugated polymers is the very fast non linear response to a probing laser pulse which makes these materials of special interest in the field of optical devices. The high non-linearity, although it is a prerequisite which seems to be a general property of conducting conjugated polymers, should be accompanied by a low absorption loss; for this reason it is important to design new molecular structures which are transparent in selected regions of the optical

159

J. Messier et al. (eds.), Nonlinear Optical Effects in Organic Polymers, 159–172.

spectrum. On the other hand the time response depends on the energy difference between the probing photon beam and the strong $\pi-\pi^*$ electronic absorption. The combination of low loss and fast and strong response narrows the range of suitable conjugated polymers. It is widely agreed that the fast non-linear response in conjugated polymers is in relation with the fast formation and decay of photoexcited species like solitons and bipolarons. In this respect then the molecular characteristics required for making good conducting polymers are the same as those required for obtaining high non-linear optical properties.

A versatile route for the preparation of conducting conjugated polymers with a controlled chemical architecture is given by the electrochemical polymerization of tailored monomer molecules. Following this method, provided that the monomer molecule has a low oxidation potential and selective polymerization sites, different polymer backbone structures may be achieved. Particularly suitable for this approach are the heteroaromatic five-member ring organic compounds like thiophene and pyrrole.

Furthermore, by means of the electro-polymerization, the polymer thin film may be obtained directly on suitable substrates in the doped (oxidized) state and by reversing the applied electric potential the undoped (reduced) state is obtained; such a process has been shown to be reversible and allows for instance to control the level of doping. A certain number of new polymeric structures have been prepared following this route[5].

From the comparative study of optical, electrical and structural properties some general conclusions on the molecular characteristics which are required in order to impart good charge transport properties have been derived. Namely:

i) the polymer backbone should have an extended π-electron conjugation and therefore should be planar and free from chemical defects (such as sp^3 hybridizations) that otherwise would interrupt the π-electron conjugation;

ii) the polymer structure should have an appropriate symmetry in order to achieve a wide bandwidth of the HOMO-band and hence to increase the mobility of charge carriers.

The charge transport in real conducting polymers is described by the superposition of three elementary transport processes: i) the intrachain transport; ii) the interchain transport; and finally iii) the interparticle transport. The first step is of crucial importance for the overall process since the real structure of conjugated polymers is far from the ideal one-dimensional infinite chain model. In reality chains have a certain distribution of finite conjugation lengths. On the other hand, in the molecular models of conducting polymers which have been investigated until now, the π-electron conjugation occurs along the direction of the polymer backbone. In this scheme, the ineluctable occurrence of defects along the direction of propagation of the polymer backbone constitutes an intrinsic obstacle to the charge transport. An alternative approach would be to obtain a bidimensional network of conjugated chains in such a way that the charge transport between adjacent chains would occur coherently rather than through hopping.

2 CHEMICAL DESIGN OF A LOW ENERGY GAP POLYMER

The band gap is an important characteristic which would be interesting to control with the aim of preparing intrinsically conducting polymers without the need of doping as well as conjugated polymers with suitable spectral windows of transparency.

In view of the above stated considerations we have designed the molecular architecture of a new monomer molecule; dithieno[3,4-b;3',4'-d]thiophene (DTT') shown in the following Figure.

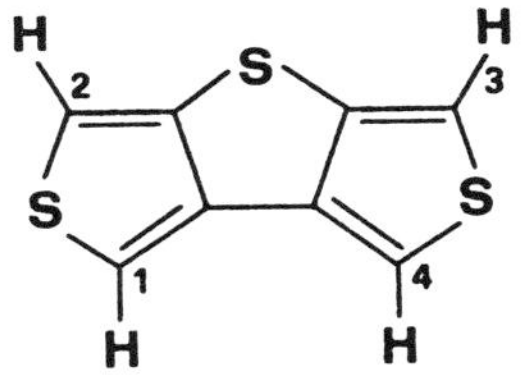

Fig. 1 Molecular structure of dithieno[3,4-b;3',4'-d]thiophene (DTT'). The positions labelled 1,2,3 and 4 represent the possible sites of propagation of the polymer chain.

The increase of the π-electron delocalization in the monomer unit is achieved by the fusion of three aromatic five-member rings in β position with respect to lateral thiophene moieties. It has been shown in fact by F. Wudl et al. [10] that the condensation of a thiophene molecule with a benzene ring in β position gives rise to a low energy gap conducting polymer, poly-ITN with an Eg of 1 eV. A theoretical investigation (MNDO-SCF-MO and VEH) of the geometry and electronic properties of poly-ITN [12] has shown that the large π−electron delocalization induced by the presence of the benzene ring attached in β position gives rise to a remarkably reduced ionization potential and a low π−π* transition energy compared to the parent polythiophene macromolecule. Similar results have been obtained also by means of an *ab initio* Hartree-Fock MO treatment [13].

Moreover the molecular structure of DTT' is such that only the α-positions are available for the linking propagation of the polymerization, reducing the possibility of the formation of chemical defects such as α-β' linkages that would otherwise interrupt the Π-electron conjugation.

An additional and very interesting feature of the DTT' molecule in relation to the consideration on the dimensionality (vide supra) is given by the two equivalent thiophene sub-units, at both sides of the molecule (see Fig. 1). In principle both the thiophene sub-units may undergo to polymerization, inducing some kind of multi-dimensionality to the resulting polymer backbone and preserving at the same time the Π-electron conjugation between the adjacent chains. The macro molecular fragment that may provide such a possibility is schematically represented in Fig. 5c.

The possibility of the occurrence of some conjugated cross-links in poly-DTT' may provide an alternative path for the charge transport of the carriers which may eventually overcome the obstacles to the propagation providing a conjugated percolating path through the macro-structure as well as increasing the Π-electron delocalization.

In this respect, the molecular architecture of DTT' is definitely more suitable than the isomer molecule dithieno[3,2-b:2',3'-d]thiophene (DTT) which, on the other hand, has been shown to give rise to an interesting conducting polymer [14], but in DTT the sites in β positions are also available for the propagation of the polymerization giving rise to possible α-β' linkages that would interrupt the conjugation.

3 RESULTS AND DISCUSSION

3.1 PREPARATION

The chemical synthesis of the DTT' molecule is described in Ref. 15. DTT' is a rather stable molecule at ambient atmosphere and room temperature.

The electro-polymerization has been performed in a two-compartment cell at room temperature with platinum electrodes at constant current density of 1 mA/cm^2 from an acetonitrile solution of the monomer (0.03M) and LiClO$_4$ as an electrolyte. During the polymerization the potential remains constant at 1.04 V vs the S.C.E.

Elemental chemical analysis shows that the degree of doping of the as-grown oxidized polymer corresponds to one counter-ion every three monomer units.

3.2 STRUCTURE

The molecular structure of the DTT' monomer has been studied by X-Ray diffraction performed by means of a Debye camera under helium atmosphere and the geometrical details are reported in the following Figure.

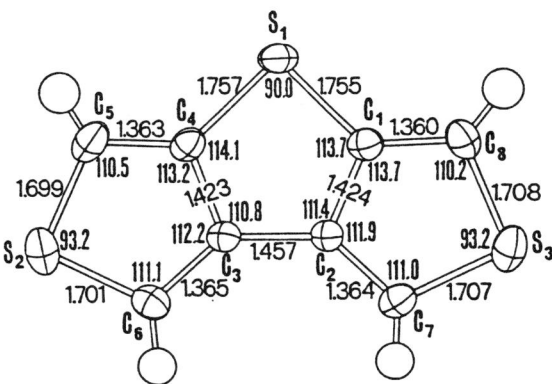

Fig.2 Bond lengths and bond angles of the DTT' monomer molecule derived from the refinement of the single crystal X-Ray diffraction study.

Moreover X-ray diffraction experiments, performed on both the doped (oxidized) and the undoped (reduced) polymers show the presence of only one broad diffraction peak. It is commonly agreed that this is the evidence of a graphite-like spacing between the polymer layers [16]. In this case the spacing is of 3.6 Angstroms which should be compared to 3.2-3.3 Angstroms in poly-DTT [6]. The absence of diffraction peaks ascribed to the ClO_4^- ions indicates that the dopant is randomly distributed in the proximity of polymer chains.

3.3 ELECTRICAL CONDUCTIVITY

The electrical conductivity of the doped poly-DTT' is measured by means of a *test pattern* with four gold contacts separated by a 33 micron gap deposited on a SiO_2 surface. The *test pattern* with short circuited contacts is used as the anode. During the electropolymerization the growing polymer bridges the gap between the electrodes. The *in situ* conductivity is measured by means of the standard four contact method; the thickness of the polymer grown between the electrodes is estimated by scanning electron microscopy (SEM) observations. The measured specific conductivity is 1 S cm^{-1} which is of the same order of magnitude as polythiophene [6] measured with the same method.

3.4 ELECTRONIC PROPERTIES

A direct estimate of the degree of the Π-electron delocalization is given by the spectrcscopic characterization of poly-DTT'.

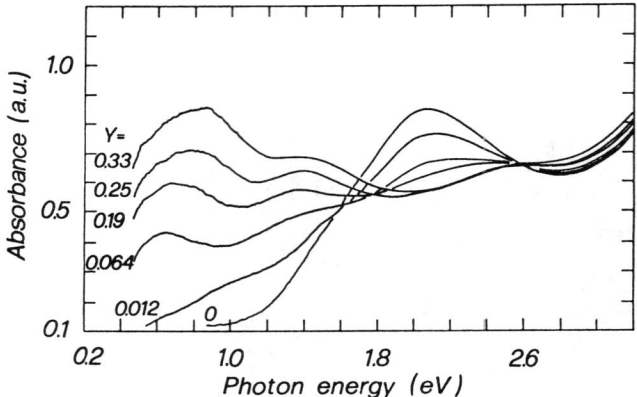

Fig. 3 *In situ* **UV-VIS-NIR spectra of polyDTT' at different degrees of ClO_4^- doping; the degree of doping is indicated on each curve in units of dopant moles per monomer unit.**

The *in situ* UV-VIS-NIR spectra of poly DTT' at various doping levels are reported in Fig. 3. A thin film of undoped (reduced) polymer grown on ITO conducting glass is placed inside a sealed electrochemical cell especially designed for spectroscopic measurements with acetonitrile and $LiClO_4$ as an electrolyte; various doping levels are achieved by applying a voltage at constant current density (0.1 mA/cm^2) for a controlled period of time, between the neutral polymer electrode (anode) and the platinum cathode.

The neutral poly-DTT' spectrum shows a strong absorption band assigned to the first $\Pi-\Pi^*$ electronic transition with a maximum at 2.1 eV (590 nm); the bandgap derived from the flection of the low energy slope of the optical absorption is about 1.1 eV (lower than polythiophene by 0.9 eV).

This makes the poly-DTT' polymer the lowest bandgap of any known conjugated polymer together with poly-ITN (Eg = 1eV) [11].

The effect of doping is shown by the gradual decrease in intensity of the $\Pi-\Pi^*$ electronic absorption and the consequent increase in intensity of two new bands on the low energy side. This behaviour is a general characteristic of conjugated conducting polymers with a non-degenerate ground state and is accounted for by the formation of two bipolaron bands.

The energy of the two bipolaron bands shifts remarkably to higher energies; the lower bipolaron band from approximately .53 eV to .83 eV and the higher bipolaron band from 1.06 to 1.39 eV, while the $\pi - \pi^*$ electronic transition shifts from 2.12 to 2.25 eV. A further evidence of the bipolaron nature of the two doping induced bands inside the gap is the fact that the sum of the energies of the two bipolaron bands is at all level of doping approximately equal to the energy of the interband transition which implies that the two new bands are symmetrical in respect to the band gap.

The drastic reduction in intensity of the interband transition upon doping reflects an interesting aspect of the optical properties of this new polymer since the $\pi - \pi^*$ interband transition lies in the middle of the visible spectral range.

The reversible switching from the undoped to the doped species, combined with the dramatic change of the optical properties in the visible spectral range, gives rise to an interesting electrochromic effect. The visual observation of the process of doping is accompanied by a switching from an almost opaque highly resistive polymer film to a colourless and semi-transparent highly conducting polymer as it is shown in Fig. 4.

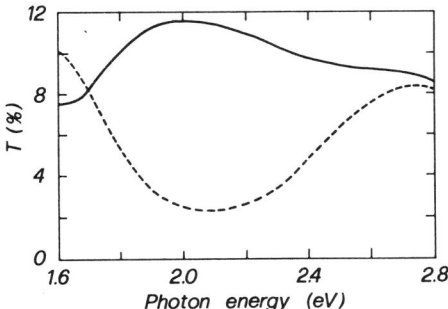

Fig.4 Visible transmission spectrum of a thin film (1.2 μm) of poly-DTT' on ITO glass. The dotted line indicates the undoped polymer and the full line indicates the doped polymer. The switching between the two forms is achieved by reversing the potential between the polymer electrode and the counter electrode. Note that the maximum change in transmission is observed in the green range of the visible spectrum corresponding to the maximum sensitivity of the eye.

The remarkable low energy of the first $\pi - \pi^*$ electronic transition of the undoped (reduced) poly-DTT' is the evidence of the high π-electron delocalization which is achieved in this new polymer backbone architecture.

3.5 POLYMER CONFIGURATION

In principle all the configurations that involve two, three and four bridgings between the α positions labelled 1,2,3 and 4 in Fig. 1, are equally possible because all the resulting configurations preserve the double bond alternation on the polymer backbone.

An additional requirement is given by the electronic spectral behaviour upon doping. The two low energy doping induced bands in the near infrared are the signatures of the two bipolaron states that require that the ground state be described with comparable weight by the aromatic and quinoid valence-bond resonance structures. In this case there are only three types of enchainment which allow for low energy quinoid-type resonance structures. These are represented in Figure 5.

On the other hand the fully bridged configuration represented in Figure 5c cannot give the major contribution because the infrared spectra of the undoped polymer show the presence of a moderately intense C-H stretching band indicating that there are unsubstituted α carbon atoms in the polymer structure.

Knowing the molecular structure of the monomer DTT' molecule from X-ray studies we have performed some molecular modelling calculations of the possible polymer structures reported in Fig. 5a and 5b by assuming that the DTT' moiety preserves its planarity in the polymer (as it is suggested by optical spectra).

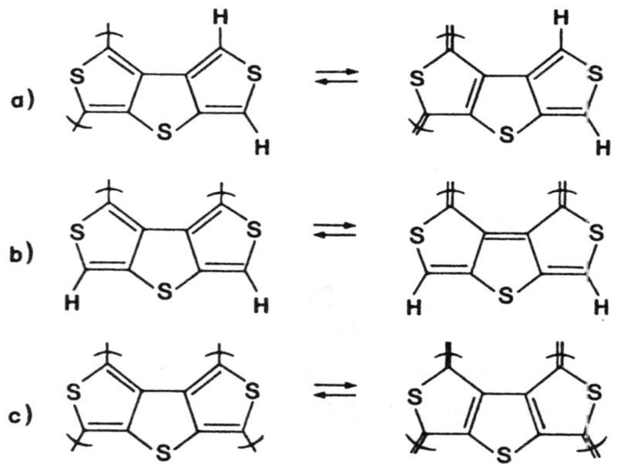

Fig.5 **Resonance valence-bond ground state structure of DTT' by assuming the several possible points of attachment of the polymerization. Note that only the α-positions with respect to the sulphur atom of the two external thiophene units are available as sites for the oxidative polymerization.**

In the case of the configuration represented in Fig.5b, the formation of acceptable intermolecular contacts requires a deviation from the mean molecular plane of at least +/- 56 degrees, with the consequence of a drastic decrease of the π-orbital overlap. Such an hypothesis is discarded because it is in contrast with the large π-electron delocalization inferred from optical properties.

The configuration represented in Fig. 5a requires, on the other hand, a deviation from the mean molecular plane of +/-10 degrees. The π-electron overlap would not be affected sizeably by such a minor deviation from planarity. For this reason we suggest that this type of enchainment is the most likely to occur. Other studies like NMR and XPS spectroscopy are needed in order to solve this problem.

3.6 PHONON PROPERTIES

The FT-IR spectra of the *as grown* polymer films on a silicon substrate in doped (oxidized) and successively undoped (reduced) forms are shown in Figure 6.

By comparing the two spectra we notice that a series of intense doping induced bands (DIB) dominates the doped polymer spectrum in the $1100-1500$ cm^{-1} spectral range. DIB are observed at 853, 1157, 1204, 1281, 1309, 1360, and 1409 cm^{-1}. A qualitatively similar behaviour is manifested by other conducting polymers based on fused thiophene rings [5,6].

The generation of positively charge carriers in the polymer backbone, which is achieved by electrochemical doping, causes a perturbation of the local backbone geometry surrounding the charge. As a consequence, there is a breakdown of the translational symmetry invariance of the polymer backbone such that the Raman active and dipole forbidden infrared vibrations of the undoped polymer become infrared active in the doped species.

The DIB are therefore *signatures* of the charge carriers which manifest themselves also with the two low energy electronic absorptions. This implies that charges are correlated in pairs in form of bipolarons. The intensity of the DIB may be related to the number of charge carriers. The relative intensity of the DIB in the $1100-1500$ cm^{-1} spectral range, compared with the C$=$C asymmetric stretching band of the undoped species at 1513 cm^{-1}, suggests that indeed the number of charge carriers in the doped polymer is of the same order of magnitude as in other thiophene based polymers [6].

On the high energy side of the IR spectrum of the doped and the undoped species we observe a relatively sharp C-H stretching band at 3095 cm^{-1} which reflects the presence of C-H bonds in the polymer backbone. This observation indicates that the polymerization does not proceed by removing all hydrogen atoms as it is expected if both the external thiophene sub-units were involved in the polymerization. In other words the configuration of the polymer is not, at least fully, represented by the ladder configuration shown in Fig. 5c.

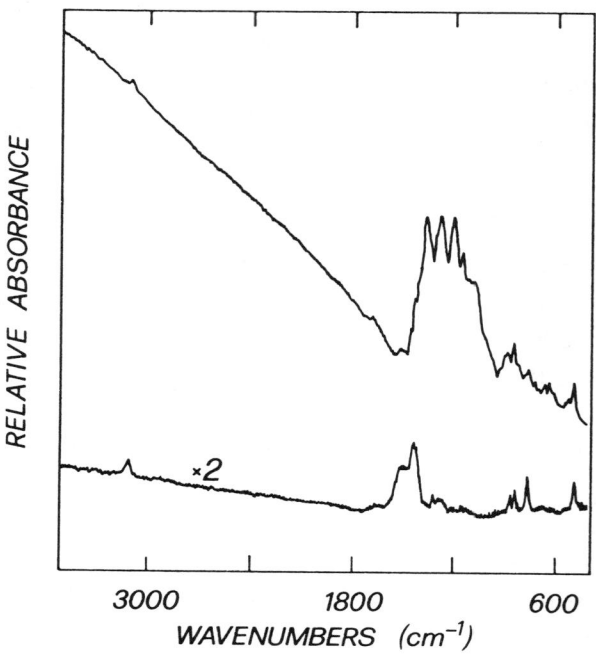

Fig.6 FT-IR spectra of the same film of poly-DTT' polymer deposited on silicon. The upper curve shows the doped polymer and the lower curve shows the undoped polymer.

3.7 PHOTOINDUCED ABSORPTION AND NON-LINEAR OPTICAL PROPERTIES

The infrared photoinduced absorption (PA) of the undoped poly-DTT' at 20K is shown in Figure 7, together with the infrared spectrum of the doped and undoped species.

The PA has been measured by means of a modified Bruker FT-IR (mod. 113 v) in which a laser beam of photon energy of 1.96eV at 50 mW/cm^2 was allowed to illuminate the sample held at low temperature by means of an Air-Liquide helium flow cryostat. The samples consisted of pressed pellets of DTT' mixed with KBr at about 0.5% wt. A fast liquid helium cooled Ge-Cu solid state device as well as liquid nitrogen cooled MCT were used as detectors and the resolution was set at 1 cm^{-1}. Consecutive interferograms of 32 scans each, with laser on and off, were stored for a total number of about 5000 scans in order to achieve an appreciable S/N ratio. The separate interferograms were then manipulated in two ways:

a) the interferograms were Fourier transformed separately and the spectra were then coadded;

b) the interferograms were first coadded and then transformed.

The PA spectrum was derived by calculating the $-\Delta T/T$ quantity and the agreement between the measurements performed following both a) and b) procedures gave us confidence on the reliability of the results.

TABLE 1. Comparison of the frequencies (cm^{-1}) of the infrared spectra of poly-DTT' with the photoinduced absorption spectrum

Infrared Absorption		Photoinduced Absorption	
Undoped	Doped		
503 w	503 m		
	645 w		
	670 w		
767 w	764 w	750	
	853 m	850	
866 w		865	bleaching
	885 w		
		945	
	1157 w	1130	
	1204 vs	1204	
	1281 vs	1270	
	1309 m		
	1360 vs	1375	
	1409 w		
1440 m		1435	bleaching
1513 m		1500	
		1600	

The PA spectrum shows intense bands at 1204, 1270, 1375, and 1500 cm^{-1} as well as weaker bands at 750, 850, and 945 cm^{-1}. Two bleachings, a moderately intense one at 850 cm^{-1} and a strong one at 1435 cm^{-1}, are also observed. The PA bands in the 1100-1500 cm^{-1} appear at approximately the same frequencies as the DIB observed in the doped spectrum and the bleaching at 1435 cm^{-1} corresponds to the C=C out of phase stretching mode while the bleaching at 865 cm^{-1} corresponds to the ring deformation mode observed at 866 cm^{-1} in the undoped polymer. A summary of infrared and PA frequencies is reported in Table 1.

The correspondence between the DIB and the PA bands in the region of 1100-1500 cm^{-1} reflects the identical nature of the charge defect states which may be indifferently generated by doping or by direct preparation of excited states (and subsequent formation of e-h pairs) by photon excitation at energies higher than the energy gap.

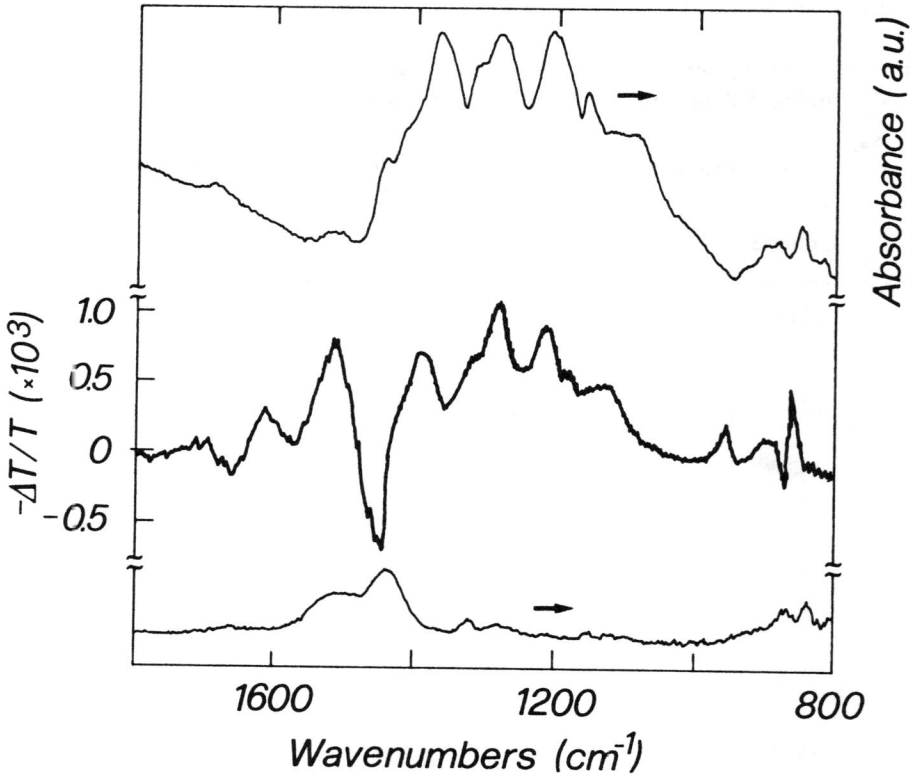

Fig 7 Infrared photoinduced absorption spectrum of poly-DTT' (middle trace); the photogeneration of charge carriers is obtained by laser excitation at 1.96 eV and the sample temperature is 20K. The infrared spectra of the doped (upper trace) and undoped species (lower trace) are shown for comparison.

The same behaviour is manifested by polythiophene[17] and poly-(3methyl)thiophene [18] and is generally agreed that this is a further evidence of the bipolaron nature of the charge carriers in the polymer backbone.

On the other hand the generation of bipolarons by means of laser illumination, with its consequent shift of the oscillator strength from the interband transition to the low energy transitions inside the gap and to the molecular vibrations, has been indicated to be of great relevance in inducing a strong non-linear response of the medium to a laser beam. In the case of poly-DTT' the low energy gap of this new polymer should extend this non-linear effect further into the near infrared spectral range compared to the other known conjugated polymers.

CONCLUSIONS

Making use of the knowledge on the relations between the chemical architecture and the electronic properties of the conjugated conducting polymers, and of the flexibility offered by organic chemistry, we have prepared and characterized a new transparent conducting polymer with good charge transport properties and a low energy gap in the undoped species.

The structure of both the monomer and the polymer has been investigated.

The optical properties that have been characterized both in the electronic and the phonon spectral range are accounted for by the general bipolaron model.

The photoexcitation at energies higher than the energy gap are shown to give intense "forbidden" IR bands; their nature is generally related to the lattice distortion around the charge defects which are equivalent to those produced by doping (i.e. bipolarons); this behaviour suggests that this new material possesses a fast non-linear optical response.

The reversibility between the doped and the undoped species gives rise to a dramatic electrochromic effect in the visible spectral range between a semi-transparent and an opaque film.

Finally some hypothesis on the possible polymer configuration has been made on the basis of molecular modelling considerations.

ACKNOWLEDGEMENTS

The authors wish to thank Mr. S. Guerri for valuable technical contribution. The authors also wish to thank Dr. M. Cozzi of W. Pabisch SpA for the use of a JASCO NIR-VIS-UV spectrophotometer.

REFERENCES

1. W.P. Su, J.R. Schriffer & A.J. Heeger, Phys. Rev. B, **22**, 2099 (1980).
2. S.A. Brazovskii & N.N. Kirova, Pisma. Zh. Eksp. Teor. Fiz. **33**, 6 (1981) [JETP Lett. **33**, 4 (1981)]
3. A.R. Bishop, D.K. Compbell & K. Fesser, Mol. Cryst. Liq. Cryst. **77**, 253 (1981)
4. J.L. Bredas, R.R. Chance & R.S. Silbey, Mol. Cryst. Liq. Cryst. **77**, 319 (1981)
5. C. Taliani in Molecular Electronics, p. 394, ed. by M. Borissov, World Scientific Publishers, Singapore 1987.
6. C. Taliani, R. Danieli, R. Zamboni, P. Ostoja & W. Porzio, Synth. Met. **18**, 177 (1987)
7. O. Wennerstrom, Macromolecules, **18**, 1977 (1985)
8. A.K. Bakhshi, J. Ladik & M. Seel, Phys. Rev. B, **35**, 704 (1987)
9. J.L. Bredas, Synth. Met. **17**, 115 (1987)
10. F Wudl, M. Kobayashi & A.J. Heeger, J. Org. Chem. **49**, 3382 (1984)
11. M. Kobayashi, N. Colaneri, M. Boysel, F. Wudl & A.J. Heeger, J. Chem. Phys. **82** 5717 (1985)
12. J.L. Bredas, B. Themans, J.M. Andre, A.J. Heeger & F. Wudl, Synth. Met. **11**, 343 (1985)
13. A.K. Bakhshi & J. Ladik, Solid State Commun. **61**, 71 (1987)
14. P. Di Marco, M. Mastragostino & C. Taliani, Mol. Cryst. Liq. Cryst. **118**, 241 (1985)
15. F. De Jong & M.J. Janssen, J. Org. Chem. **36**, 1645 (1971)
16. S. Bruckner & W. Porzio, Makromol.Chem. **189**, 961 (1988)
17. H.E. Schaffer & A.J. Heeger, Solid State Commun. **59**, 415 (1986); Z. Vardeny, E. Ehrenfreund, O. Brafman, A.J. Heeger & F. Wudl, Synth. Met. 18, 183 (1987)
18. Y H. Kim, S. Hotta & A.J. Heeger, Phys. Rev. B, **36**, 7486 (1987)

SYNTHETIC APPROACHES TO STABLE AND EFFICIENT POLYMERIC FREQUENCY DOUBLING
MATERIALS. SECOND-ORDER NONLINEAR OPTICAL PROPERTIES OF POLED,
CHROMOPHORE-FUNCTIONALIZED GLASSY POLYMERS

C. YE, N. MINAMI, and T. J. MARKS, Department of Chemistry and the
Materials Research Center; J. YANG and G. K. WONG, Department of Physics
and the Materials Research Center, Northwestern University, Evanston, IL
60208, USA

1. INTRODUCTION
    Materials exhibiting highly nonlinear optical (NLO) characteristics
are currently the focus of intense scientific and technological interest
(1-5). While such materials have traditionally been inorganic substances
(e.g., KDP, KTP, LiNbO$_3$, etc. (5)), growing evidence now indicates that
optical materials composed of organic π-electron chromophores hold considerable promise. In particular, polymeric NLO materials of this type
would be ideally suited as the active components of numerous guided-wave
and integrated optics devices (1-4). The advantages of such materials
include the structure/performance tailorability inherent in organic
polymeric structures and ease of processing to fabricate thin films,
fibers, waveguide structures, etc. Glassy polymer systems would offer, in
addition, high optical quality, low dielectric constants and dielectric
loss, and a large synthesis/processing information base.
    Crucial design strategies for polymeric NLO materials must include
appropriate choice of chromophore (6) and effective chemical means of
incorporating the chromophore into a polymeric structure. The latter task
must achieve high chromophore densities (to maximize NLO performance)
without phase separation, and high environmental stability with regard to
chromophore detachment or decomposition. In addition, strategies for
second harmonic generation (SHG) materials must bring about (e.g., by
electric field poling (7-9)) significant and persistent acentricity.
Understanding and controlling the temporal aspects of induced acentricity
in a polymeric material is a particularly challenging problem (10).
    We recently demonstrated that covalently linking NLO chromophores to
the backbones of glassy polymers, followed by electric field poling,
offers a new and efficacious approach to polymeric SHG materials (11-14).
The present contribution extends these efforts to glassy, high-T$_g$ polystyrene systems having the possibility of higher levels of chromophore
functionalization (11) and structure reinforcement/chromophore immobilization by hydrogen bond networks.

2. EXPERIMENTAL
    The functionalization of poly($\underline{p}$-hydroxystyrene) ((PS)OH, $M_W \approx 6,000$)
was carried out as shown in Scheme I. The chromophores NPP (N-(4-nitro-
phenyl)-L-prolinol) and Disperse Red 1 (4-(4-nitrophenylaza)(N-ethyl)(2-
hydroxyethyl)aniline) (DR) were converted to the corresponding tosylates
by standard procedures. The tosylates were then subjected to nucleophilic
substitution by (PS)O$^-$ using unexceptional methodology to yield the
chromophore-substituted polymers. Chromophore functionalization levels up
to 60% of the phenolic rings can be readily achieved. The functionalization of poly(vinylbenzyl chloride) (ca. 60/40 3-/4- isomers) was carried
out with the anion of NPP as shown in Scheme II. All new products were

J. Messier et al. (eds.), Nonlinear Optical Effects in Organic Polymers, 173–183.
© *1989 by Kluwer Academic Publishers.*

174

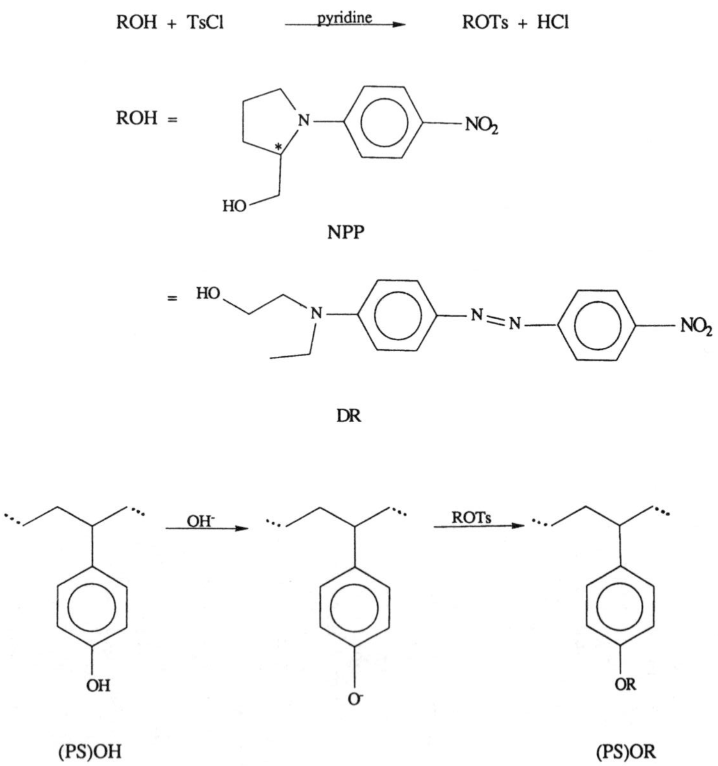

SCHEME II

ROH + NaH $\xrightarrow{\text{DMF}}$ RO⁻Na⁺ + H₂

ROH =

NPP

(PS)CH₂Cl $\xrightarrow{\text{RO}^-}$ (PS)CH₂OR

SCHEME I

ROH + TsCl $\xrightarrow{\text{pyridine}}$ ROTs + HCl

ROH =

NPP

DR

= HO

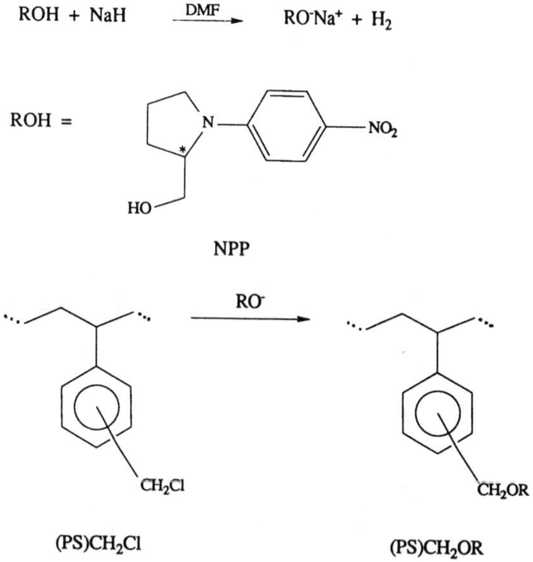

(PS)OH $\xrightarrow{\text{OH}^-}$ (PS)O⁻ $\xrightarrow{\text{ROTs}}$ (PS)OR

characterized by 400 MHz ^{1}H NMR, elemental analysis, DSC, and optical spectrophotometry.

In a class 100 laminar-flow clean hood, thin films (1-4 μm as measured by a Tencore Alpha-Step Profiler) of (PS)O-NPP,DR and (PS)CH$_2$-NPP were cast onto ITO-coated conductive glass from multiply filtered solutions of the functionalized polymers. THF and ClCH$_2$CH$_2$Cl were used as the solvents for casting these functionalized PS(OH) and PS films, respectively. The films were then annealed at 100°C for 2 hrs prior to dc poling. The effect of annealing is to enhance the ultimate poling fields which can be achieved and the temporal stability of the SHG capacity (vide infra). The cause appears to be removal of traces of solvent (THF or ClCH$_2$CH$_2$Cl) and other volatiles (verified by FT-IR) which may plasticize the material and enhance macromolecule mobility (supported by DSC data), changes in the (PS)OH/(PS)OR hydrogen bond network (verified by FT-IR), and changes in film morphology (suggested by SEM) which may seal imperfections which lead to dielectric breakdown during poling.

Annealed films were covered with an aluminum electrode, heated to 80°C for (PS)O-NPP,DR films and 110°C for (PS)CH$_2$-NPP films, and poled by incrementally increasing the field with online current monitoring (to avoid breakdown processes). The film was held at the maximum field for 0.5 hrs and then cooled in the presence of the field. Careful film fabrication and annealing allowed poling at dc fields as high as 1.8 MV/cm for functionalized PS(OH), which appears to be near the limit of what is practicable for typical polymer thin films (15). In contrast, films of (PS)CH$_2$-NPP were more susceptible to breakdown during poling. The second harmonic generating capacities were also lower than those of (PS)O-NPP and -DR films having similar functionalization levels and poling conditions. The differences in behavior may be due to impurities in (PS)CH$_2$-NPP. Further purification efforts are in progress. In order to directly compare merits of covalently functionalized versus chromophore-doped polymer systems (8-10), experiments were attempted with PS and (PS)OH films which were simply doped with NPP. The results indicated phase separation (and opacity) at chromophore concentrations as low as 5 mol%, frequent breakdown at low poling fields (presumably reflecting greater mobility), and weak, short-lived SHG performance (also presumably reflecting greater chromophore mobility). Only a small number of NPP/PS films were of sufficient quality for SHG measurements.

3. RESULTS

A poled glassy film is uniaxial and belongs to the point group ∞mm. Based on spatial symmetry considerations, the second harmonic polarization can be written in terms of three independent nonzero second harmonic coefficients d_{31}, d_{15}, and d_{33} (5,16), where the 3-axis is taken to be the uniaxial axis defined by the direction of poling field (eqs.(1)-(3)),

$$P_x^{2\omega} = 2d_{15}E_xE_z \tag{1}$$

$$P_y^{2\omega} = 2d_{15}E_yE_z \tag{2}$$

$$P_z^{2\omega} = d_{31}E_x^2 + d_{31}E_y^2 + d_{33}E_z^2 \tag{3}$$

where the conventional contracted notation has been used (16). Assuming Kleinman symmetry (17), one can further set $d_{31} = d_{15}$. In general, the transmitted second harmonic power generated by the nonlinear polarization in eqs.(1)-(3) is given for radiation incident on polymer films from the glass side by eq.(4) (18)

$$p^{2\omega} = (512\pi^3/A)t_g^4 t_\omega^4 T_{2\omega} d^2 p^2 P_\omega^2 [1/(n_\omega^2 - n_{2\omega}^2)^2] \sin^2\Psi(\theta), \tag{4}$$

where $P^{2\omega}$ is the transmitted second harmonic power, P^ω is the incident fundamental power, A is the laser beam area, dp is the product of appropriate second harmonic coefficients with angular factors resulting from projection of the nonlinear polarization components onto the direction of $E_{2\omega}$ in the film, t_ω and $T_{2\omega}$ are Fresnel-like transmission factors, t_g is the Fresnel transmission factor of the fundamental light through the glass substrate, and the n's are the refractive indices at the indicated frequencies. Ψ is an angular factor resulting from interference between free and bound waves. For films with thickness small compared with the coherence length, $l_c = \lambda/4(n_\omega - n_{2\omega})$, $\sin^2\Psi$ is given approximately by eq.(5),

$$\sin^2\Psi(\theta) \sim \left[\frac{\pi}{2} \frac{1}{l_c} \frac{n}{(N^2-\sin^2\theta)^{1/2}} \right]^2 \tag{5}$$

where l is the film thickness, θ is the incident angle, $\bar{n} = (n_\omega + n_{2\omega})/2$ and $N^2 = \bar{n}^2 + [(n_\omega - n_{2\omega})/2]^2$. Since $(n_\omega - n_{2\omega}) \ll \bar{n}$, the second harmonic intensity is independent of the coherence length and depends quadratically on l. When the incident laser light and the second harmonic light are both p-polarized (p-p), dp can readily be shown to be given by eq.(6).

$$dp = (d_{33}\sin^2\theta_\omega + d_{31}\cos^2\theta_\omega)\sin\theta_{2\omega} + 2d_{31}\cos\theta_\omega\sin\theta_\omega\cos\theta_{2\omega} \tag{6}$$

With the incident laser light s-polarized and the second harmonic light p-polarized (s-p), the appropriate dp is then given in eq.(7)

$$dp = d_{31} \sin\theta_{2\omega} \tag{7}$$

Thus, p-p and s-p measurements together can determine both d_{33} and d_{31}.

Second harmonic coefficients of the polymer films synthesized in this study were measured at 1.064 μm using the instrumentation and calibration techniques described previously (11-14). Both p-p and s-p measurements were carried out on selected films. Data for a typical (PS)O-NPP film are shown in Fig. 1. For films poled in moderate dc fields (≤ 0.6 MV/cm), the results are consistent with $d_{33} = 3d_{31}$ as predicted by the statistical average calculated using a Boltzmann-like orientational function to describe alignment of chromophores in the poling field. For most films, only p-p measurements were carried out and the relation $d_{33} = 3d_{31}$ was assumed in calculating d_{33}. Representative d_{33} data are set out in Table 1. It can be seen that the (PS)O-NPP second harmonic coefficients are rather large, exceeding the corresponding d_{36} of KDP (1.1 x 10^{-9} esu) and equalling or exceeding d_{31} of LiNbO3 (14.2 x 10^{-9} esu). In a noninteracting oriented molecular gas model, the steady-state d_{33} that can be achieved is given by eqs.(8) and (9) (8,9), where it is generally assumed

$$d_{33} = Nf^{2\omega}f^\omega f^\omega \, \beta_{\zeta\zeta\zeta} L_3(p) \tag{8}$$

$$p = \frac{f^0 \mu E_p}{kT} \tag{9}$$

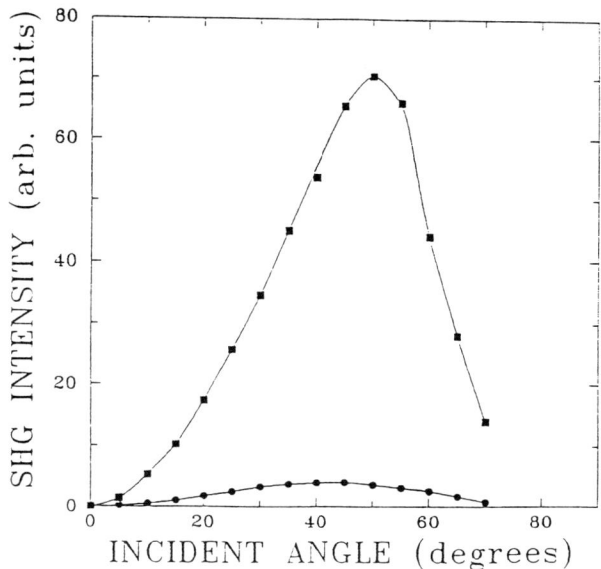

FIGURE 1. Second harmonic intensity at 0.532 μm for a (PS)O-NPP film (25% functionalization level) as a function of incident angle for p-p (■) and s-p (●) geometries.

TABLE 1. Second Harmonic Coefficients for Chromophore-Functionalized Polystyrenes[a]

Material	Functionalization Level (% Phenyl Rings)	Poling Field (MV/cm)	τ_2 (days)[b]	d_{33} $(10^{-9}$ esu)[c]
(PS)CH$_2$-DR	12.5	0.3		2.7[d]
(PS)CH$_2$-DASP	4.5	0.3		0.12[d]
(PS)CH$_2$-NPP	36.0	0.7		3.8(8.9)
(PS)O-NPP	15.0	0.7	313	5.1(3.7)
(PS)O-NPP	25.0	0.3	195	3.0(2.6)
(PS)O-NPP	48.0	0.6		11.6(10.1)
(PS)O-NPP	48.0	1.6	42	18.0

[a]Measured within 0.5 h of poling; λ = 1.064 μm.
[b]Long-term SHG decay lifetime from fitting to eq.(10).
[c]Experimental SHG coefficients. Theoretical Values calculated from eq.(8) are given in parentheses.
[d]From reference 11. DR = 4-(4-nitrophenylaza)(N-ethyl)(2-0-ethyl)aniline; DASP = 4-(4-N,N-dimethylaminostyryl)pyridinium iodide.

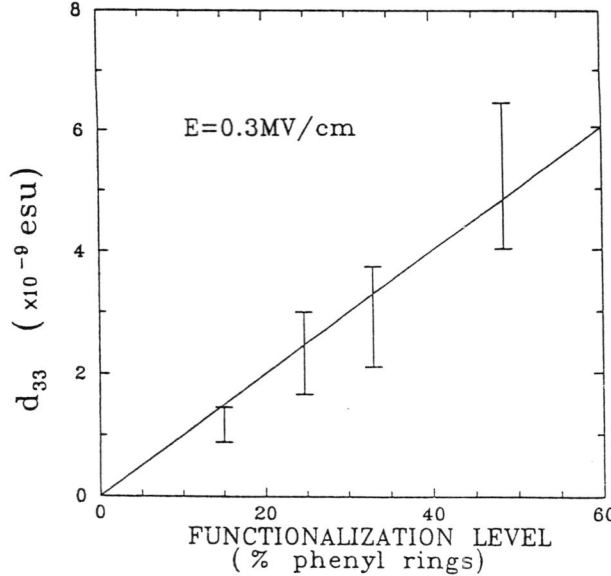

FIGURE 2.   Second harmonic coefficients of (PS)O-NPP films poled at 0.3 MV/cm as a function of chromophore functionalization level for 1.064 μm incident radiation.

that the chromophore is rod-like with the dipole moment along the molecular axis and $\beta_{\zeta\zeta\zeta}$ (ζ refers to the molecular axis) the only nonzero molecular hyperpolarizability tensor elements.   Here N is the chromophore number density, μ the chromophore molecular dipole moment, E_p the poling field. $L_3(p)$ the third order Langevin function, and the f's the corresponding local field factors at the frequencies indicated.

The present results indicate that the above Langevin model works quite well for functionalized (PS)OH.   Thus, it can be seen in Fig. 2 that there is a fairly linear dependence of d_{33} (measured immediately after switching off the poling field) on NPP functionalization level (which should approximately scale as the chromophore density) up to the highest chromophore densities, as predicted.   The dependence of d_{33} on dc poling field (measured immediately after the poling field was switched off) is shown in Fig. 3.   It can be seen that at low fields, d_{33} varies linearly with E_p, in agreement with the expected behavior of $L_3(p)$ for small p (eqs. (8), (9)).   Furthermore, saturation of $L_3(p)$ is observed at high fields (Fig. 3), as expected when $\mu E_p \geq kT$.   To our knowledge, this is one of the few instances in which such saturation effects have been directly observed for an NLO polymer.   This observation lends further support to the validity of the oriented gas, Langevin model.   Table 1 compares experimental, steady-state d_{33} values for (PS)O-NPP films with those calculated via eq. (8) taking $\mu\beta = 307 \pm 66 \times 10^{-30}$ cm^5 D/esu for NPP (19). It can be seen that there is acceptable agreement.   For the (PS)CH$_2$-NPP polymers, experimental d_{33} values are consistently much below those calculated via eq. (8), possibly because of the purity problems noted earlier.

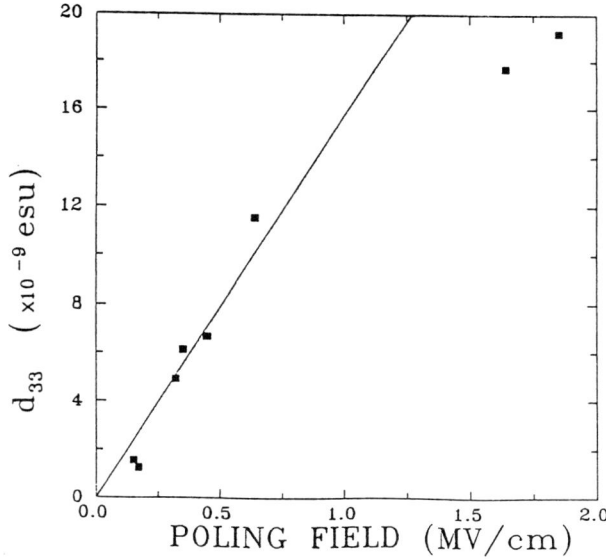

FIGURE 3. Second harmonic coefficient of (PS)O-NPP films (48% chromophore functionalization level) as a function of poling field for 1.064 μm incident radiation.

The temporal stability of poling-induced chromophore alignment is of great practical importance in all polymeric SHG materials. The decay of this preferential alignment and the concurrent loss of SHG efficiency is intimately connected with physical aging of the polymer (20), and reflects a variety of complex reorientational processes by which the system relaxes to thermodynamic equilibrium and minimum free volume. For polymeric SHG systems in which NLO chromophores are simply doped into a host polymer (8,9), these relaxation processes are generally rapid (10). It was anticipated that connecting the NLO chromophores to the backbone of a relatively high-T_g, high-T_β polymer would impede chromophore reorientation and thus help to preserve SHG efficiency. In addition, the presence of hydrogen bonding networks as in the (PS)OH/(PS)OR materials would be expected to effect further macromolecule immobilization. Gratifyingly, such a strategy is found to be effective.

Typical SHG decay curves for (PS)O-NPP films are shown in Figs. 4 and 5. Impressive SHG signal lifetimes are observed, which are further enhanced by annealing the films. FT-IR spectra show that annealing removes traces of THF (ν = 1050, 885 cm^{-1}) remaining from the film fabrication process. That THF acts as a plasticizer and promotes facile chromophore realignment is supported by the observation that the T_g of (PS)O-NPP samples increases by more than 50° on annealing. The FT-IR spectra also reveal concurrent changes in the ν_{OH} region ($\Delta\nu_{OH} \approx +25$ cm^{-1}) which are suggestive of changes in the hydrogen bonding network (21). While THF removal will likely affect some hydrogen bonding interactions within the polymer, changes in interactions involving the NPP chromophore are difficult to identify since nitro group vibrational frequencies are not particularly sensitive to hydrogen bonding (22).

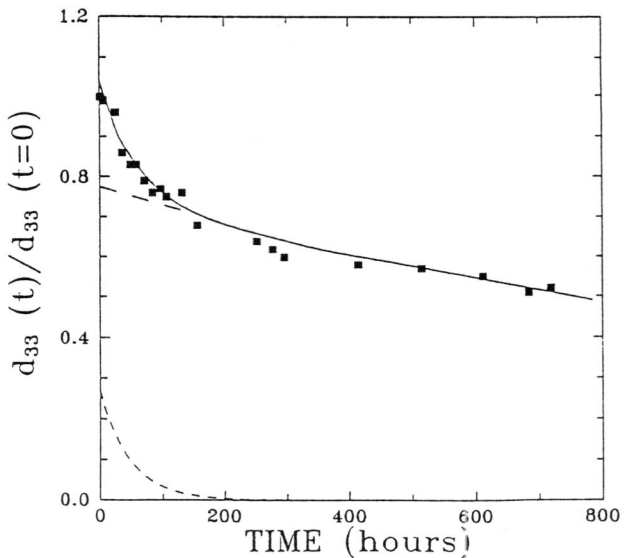

FIGURE 4. Temporal behavior of the second harmonic coefficient of an unannealed (PS)0-NPP film (25% functionalization level) at room temperature. The solid line shows a two exponential fit (eq.(10)) to the data points while the two dashed lines represent the component functions.

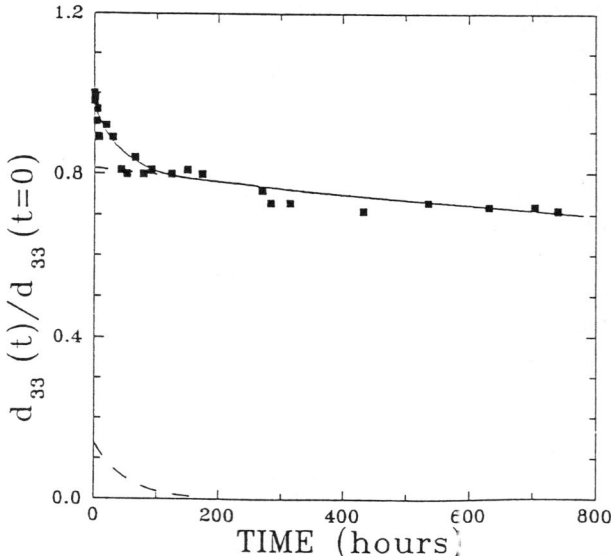

FIGURE 5. Temporal behavior of the second harmonic coefficient of an annealed (PS)0-NPP film (25% functionalization level) at room temperature. The solid line shows a two exponential fit (eq.(10)) to the data points while the two dashed lines represent the component functions.

The SHG decay functions such as those in Figs. 4 and 5 cannot be fit to a simple single exponential or Williams-Watt stretched exponential (23). Fits to a two-exponential model (eq.(10)) are more satisfactory (Figs. 4

$$d_{33} = Ae^{-t/\tau_1} + Be^{-t/\tau_2} \qquad (10)$$

and 5) and suggest several relaxation processes. Derived τ_2 values for the long-term decay process are given in Table 1, with the maximum τ_2 observed being a very large 313 days. We find that the amplitude of the short-term process is more sensitive to the presence of THF and can be greatly diminished by annealing. This argues that the short-term process may involve rapid chromophore reorientation in THF-rich (high local free volume) microenvironments. It can also be seen in Table 1 and in Fig. 6 that the long-term decay rate increases significantly upon increasing levels of NPP functionalization. Increasing chromophore functionalization (0%→48%) is also accompanied by a shift of ν_{ArOH} to higher frequency ($\Delta\nu_{OH} \approx +50$ cm^{-1}), suggesting changes (probably diminution) in ArOH-ArOH hydrogen bonding as the concentration of phenolic groups decreases. Fig. 6 also shows that the lifetime of the 60% functionalized (PS)O-DR film can be roughly extrapolated from the (PS)O-NPP data. It can also be seen that the lifetime of the (PS)CH$_2$-NPP films falls significantly below the (PS)O-NPP data. While the chromophore concentration of the doped NPP/PS film is very low, and the SHG lifetime is below that of the best (PS)O-NPP samples, it does appear that the chromophore orientation is more stable than that of many other doped glassy polymers (10). These results

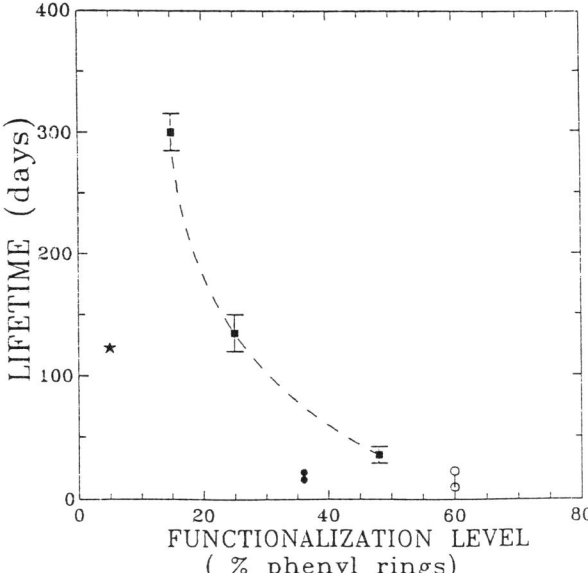

FIGURE 6. Dependence of the second harmonic coefficient exponential lifetime (τ_2 in eq.(10)) on functionalization level for ■ (PS)O-NPP; o (PS)O-DR; ● (PS)CH$_2$-NPP; * NPP doped into PS.

raise the interesting question of whether chromophore aggregation by, for example, hydrogen bonding might offer a mechanism to restrict chromophore mobility. This possibility clearly warrants additional investigation.

4. CONCLUSIONS

The information presented here considerably extends what is known about the NLO characteristics of chromophore-functionalized polystyrenes. In particular, it demonstrates that systems in which NLO chromophore molecules are covalently linked to the polymer matrix offer a number of advantages over simple doped systems. Notably, the chromophore-functionalized systems can support much higher chromophore densities without deleterious phase separation and opacity, appear to withstand much higher poling fields with far greater reliability, and generally exhibit longer-lived SHG performance. The net result is a family of processable, polymeric SHG materials which exhibit persistent second order nonlinearities equal to or higher than those of the common inorganic SHG materials.

Nevertheless, an important goal for future efforts must be to further improve the temporal stability of SHG characteristics. A major part of this effort must be to better understand the microdynamics of the chromophore reorientation processes and to synthesize materials which test our current understanding of such phenomena. Building in more extensive hydrogen bonding networks and other types of cross-linking mechanisms (24) are two of many interesting approaches.

ACKNOWLEDGMENTS

This research was supported by the NSF-MRL program through the Materials Research Center of Northwestern University (Grant DMR85-20280) and by the Air Force of Scientific Research (Contract AFOSR-86-0106).

REFERENCES

1. Prasad, P.N.; Ulrich, D.R. (eds): "Nonlinear Optical and Electroactive Polymers," Plenum Press: New York, 1988.
2. Heeger, A.J.; Orenstein, J.; Ulrich, D.R. (eds): "Nonlinear Optical Properties of Polymers," Mats. Res. Soc. Symp. Proc. 1988, 109.
3. Chemla, D.S.; Zyss, J. (eds): "Nonlinear Optical Properties of Organic Molecules and Crystals," Vols. 1,2; Academic Press: New York, 1987.
4. Khanarian, G. (ed): "Molecular and Polymeric Optoelectronic Materials: Fundamentals and Applications," SPIE 1986, 682.
5. Shen, Y.R.: "The Principles of Nonlinear Optics," Wiley: New York, 1984.
6. For an efficient theoretical approach, see: Li, D.; Ratner, M. A.; Marks, T. J.: J. Am. Chem. Soc. 1988, 110, 1707-1715, and references therein.
7. Meredith, G.R.; Van Dusen, J.G.; Williams, D.J.: Macromolecules 1987, 15, 1385-1389.
8. Singer, K.D.; Sohn, J.E.; Lalama, S.J.: Appl. Phys. Lett. 1986, 69, 248-250.
9. Singer, K.D.; Kuzyk, M.G.; Sohn, J.E.: J. Opt. Soc. Am. B. 1987, 4, 968-975.
10. Hampsch, H. L.; Yang, J.; Wong, G. K.; Torkelson; J. M.: Macromolecules 1988, 21, 526-528.

11. Ye, C.; Marks, T. J.; Yang,Y.; Wong, G. K.: <u>Macromoelcules</u> 1987, <u>20</u>, 2322-2324. Functionalization levels no higher than 12.5% were achieved in early studies with (PS)CH$_2$X materials.
12. Ye, C.; Minami, N.; Marks, T. J.; Yang, J.; Wong, G. K.: <u>Mats. Res. Sympos. Proc.</u> 1988, <u>109</u>, 263-269.
13. Li, D.; Yang, J.; Ye, C.; Ratner, M. A.; Wong, G. K.; Marks, T. J.: in reference 1a, pp. 217-228.
14. Ye, C.; Minami, N.; Marks, T. J.; Yang, J.; Wong, G. K.: submitted for publication.
15. Mathes, K. N.: in "Encyclopedia of Polymer Science and Engineering"; Wiley: New York, 1986; Vol. 5, pp. 512-593.
16. Nye, J. F.: "Physical Properties of Crystals, Clarenden: Oxford, 1967.
17. Kleinman, D. A.: <u>Phys. Rev.</u> 1962, <u>126</u>, 1977-1979.
18. Jerphagnon, J.; Kurtz, S. K.: <u>J. Appl. Phys.</u> 1970, <u>41</u>, 1667-1681.
19. Barzoukas, M.; Josse, D.; Fremeux, P.; Zyss, J.; Nicoud, J. F.; Morley, J. O.: <u>J. Opt. Soc. Am</u> 1987, <u>B1</u>, 977-986.
20. Struik, L. C. E.: "Physical Aging in Amorphous Polymers and Other Materials"; Elsevier: Amsterdam, 1978; Chapts. 2, 3, 9.
21. Moskala, E. J.; Varnell, D. F.; Coleman, M. M.: <u>Polymer</u> 1985, <u>26</u>, 228-234, and references therein.
22. Bellamy, L. J.: "Infrared Spectra of Complex Molecules"; Second Ed.; Chapman and Hall: London, 1980; pp. 264-265, and references therein.
23. Hodge, I. M.: <u>Macromolecules</u> 1983, <u>16</u>, 898-902.
24. Hubbard, M. A.; Marks, T. J.; Yang, J.; Wong, G. K.: manuscript in preparation.

LANGMUIR-BLODGETT FILMS OF RIGID ROD POLYMERS WITH CONTROLLED LATERAL ORIENTATION

C. Bubeck, D. Neher, A. Kaltbeitzel, G. Duda
T. Arndt, T. Sauer, G. Wegner

Max-Planck-Institut für Polymerforschung,
Postfach 3148, D-6500 Mainz, West Germany

ABSTRACT

Some new concepts for the preparation of ultrathin polymer films with controlled lateral orientation are briefly reviewed. Rigid rod polymers such as phthalocyaninato polysiloxanes or polyglutamates that are substituted with alkyl chains can be used to built up multilayers with good transfer properties from the air-water interface to solid substrates. During the film transfer the long axes of the rigid rods are oriented parallel to the dipping direction. Investigations of nonlinear optical properties of substituted phthalocyaninato-polysiloxane films are presented.

1. INTRODUCTION

Langmuir-Blodgett (LB) films of materials with nonlinear optical properties have attained considerable interest in recent years because they might be used to construct integrated optical devices (1-3). There is strong need for thin films with well defined thickness, refractive index and orientation of the molecules. The control of the molecular orientation is crucial with respect to two aspects: the orientation perpendicular to the layer plane is important to obtain noncentrosymmetric structures whereas a control of the lateral orientation might be of considerable value to obtain an anisotropic refractive index for phase matching of guided waves.

However, several inherent problems of LB films are well recognized but nevertheless hard to overcome. The most important ones are the poor mechanical and thermal stability of films made from low molecular weight compounds and their tendency to form microcrystalline domains that can scatter light strongly (4). Polymeric LB films with improved mechanical stability can be prepared using two possible routes: either the LB film is made from amphiphilic, reactive monomers and polymerized by UV-light or other high energy radiation (5,6), or the film is prepared from preformed polymers (7-9). Fig. 1 shows a comparison of the structure of the different classes of LB-films. The problems of the polymerization in the LB multilayer are possible

185

J. Messier et al. (eds.), Nonlinear Optical Effects in Organic Polymers, 185–193.
© *1989 by Kluwer Academic Publishers.*

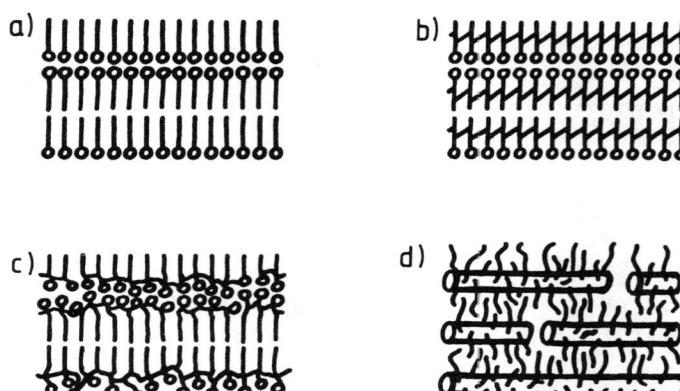

FIGURE 1. Comparison of the structure of Langmuir-Blodgett layers of different compounds (roughly simplified; view parallel to the layer plane):
a) classical amphiphiles
b) reactive amphiphile after solid-state polymerization
c) prepolymerized amphiphilic polymer
d) rigid rod polymer substituted with flexible side chains.

photodegradation and restrictions due to the packing requirements for the solid state reactivity reviewed recently (10). The problems with the preparation of LB films of preformed polymers lie in the film transfer from the air water interface to the substrate which is not always complete. Quite frequently a change of the transfer ratio at higher numbers of dipping cycles is observed, allowing the formation of thicker films with a few systems only.

The emphasis of this contribution is twofold. First we summarize some recent progress in the formation of films of unconventional polymers for the LB technique. Their molecular structure differs significantly from the classical principles of the molecular design of molecules applicable for the LB technique. In the second part we report on nonlinear optical properties of phthalocyaninato-polysiloxane LB films for the first time.

2. PREPARATION CHARACTERISTICS AND LINEAR OPTICAL PROPERTIES OF PHTHALOCYANINATO POLYSILOXANE LB FILMS

The synthesis of phthalocyaninato polysiloxane was described recently (11,12). The structure is shown in scheme 1. Since the Si-O-Si bond (3.33 Å) is smaller than the intermolecular Van-der-Waals distance of two adjacent carbon atoms ($\approx$ 3.6 Å), this polymer has an extremely stiff chain. The polymer is

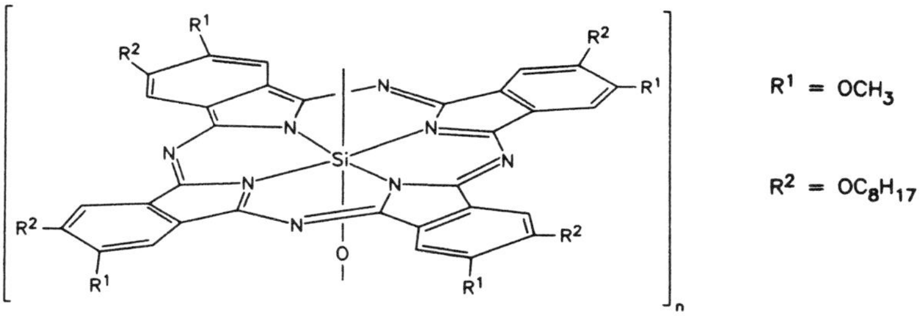

Scheme 1: Chemical structure of Phthalocyaninato-polysiloxane

$R^1 = OCH_3$

$R^2 = OC_8H_{17}$

unsymmetrically substituted with R1 and R2 to make it soluble in common organic solvents. For spreading at the air water interface chloroform can be used. At constant pressure (25 mN/m) multilayers can be transferred to hydrophobic substrates with y-type deposition (13). The transfer ratio is 100% and stays constantly at this value up to high numbers of transfer cycles.

The absorption spectra of a LB film of the polymer are shown in Fig. 2. The absorption maximum of the polymer is observed at 560 nm. As compared to the monomer absorption at 680 nm (12) it is considerably broadened and blue shifted due to coupling of the electronic transition dipole moments of neighbouring ring systems. Strong dichroism of the polymer absorption can be seen in Fig. 2. The dichroitic ratio is approximately 2.5 as observed earlier (13). This ratio can be improved by a factor of two by anealing of the film at elevated temperatures. To investigate this effect, the film was stored subsequently at higher temperatures for 30 min. Thermal treatment at $50^{\circ}C$ does not change the spectrum. At temperatures between $50^{\circ}C$ and $75^{\circ}C$ the dichroitic ratio begins to increase without a change of the shape of the spectrum. At anealing temperatures above $150^{\circ}C$ and longer treatment times the absorption spectrum broadens further and decreases in intensity. The relative influence of anealing time and temperature needs further examination.

During laser irradiation the films show considerable light scattering with a speckle type spatial distribution. This light scattering can not be reduced by the thermal treatment at various temperatures described above.

The following conclusions can be drawn from these observations:
Since the transition dipole moment is located preferentially parallel to the plane of the phthalocyanine ring, the strong dichroism indicates an orientation of the long axis of the rigid rod polymer parallel to the transfer direction of the substrate. The reason for this orientation phenomenon is thought to originate from shear forces during the transfer process. The flexibility of the substituents obviously enables this orientation process.

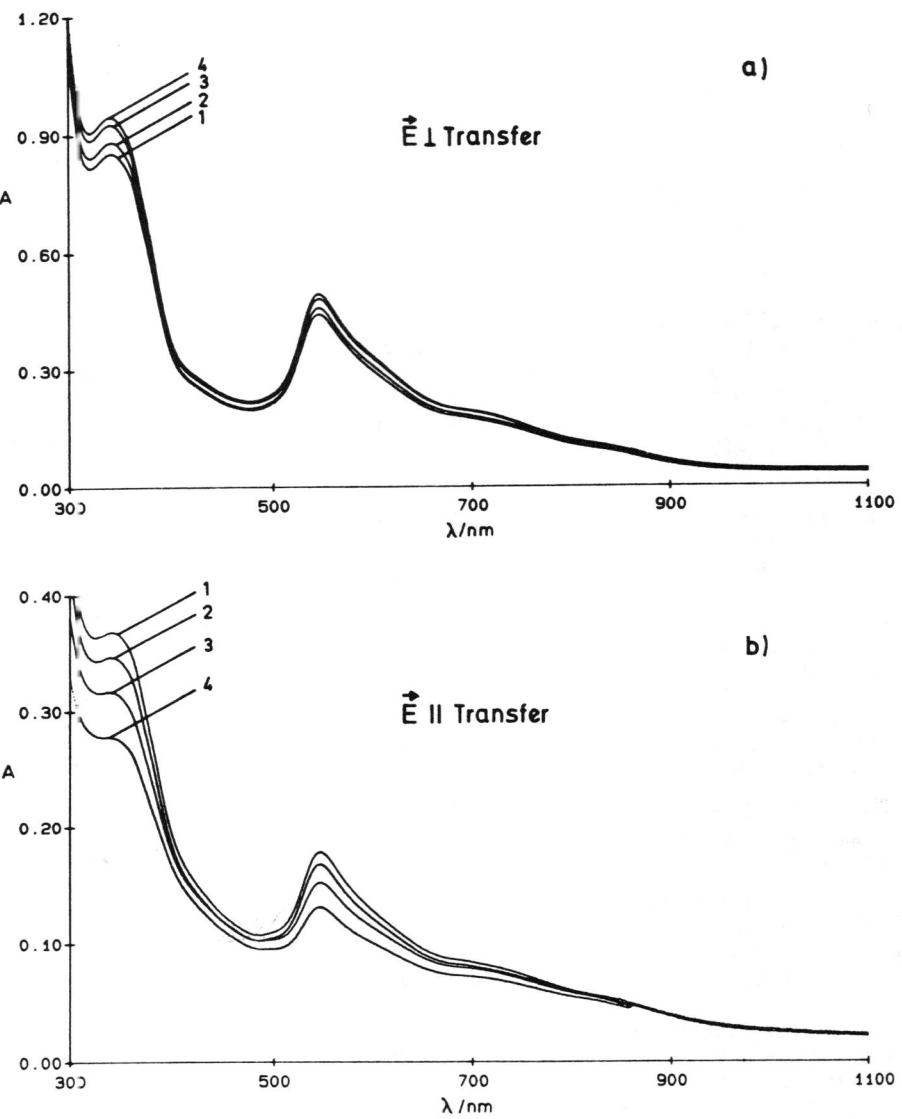

FIGURE 2. Polarized optical absorption spectra of phthalo-cyaninato-polysiloxane LB film (100 layers) with polarizer orientations vertical (a) and parallel (b) to the transfer direction of the substrate. The spectra are measured at room temperature after subsequent anealing of the sample for 30 min at elevated temperatures:

1. untreated, 2. 75°C, 3. 100°C, 4. 125°C.

The polymer is presumably in a liquid crystalline state (13). This might explain the possibility to improve the molecular orientation by thermal treatment and is responsible for the considerable light scattering of these films. Ways to circumvent this problem are under active investigation.

3. LANGMUIR-BLODGETT FILMS OF POLYGLUTAMATES

Substituted polyglutamates are a second example of a rigid rod polymer that can be used to form oriented LB films (14). A series of copolymers of poly (γ-methyl-L-glutamate-co-n-alkyl-L-glutamate) was synthesized following the method of Watanabe (15). These copolyglutamates could be transferred onto hydrophobic substrates at 20 - 25 mN/m as y-type multilayers. The transfer ratio was 100% and constant up to 200 layers.

The orientation of the rigid polymer chain was investigated with polarized infrared transmission spectroscopy. The spectra show the typical amide I and amide II bands. This indicates that the α-helical conformation of the polymer backbone is preserved. The α-helices are oriented with the main axes preferentially parallel to the transfer direction.

The molecular order and orientation of the alkyl chains can be investigated with grazing incidence IR-spectroscopy (16-18). Inspection of the CH-stretching bands clearly indicates that the alkyl substituents are oriented randomly.

Recently it was demonstrated that oleophilic dyes can be cospread with these alkyl substituted polyglutamates onto the air-water interface (19). LB-assemblies can be built up in which the dyes such as β-carotene are dissolved in the liquid like phase provided by the mixture of the long and short n-alkyl chains attached to the helical polypeptid backbone. The dichroic absorption of the dye-containing LB assemblies indicates that the dyes are not isotropically distributed in the layers but are located with a preferential orientation of the molecular axis with regard to the director axis of the polypeptide.

4. COMPARISON OF THE FILMS PREPARED FROM ALKYL-SUBSTITUTED RIGID ROD POLYMERS WITH CLASSICAL LB FILMS

The structural model of the LB films of the rigid rod polymers is presented in Fig 1d. First of all it is very surprising that these polymers form LB layers at all because their molecular structure differs considerably from the classical amphiphiles with a hydrophilic head group and a hydrophobic tail. The molecular order in the range of the alkyl chains is relatively high in the classical LB films, however, it is nearly absent in the case of rigid rod polymer films. This flexibility of the side groups originates from their substitution pattern. The liquid crystalline nature of these rigid rod polymers enables the shear forces that are present during the layer transfer to solid substrates to orient the rigid rods with their long axis parallel to the transfer direction.

The layer stabilization is achieved in the classical LB films by Van-der-Waals and ionic interaction energies of the solid analog phases with crystalline packing of the alkyl chains. In the case of the rigid rod polymers their liquid crystalline nature already tends to form ordered structures. The layer forming principle is based on the stiffnes of the macromocules.

5. THIRD HARMONIC GENERATION IN PHTHALOCYANONATOPOLYSILOXANE LB FILMS

Third harmonic generation (THG) measurements have been performed with an active/passive mode-locked Nd:YAG laser working at a repetition rate of 10 Hz. A single pulse was selected from the pulse train giving an energy of 0.4 mJ per pulse with a length of 20ps at 1064nm. The IR-beam was focused on the sample by a f = 350 mm lens. All measurements were performed with the focal region in air. The THG-signal was selected by a monochromator and detected with a photomultiplier. The IR-intensity on the sample was adjusted with calibrated attenuation-filters giving a THG-signal of approximately 100 photons/pulse. The polymer films were prepared by the LB technique as described recently (13). The sample was mounted on a rotation stage with the rotation axis parallel to the incident beam polarisation. The THG-signal was measured as function of the angle α betwen the beam direction and the surface normal of the sample.

THG signals as a function of the angle α are shown in Fig. 3. The intensities are corrected with respect to a fused silica reference sample. The anisotropy of the nonlinear properties of the LB film can be clearly seen by a comparison of Fig. 3b and 3c. The measurements are fitted using formulas derived by Messier et al. (20, 21) including air contribution and the absorption of the fundamental and harmonic light. For an exact determination of $\chi^{(3)}$ the knowledge of the refractive indices $n(\omega)$ and $n(3\omega)$ is necessary. Since we do not have these values yet we present the ratio $\chi^{(3)}/n^{3/2}(\omega) \cdot n^{1/2}(3\omega)$ in Tab. 1 together with the measured absorption coefficients (see Fig.2).

	$\alpha(3\omega)$ [cm^{-1}]	$\alpha(\omega)$ [cm^{-1}]	$\chi^{(3)}/n^{3/2}(\omega) \cdot n^{1/2}(3\omega)$ [esu]
E ∥ T	$3,8 \cdot 10^4$	$1,5 \cdot 10^3$	$0,5 \cdot 10^{-12}$
E ⊥ T	$8,5 \cdot 10^4$	$4,0 \cdot 10^3$	$1,0 \cdot 10^{-12}$

TABLE 1. Absorption coefficients α and ratio of $\chi^{(3)}$ with respect to the refractive indices of a phthalocyaninato-polysiloxane LB film for the polarizations of the incident beam parallel and perpendicular to the transfer direction T.

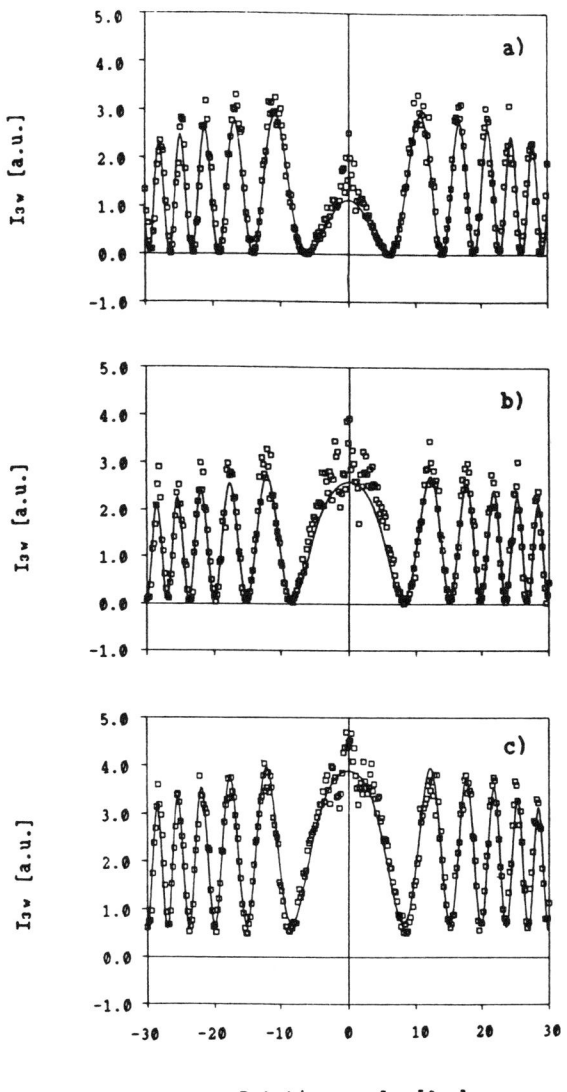

FIGURE 3. Third harmonic intensity as a function of the angle
of incidence for:
a) glass substrate without film
b) phthalocyaninato-polysiloxane LB films on the back side of
   a glass substrate. The polarization of the incident beam is
   parallel to the dipping direction (E || T)
c) same as b), but E ⊥ T
The squares are measured and the theoretical curve is obtained
by a least square fit.

If we assume the values $n(\omega) = 1,7 \pm 0,3$ and $n(3\omega) = 2,3 \pm 0,7$ typical for organic materials near resonance, we can estimate $\chi^{(3)}$ to be of the order of $(1-3)$ 10^{-12} esu. By inspection of the absorption spectrum shown in Fig. 2, we see that this estimated value of $\chi^{(3)}$ may contain resonant contributions. Experiments to determine the wavelength dependence of $\chi^{(3)}(-3\omega, \omega, \omega, \omega)$ are in preparation.

Recently the third harmonic generation in monomeric phthalocynanine polycrystalline films was reported by Ho et al. (22). Our estimated value of $\chi^{(3)}$ is about one order of magniture lower than that of the monomeric phthalocyanine. We assure that the shift of the electronic transitions of the polymer to shorter wavelengths accounts for the reduced hyperpolarizability. However, the well defined structure of the LB films and the possibility to vary the chemical structure of polymeric phthalocyanines still make them interesting candidates for further nonlinear optical investigations.

ACKNOWLEDGEMENT

This work is supported by the Bundesministerium für Forschung und Technologie under the project "Ultrathin Layers of Polymers" 03M4008E9. The authors thank H.J. Menges and W. Scholdei for considerable technical support.

REFERENCES
1. D. J. Williams, Angew. Chemie 96, 637 (1984)
2. J. Zyss, J.Mol. Electronics 1, 25 (1985)
3. G.I. Stegeman and C. T. Seaton, J.Appl. Phys. 58, R 57 (1985)
4. G. Lieser, B. Tieke and G. Wegner, Thin Solid Films 68, 77 (1980)
5. B. Tieke, Adv. Polym. Sci. 71, 79 (1985)
6. A. Laschewsky, H. Ringsdorf and G. Schmidt, Thin Solid Films, 134, 153 (1985)
7. R. H. Tredgold and C. S. Winter, J. Phys. D. 15, L55 (1982)
8. R. H. Tredgold, Thin Solid Films 152, 223 (1987)
9. S. J. Mumby, J.D. Swalen and J. F. Rabolt, Macromol. 19, 1054 (1986)
10. C. Bubeck, Thin Solid Films 159, (1988) in press
11. E. Orthmann and G. Wegner, Makromol. Chem. Rapid Commun. 7, 243 (1986)
12. C. Sauer and G. Wegner, Makromol.Chem.Symp., (1988) in press
13. E. Orthmann and G. Wegner, Angew. Chem. Int. Ed. Engl. 25, 1105 (1986)
14. G. Duda, A. J. Schouten, T. Arndt, G. Lieser, G. F. Schmidt, C. Bubeck and G. Wegner, Thin Solid Films (1988), in press
15. J. Watanabe, H. Ono, I. Uematsu and A. Abe, Macromolecules, 18, 2141 (1985)

16. J. D. Swalen and J.F. Rabolt in "Fourier Transform Infrared Spectroscopy", Vol 4, J. R. Ferraro, L. J. Basile (eds) Acad. Press, New York, Chap.7 (1985)
17. T. Arndt and C. Bubeck, Thin Solid Films (1988), in press
18. T. Arndt, C. Bubeck and G. Wegner, Langmuir, submitted
19. G. Duda and G. Wegner, Makromol. Chem. Rapid Commun., submitted
20. F.Kajzar and J. Messier, Phys. Rev. A 32, 2352 (1985)
21. P.A. Chollet, F. Kajzar and J. Messier, Thin Solid Films 132, 1 (1985)
22. Z.Z. Ho, C.Y. Ju and W.U. Hetherington, J. Appl. Phys. 62, 716 (1987)

POLYMERS AND MOLECULAR ASSEMBLIES FOR SECOND-ORDER NONLINEAR OPTICS

D. J. Williams,* T. L. Penner, J. J. Schildkraut, N. Tillman, A. Ulman, and C. S. Willand

Corporate Research Laboratories, Eastman Kodak Company, Rochester, New York 14650

1. INTRODUCTION

In the past decade considerable effort has gone into the design, synthesis, and characterization of new materials for second-order nonlinear optics [1,2]. The driving force for much of this effort has been the recognition that extremely large hyperpolarizabilities were associated with certain aromatic π electronic systems [3-5] which could lead to significant applications for new materials. The origin of the hyperpolarizability, β, has been attributed to charge transfer resonances involving appropriately substituted π-electronic systems. Early theoretical efforts to account for these properties assumed that the molecule was a two-level system with a nonpolar ground state and highly polarized first excited state [4]. The action of the optical field on the molecule was described by time-dependent perturbation theory [6]. This approach requires calculating matrix elements of the dipole moment operator and leads to a straightforward prediction that β is proportional to the oscillator strength of the optical transition between the states, the difference between ground and excited state dipole moment. Electronic dispersion was also accounted for. While this relatively simple approach provided qualitative agreement with many experimental observations of β [4,7], it was apparent that refinements were needed to account more accurately for these properties. Thus the use of SCF approaches to calculate more accurate ground state wavefunctions, the generation of more representative excited states by extensive configuration interaction, and the inclusion of elements of the dipole moment operator between ground and excited states and between pairs of excited states leads to greater understanding of observations and to a new level of confidence in predictive capabilities [8-10].

In addition to the molecular requirements for large second-order nonlinear responses, the orientation of the molecular units in the condensed state determines the efficiency of nonlinear processes in the medium. In the dipolar approximation, which is valid when the size of the polarizable unit is small compared to the wavelength of light, the minimal requirement for bulk nonzero second-order nonlinearity is that the medium be noncentrosymmetric. The ability to alter substituent patterns and tinker with attractive and repulsive interactions including chirality [11], dipole-dipole interactions [12], and hydrogen bonding [13] led to a variety of efforts to prepare crystals in noncentrosymmetric space groups

J. Messier et al. (eds.), Nonlinear Optical Effects in Organic Polymers, 195–218.
© 1989 by Kluwer Academic Publishers.

for the study of their second-order nonlinear optical properties. Noncentrosymmetry is, however, only a minimal requirement for a second-order response. The merocyanine chromophore
which exhibits an extremely large β [14,15], when incorporated
into a crystal as the salt of d-camphor sulfonic acid, showed
an extremely small powder harmonic generation signal [16] due
to pseudo symmetry in the unit cell. Considerable progress in
the understanding of optimal molecular packing in noncentrosymmetric structures was made by Ouder and Zyss [17,18] who
derived general expressions for macroscopic nonlinear tensor
components from the molecular tensor components, using appropriate transformations to account for molecular orientation
and symmetry. Significant progress has since been made in the
tailoring of new molecular structures for optimized crystal
growth [19] and growing crystals in formats of interest for
waveguide nonlinear optics [20,21].

It was recognized in the early 1980's [22] that a polymeric
thin-film-based approach to second-order nonlinear materials
might offer significant fabrication advantages relative to
single crystals if the orientational averaging of nonlinear
chromophores incorporated into polymeric film could be
removed. The initial demonstration of electric field poling
of a thermoplastic doped with a nonlinear chromophore, as well
as the statistical thermodynamic analysis of the relationships
between electric field-induced nonlinear tensor elements, local
potentials such as those associated with liquid crystallinity,
and the poling field, was done by Meredith et al. [22,23] and
Williams [24]. This approach has the potential advantage of
being compatible with planar fabrication techniques and integration of active films onto Si or GaAs substrates as well as
compatibility with fabrication of active fiber optic based
devices. It has received considerable research attention in
recent years [25] and may have significant commercial potential
[26]. Efforts to enhance poling responses in polymers through
cooperative effects in polar polymer chains have been undertaken in an effort to achieve higher nonlinearities at lower
poling fields [27,28]. Preliminary efforts to understand the
potential as well as problems associated with this approach
are described in section 2.

Given the considerable effort and progress in crystal design
and growth of poled polymers, molecular assembly techniques
might be viewed as a third generation approach to materials
for nonlinear optics. They offer, at least conceptually, the
ultimate in design flexibility for linear and nonlinear optical
properties for thin film and waveguided devices.

The first molecular assembly approach to be investigated in
this regard is the Langmuir-Blodgett (L-B) technique. Here,
chromophores substituted with fatty acid tail groups and polar
head groups are deposited from a compressed film on a water
surface onto the surface of a substrate. The balance of a
variety of forces determines the orientation and packing of
the film, its thermal stability, and the orientation and
packing of subsequently deposited layers [29].

A variety of recent studies of second-order nonlinear
optical properties of L-B films as monolayers [30-34],
multilayers [35-38], and films made from preformed polymers

[39,40] have been reported recently. Studies of films based on the merocyanine chromophore show good SHG efficiency but a tendency to undergo protonation to a less active form [30]. Multilayers of the merocyanine and hemicyanine chromophores exhibit y-type deposition with themselves to produce centro-symmetric nonactive or weakly active films. However, when alternately deposited with inert layers such as ω-tricosenoic acid, multilayers with additive nonlinearities can be obtained [35,36]. The predicted intensity dependence on the square of the number of layers is not observed in these systems, presumably indicating disorder and instability. An exception to this general subquadratic dependence on layer number is the study by Neal et al. [37]. Here, an apparent enhancement in bilayer structures formed from hemicyanine and nitrostilbene-based amphiphiles was observed. In these bilayers the polar and nonpolar regions of the molecules interact in the conventional Y sense, but chromophore directions are reversed in the molecule so that nonlinear responses are additive.

A third approach, which holds the promise of increased stability in deposited layers, is the preformed polymer approach for which two preliminary studies have appeared [39,40].

In section 3 we describe recent efforts in our laboratory to explore the use of photopolymerization in L-B films to increase the stability of L-B monolayer structures. The nature of the photopolarizable chromophore, its position in the chain, and the number per chain have been systematically varied in order to establish a chemical approach compatible with the incorporation of nonlinear chromophores. We also describe recent efforts in our laboratory to understand the role of inter-chromophore interactions on nonlinear responses within a monolayer.

A second molecular assembly approach, which has not been discussed in the literature in the context of second-order nonlinear optics, is the self-assembled monolayer approach (SAM) [41]. This approach bears some relationship to the L-B approach but has many distinctive differences and potential advantages. SAM's generally consist of a head group that is reactive towards a specially prepared substrate (a long alkyl chain possibility containing a chromophore) and a reactive tail group that can be chemically converted and activated so as to combine the deposition process of subsequent layers. Typically, the head group is the trichlorosilyl group which reacts with hydroxyl group-containing surfaces such as Si to form a covalently bonded polymerized monolayer. Reduction of the surface methyl ester with $LiAlH_4$ creates a new hydroxyl-ated surface which can be subsequently deposited with a new layer.

We have performed extensive studies of orientation, order, and packing in SAM's containing aromatic chromophores as well as the dependence of these factors on the placement of the chromophore in chains of various lengths. We have also investigated the conditions for multilayer formation [42] and have shown that highly ordered multilayers can be obtained [43]. Extensive physical characterization of this process was conducted and is reported in section 4.

2. POLAR POLYMERS

In the poling process, an electric field is applied to a polymer film in a softened state, and molecular dipoles are subjected to a force tending to align them in the direction of the field. This orienting force is opposed by forces associated with thermal fluctuations in the system tending to randomize the alignment. A certain amount of alignment will be retained when the system is cooled and hardened, leading to partial removal of the orientational averaging and introduction of uniaxial ($C_{\infty v}$) symmetry. Assuming Kleinman symmetry, only two independent $\chi^{(2)}$ tensor components exist [22-24]. The expressions for second-harmonic generation are

$$\chi^{(2)}_{zzz} = N\beta F(\omega)^2 F(2\omega) <\cos^3\theta> \tag{1}$$

and

$$\chi^{(2)}_{zxx} = N\beta F(\omega)^2 F(2\omega)(<\cos\theta> - <\cos^2\theta>) \tag{2}$$

where N is the concentration of polarizable molecules; β is the vector part of the molecular hypolarizability tensor, parallel to the molecular dipole; $F(\omega_i)$ is the local field factor at frequency i; and θ is the angle between the poling field and the molecular dipole. The poling field dependence of the nonlinear coefficients can be readily accounted for if the orientational averages can be determined. This is readily done from the Boltzman distribution law:

$$<\cos^3\theta> = \frac{\int_0^\pi F(\theta)\cos^3\theta\sin\theta d\theta}{\int_0^\pi F(\theta)\sin\theta d\theta} \tag{3}$$

where $F(\theta)$ is the orientational distribution function defined as

$$F(\theta) = e^{U(\theta)/kT} \tag{4}$$

$U(\theta)$ is the potential energy of the molecular dipole and is composed of a contribution from the local environment $U(\theta)_t$ and the contribution from the poling field $\vec{E}$:

$$U(\theta) = U(\theta)_t - F(0)\vec{E}\cdot\vec{\mu} \tag{5}$$

where $\vec{\mu}$ is the dipole moment and $F(0)$ the local field factor at $\omega = 0$. In isotropic media such as a liquid or glassy polymer, $U(\theta)_t$ should be approximately constant, independent of (θ), and related to the local viscosity of the medium. In anisotropic media such as liquid crystalline systems, $U(\theta)_t$ can make a significant contribution to $U(\theta)$ and influence the poling process. A variety of polymeric systems exhibiting behavior governed by these expressions has been observed.

The effects of $U(\theta)_t$ were investigated by Meredith et al. [22,23] in molecularly doped liquid crystalline thermoplastics. Molecular doped amorphous thermopolastics were investigated by Lalama et al. [44] and Singer et al. [25]. The magnitude of the electric field-induced orientation that can be achieved in these systems is limited by two factors. The first is the size of the dipole moment of the chromophore, and the second is dielectric breakdown which limits the value of the poling field to ~2 MV/cm.

A closer look at Eq. (1) leads to additional insights into this problem. Substituting the solution to Eq. (3) into Eq. (1) leads to the following expressions for $\chi_{zzz}^{(2)}$ and $\chi_{zxx}^{(2)}$ [22]:

$$\chi_{zzz}^{(2)} = N\beta F(\omega)^2 F(2\omega) L_3(F(0)\mu E/kT) \qquad (6)$$

$$\chi_{zxx}^{(2)} = \frac{1}{2} \left[\frac{L_1(F(0)\mu E/kT}{L_3(F(0)\mu E/kT)} - 1 \right] \chi_{zzz}^{(2)} \qquad (7)$$

where $L_n(X)$ (the nth order Langevin function) gives the orientational average of $\cos^n\theta$ as a function of X and $f(\omega_i)$ is the local field factor at ω_i. These quantities are defined as

$$L_1(X) = \coth X - 1/X$$
$$= X/3 - X^3/45 + 2X^5/945 + \ldots \qquad (8)$$

and

$$L_3(X) = L_1(X)(1 - 6X^2) - 2/X$$
$$= X/5 - X^3/105 + \ldots \qquad (9)$$

where $X = F(0)\mu E/kT$. A plot of $L_3(X)$ versus electric field is shown in Fig. 1. A glance at Fig. 1 shows that only about 20% of the nonlinearity can be achieved for a chromophore of 5D for a poling field of 2 MV/cm. Local field effects, space charge effects, depolarization effects, etc., could reduce this further. On the other hand, poling a system with a 100D dipole moment at a field of 0.5 MV/cm would give 65% of saturation alignment. For a moment, 1000D over 90% alignment would be obtained under these conditions.

We have recently begun to explore an approach for achieving high degrees of dipolar alignment at lower poling fields based on cooperative effects in polar polymer chains. Levine and Bethea [45] demonstrated that poly-γ-benzyl-L-glutamate in ethylene chloride exhibited a substantial enhancement in susceptibility due to incorporation into the rigid α-helical polymer chain. We have pursued an approach based on polar condensation polymers to achieve enhancements in more useful polymer systems [28].

In order to understand the proposed enhancement mechanism more fully, consider the model for a polar polymer chain shown in Figs. 2a and 2b. The polymer chain consists of n repeat units each having an associated vector r in the polymer chain

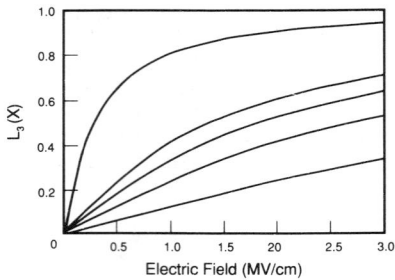

FIGURE 1. Plot of the third order Langevin function $L_3(X)$ versus electric field.

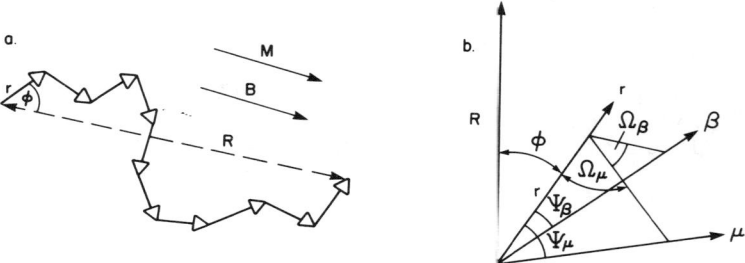

FIGURE 2. Schematic representation of (a) a polymer chain consisting of vectors r parallel to the chain with an end-to-end distance R and an angle ϕ between r and R, and (b) vectors β and μ making polar and azimuthal angles ψ_β, ψ_μ and Ω_β, Ω_μ with respect to r. M and B are defined in the text.

direction oriented at some angle ϕ with respect to the end-to-end vector R. Associated with each repeat unit is a dipole moment vector μ and a hyperpolarizability vector β. In general, neither μ nor β need to be parallel to r or to each other. The relationships between μ, β and r are shown in Fig. 2. The azimuthal angle between the planes defined by (r,β) and (r,μ) as well as the angles ψ_μ and ψ_β are determined by the nature of the chromophores and bonding in the repeat unit. For the purpose of simplifying and discussing the model we will assume a random distribution of angles between (r,R) and (μ,R). In Fig. 2a two additional vectors, M and B, are shown. These are the dipole moment and hyperpolarizability of the polymer chain, respectively, and are defined as

$$M = \sum_{i=1}^{n} \mu = n\mu <\cos\phi> \cos\psi_n \qquad (10)$$

and

$$B = \sum_{i=1}^{n} \beta = n\beta <\cos\phi>\cos\psi_{\beta} \qquad (11)$$

where < > indicates the average over the polymer chain. Since
the enhancement will be manifested as increased polar alignment
at low fields, we consider the low field limit to be the solutions of Eqs. (6) and (7).

$$\chi_{zzz}^{(2)} = NF(\omega)^2 F(2\omega) \frac{F(0)\mu\beta E_0}{5kT} \qquad (12)$$

and

$$\chi_{zxx}^{(2)} = 1/3 \; \chi_{zzz}^{(2)} \qquad (13)$$

For a chromophore in the polar polymer chain of the model we
can define the second order susceptibility as

$$\chi_{pzzz}^{(2)} = \frac{N}{n} F(\omega)^2 F(2\omega) \frac{F(0)MBE}{5kT} \qquad (14)$$

If we define the "monomer susceptibility" as $\mu\beta$, then the
monomer susceptibility as a result of being in the polymer
chain is MB/n. Taking the ratio of $\chi_{pzzz}^{(2)}$ to $\chi_{zzz}^{(2)}$ defines an
enhancement factor G:

$$G = \frac{M_z B_z}{\dfrac{n}{\mu\beta}} \qquad (15)$$

From the definitions of M and B, Eqs. (10) and (11), it is
clear that extended polymer chains will have a large enhancement relative to an equivalent concentration of monomer. On
the other hand, a tightly coiled polymer chain could have a
substantially lower nonlinearity than an equivalent concentration.

In order to examine these predictions we have synthesized a
chromophore based on p-oxy-α-cyanocinnamate (I) and incorporated it into a polymer chain (II) [27].

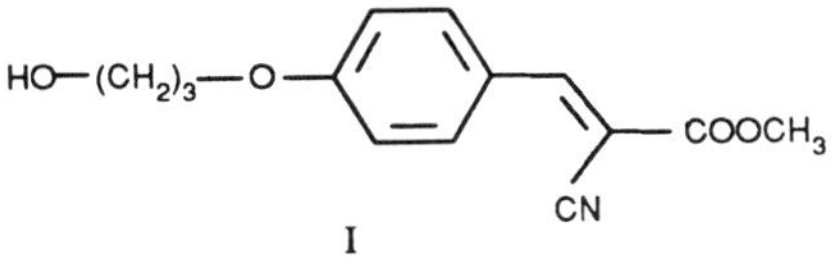

I

II

Values of the susceptibilities were determined in solution by
the EFISH technique [18] using the 1.06 μ output of a 10 pps,
Q-switched Nd:YAG laser and approximately 6 kV/cm and 100 μs
duration pulses from a pulsed high voltage power supply on the
EFISH cell [49]. The values obtained for the susceptibilities
and enhancement factors are shown in Table 1.

TABLE 1. Average monomer susceptibility and enhancement
factors for monomer and polymers.

Compound	Molecular weight, $\overline{M}_n$	n	MB/n (10^{-48} esu)	G
I		1	57	1
IIa	17,000	37	850	15
IIb	70,000	152	1140	20

It is clear that the polymers exhibit a substantial enhancement
relative to the monomer and that the enhancement effect appears
to saturate in the 17,000 M_n to 70,000 M_n region. It is not
clear at this point why polymers with large flexible chains
should have such extended rather than random coil conformations. One possibility is that the polymers are not random
but have significant blocks of the comonomers. Further studies
are planned to investigate this point.
   In order to determine the extent to which the advantage
realized in the polymer chain relative to monomers might be
obtained in films, a series of experiments on poled films was
conducted. Films of IIb were cast from chloroform solution
onto a chromium-plated optical flat. A 150 μm spacing between
poling electrodes was provided to give a uniform transverse
poling field between the electrodes. Since T_g of this polymer
is below room temperature, the temperature was controlled by
passing dry N_2 gas into a sealed chamber containing the sample
and monitoring with a thermocouple. Laser light polarized
parallel to the poling field was focused with a 7.5 cm lens
onto the sample, and the pulse energy maintained below 15
μJ/pulse in order to avoid laser-induced thermal effects. The
cast film was maintained above its T_g and a DC field applied
until a steady state temperature was reached. The film was
cooled well below T_g before the DC field was removed. The
second harmonic signal was then monitored as a function of
time and temperature.

A typical experimental result is shown in Fig. 3. On the left the second harmonic intensity is plotted as a function of temperature, and on the right the heat capacity as measured by DSC. The DSC trace indicates that T_g is slightly below room temperature and hence the poling was done at ambient temperature. Cooling to 240 K and removal of the poling field gave a signal as indicated. The signal was stable until warming above 255°C where a gradual loss occurred until the 285°C region where it precipitously decreased. The signal was not recovered by cooling unless another poling cycle was performed. The lower temperature gradual loss in signal appears to correlate with sub T_g motions and the precipitous loss in the randomization accompanying T_g.

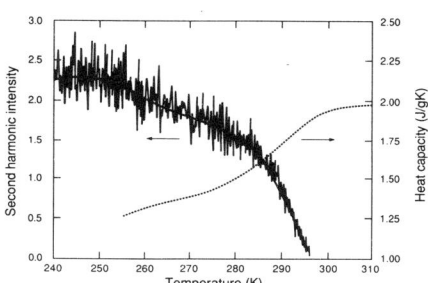

FIGURE 3. Plot of second harmonic intensity versus temperature following poling at 300 K and cooling to 240 K before removing field and DSC trace.

The field dependence of second harmonic intensity was investigated in order to gain insight into the enhancement mechanism in the rubbery state and is shown in Fig. 4. Also plotted is a nonlinear least squares fit of $L_3(X)$ (as defined in Eq. (9)) with a value for the dipole moment of 30 ± 5 D/F(0), seeming to indicate dipolar correlation of 3 to 4 monomer units in the polymer chain. Rewriting Eq. (6) for the polymer chain, we obtain for second harmonic generation using Onsager local field factors [46]

$$B_z = \frac{n \chi_{zzz}^{(2)}}{N\left(\dfrac{\eta^2 + 2}{3}\right)^3 L_3\left(\dfrac{F(0)E_0 M_z}{kT}\right)}$$

Using the measured values of $\chi_{zzz}^{(2)}$ (relative to quartz) of 5 x 10^{-9} esu, L_3 from the fit in Fig. 3, n = 152 for the polymer, ϵ = 2.6, and η = 1.6, we obtain B_z = 182 x 10^{-30} esu.

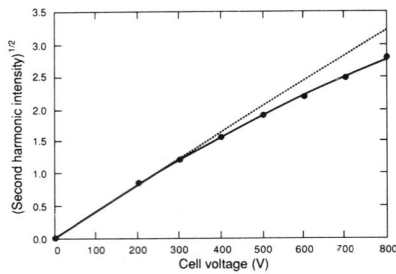

FIGURE 4. Second harmonic intensity versus cell voltage for films at 300 K; extrapolation to low field region — — —, and $L_3(X)$ versus voltage as described in text.

The enhancement factor in the film computes to be 0.4 relative to the solution value of 20. We note that it was not possible to obtain B and M independently in solution since under the conditions of the EFISH experiment only the linear part of the SHG curve could be measured, which precludes accurate determination of $L_3(X)$. We, therefore, cannot compare B and M directly to the film values--only their product. Since the enhancement is less than unity, however, we conclude that some penalty is being paid for incorporation of the monomer into the polymer chain in the film. It is tempting to conclude that the chain is substantially less extended than in solution, but because of questions regarding the copolymer sequence distribution and whatever effect this might have on the results, we are hesitant to do so at this point.

In conclusion, we have investigated electric field-induced harmonic generation in asymmetric condensation polymer systems. A model was developed within which to interpret enhancement factors and to extract details regarding the nature of the polymer chain. The large enhancement factors observed in solution were not observed in films. Further experiments on asymmetric condensation polymers, where non-alternating sequences are precluded and where chain lengths rang from oligomers to high molecular weight polymers, are in progress and should shed further light on the poling behavior of polar polymers in films as well as the practicality of this concept.

3. RECENT EXPERIMENTS WITH L-B FILMS

Langmuir-Blodgett films offer a conceptually attractive approach for the preparation of noncentrosymmetric structures for second-order nonlinear optics. In the introduction a number of recent studies on L-B films were mentioned. Two problems have surfaced from these studies that may require significant fundamental work before useful films can be fabricated by this technique. The first of these is ruggedness or stability. The second is the significant deviation of the dependence of SHG efficiency on the number of layers as well

as the concentration of chromophores within a layer from quadratic. With respect to the first issue, significant recent
progress has been made in imparting stability to L-B films
through photopolymerization. This process was studied in
detail [47,48] in L-B films of 2-chain molecules with phospho-
nate head groups III and in related

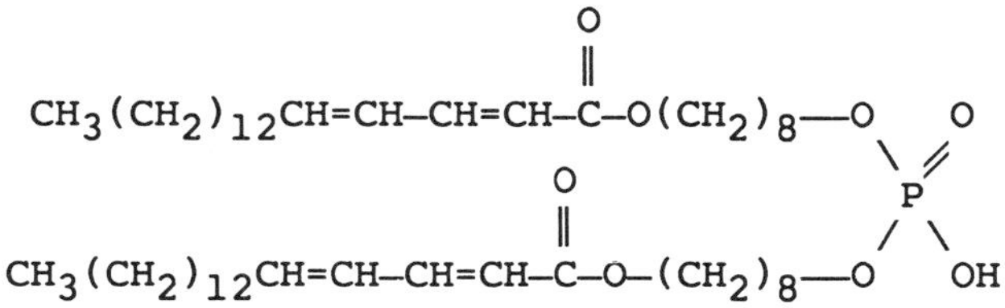

$$
\begin{array}{c}
\text{O} \\
\parallel \\
CH_3(CH_2)_{12}CH=CH-CH=CH-C-O(CH_2)_8-O \\
\end{array}
$$

CH_3(CH_2)_{12}CH=CH–CH=CH–C–O(CH_2)_8—O   O
                                                \   //
                                                 P
            O   / \
            ‖   /   \
CH_3(CH_2)_{12}CH=CH–CH=CH–C–O–(CH_2)_8—O   OH

III

structures containing 1 chain with 2 double bonds and 2 chains
with 1 double bond in each chain. It was found that a 2-chain
2-double bond system gave extensive polymerization as evidenced
by its physical properties and spectroscopic characteristics.
   In the remainder of this section we would like to discuss
recent results from our laboratory that may explain recently
reported significant deviations of SHG efficiency from quadratic in mono [31] and multilayer [38] films formed from an
amphiphilic hemicyanine dye IV diluted with arachidic acid V
and the fact the SHG is more efficient in mixed films than in
the pure dye.

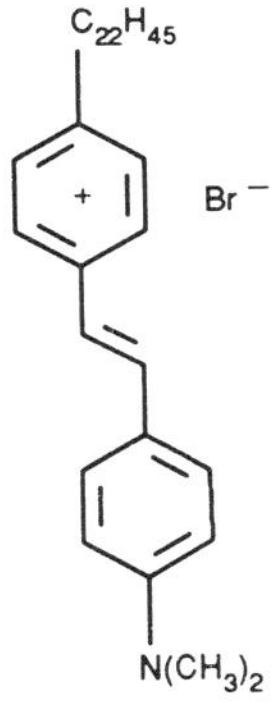

C_{22}H_{45}

Br^-

CH_3
|
(CH_2)_{18}
|
COOH

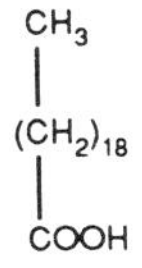

V

N(CH_3)_2

IV

   In this study we measured the absorption spectrum and SHG
efficiency in both pure IV and in a mixture of IV and V. It
will be shown that in the pure dye film the dye is almost
exclusively in the H-aggregate form, whereas in the mixed
system the dye is almost exclusively monomeric.
   The L-B films were prepared according to the following
procedure. Dye IV or mixtures of IV and V were spread from a
chloroform solution onto a 4 mM sodium borate (pH 9.2) subphase. After 15 min the film was compressed to a surface
pressure of 30 mN/in. The surface pressure was held constant

for 45 min and the film was deposited onto a fused silica substrate at a rate of 2 mm/min. Absorption spectra were measured on a Cary 2290 spectrophotometer, and SHG measurements were made using a Nd:YAG dye laser H_2-Raman-shifted with pulse energies of approximately 1 mJ incident on the sample.

The absorption spectrum of a 5.1 μm chloroform solution of the dye is shown in Fig. 5, and the absorption spectrum of a pure film of IV and 1:4 mixture of IV with V is shown in Fig. 6. In solution the dye shows 2 bands centered at 263 and 494 nm. Both the pure dye film and the diluted dye film exhibit a peak at around 260 nm but differ significantly at longer wavelengths, with the pure dye having a band at 350 nm and the diluted dye at 477 nm.

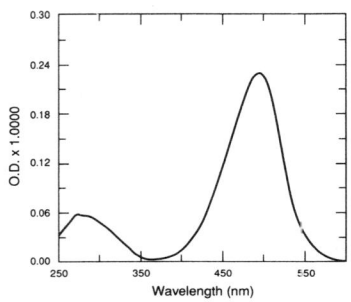

FIGURE 5. Optical absorption spectrum of hemicyanine dye in 5.1 μm chloroform solution.

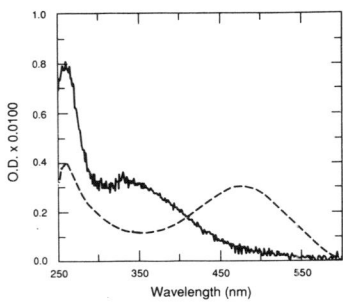

FIGURE 6. Optical absorption spectrum of pure dye film (solid line) and mixed 1:4 dye-arachidic acid film (dashed line).

SHG measurements of the two films with p- and s-polarized 1064 nm and p-polarized 1217 nm beams were made in the transmission mode. The output beams at 532 nm and 608.5 nm were almost completely p-polarized. An SHG fringe pattern due to interference of SHG signals emanating from the two faces of the substrate is shown in Fig. 7. The results of these

experiments are summarized in Table 2. The numbers in paren-
thesis have been normalized for the lower dye concentrations
in the mixed films which are 30% of the pure dye films. In
Table 3 several other polarization ratios are shown for the
pure and mixed dyes for 1064 nm input light.

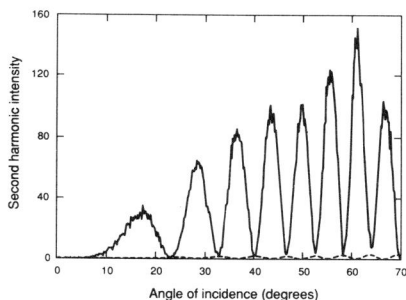

FIGURE 7. SHG of a pure dye film (dashed line) and a mixed
1:4 dye-arachidic acid film (solid line). Both the input beam
at 1064 nm and the output beam at 532 nm are p-polarized.

TABLE 2. Second harmonic intensities for the mixed and pure
dye films.

Ratio	1064 nm Input	1217 nm Input
$\dfrac{I_{p \to p} \text{ mixed}}{I_{p \to p} \text{ pure dye}}$	44	19
$\dfrac{\chi^{(2)} \text{ mixed}}{\chi^{(2)} \text{ pure dye}}$	6.6(22)	4.3(14)

The results obtained above can be explained if we assume
that the 150 nm blue shift in the dye absorption band is
associated with H-aggregate formation in the pure dye film.
Clearly, this would produce a difference in resonance enhancement of the second harmonic signals in two systems. This is
evidenced by the 19X enhancement of 1217 nm relative to the
44X enhancement at 1.06 μ. In addition to the resonance
enhancement effect, there could also be a partial cancellation
of second harmonic intensity due to the arrangement of dye
molecules in the aggregates or a reduction in hyperpolarizability due to the electronic structure of the dye molecules
in the aggregate. Taking into account dispersion using a
two-level model, we would expect a 15X and 5X enhancement in
SHG intensity at 1064 nm and 1217 nm, respectively. This is
approximately a factor of 3 less than the observed value in

TABLE 3. Second harmonic polarization ratios at 532 nm.

Ratio	Pure dye film	Mixed film
$\dfrac{I_{p \rightarrow s}}{I_{p \rightarrow p}}$	$<10^{-2}$	$<10^{-2}$
$\dfrac{I_{s \rightarrow s}}{I_{s \rightarrow p}}$	$<10^{-2}$	$<10^{-2}$
$\dfrac{I_{p \rightarrow p}}{I_{s \rightarrow p}}$	5.9	6.4

each case. The fact that $I_{p \rightarrow p}/I_{s \rightarrow p}$ is virtually the same for the pure dye and mixed films indicates that the orientation of the chromophores relative to the perpendicular direction is unchanged. We therefore conclude that both dispersion and differences in electronic structure between dye and aggregate contribute to the observed behavior.

It was mentioned above that the arrangement of dipoles in the aggregate might result in partial cancellation of signal. Antiparallel alignment of dipoles in the aggregate might account for this. We have illustrated possible local structural differences between monomeric dye and aggregated dye in Fig. 8.

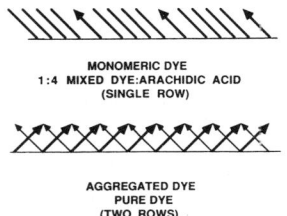

MONOMERIC DYE
1:4 MIXED DYE:ARACHIDIC ACID
(SINGLE ROW)

AGGREGATED DYE
PURE DYE
(TWO ROWS)

FIGURE 8. Hypothetical molecular dipole arrangement.

These observations verify the results reported by Girling et al. [31] and offer a plausible explanation for the observation by Hayden et al. [38] that the SHG signal does not increase quadratically with the number of layers. This latter observation may be partially due to differing relative amounts of aggregate and monomer in each dye layer. The instability of Z-type layered structures is also expected to contribute to this.

4. SELF-ASSEMBLED MONOLAYERS

An alternative approach to forming tailored multilayer assemblies is the self-assembled monolayer method SAM pioneered by Sagiv and coworkers [49-53]. In this method monolayers form spontaneously at a solid-liquid interface, resulting in close-packed monolayers chemically bonded to the surfaces. Most of the fundamental work to date has been done on derivatives of trichlorosilane and, in particular, octadecyltrichlorosilane (OTS). An advantage of trichloroalkyl silanes is that they

$$CH_3$$
$$|$$
$$(CH_2)_{17}$$
$$|$$
$$SiCl_3$$

VI

can be terminated with a functional group (e.g., C=C or an ester group) that can be activated chemically to form a fresh hydroxyl surface for subsequent deposition. This technique, therefore, has great potential utility in the construction of macromolecular structures, or artificial lattices, with order at the molecular level and with tailored optical, electronic, or nonlinear optical properties at the bulk level.

Our intention is to determine if structures of the type illustrated schematically in Fig. 9 can be developed. A number of questions emerge that can only be answered by systematic fundamental studies of SAM's. The first of these is the effect of insertion of an aromatic chromophore on packing and orientation within a layer as well as its effect on surface order and orientation. The latter may be particularly important for developing rapid quantitative surface chemistries required for multilayer formation. Secondly, the optimum position of the chromophore in the chain as well as optimum chain lengths must be determined. Thirdly, the whole question of multilayer formation and order must be addressed since one published study [57] noted that order and packing degraded seriously after the second layer for methyl 23-tri-chlorosilyltridecanoate. Finally, we might anticipate some

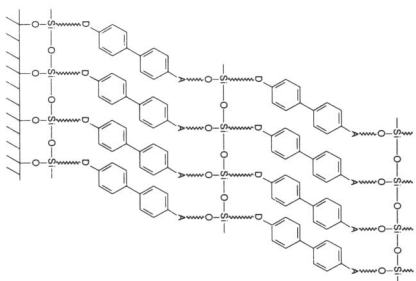

FIGURE 9. Conceptualization of a section of a film constructed from self-assembled monolayers.

impact of dipole-dipole repulsion in the monolayer or its stability. We have made significant progress to date on the first three of these issues and describe recent results in this section.

To date, there have been no publications on the incorporation of useful chromophores into a trichloroalkyl silyl chain itself. As a first step in our study we have incorporated a phenoxy group into long-chain trichloroalkyl silanes and studied the effects of its inclusion and position in the chain on the resultant monolayer properties [42]. The molecules we synthesized and studied are listed below.

	m	n	m+n
VII	8	11	19
VIII	8	4	12
IX	4	8	12
X	1	11	12

We have used analytical techniques including ellipsometry and FTIR (in both the grazing angle external specular reflection and ATR modes) to examine monolayer quality and group orientation in these systems. Surface structure and order were examined through the measurement of contact angles for various liquids.

Before describing our results on these materials, several comments are in order regarding these characterization techniques. The external specular grazing angle FTIR method relies on the existence of an electric field vector perpendicular to the surface, which is obtained by using high incidence angle light (grazing angle). The s-polarized component is strongly damped by the metal surface. In our case we use aluminized Si surfaces for these measurements. Consequently, vibrational modes having transition dipoles oriented perpendicular to the surface show relatively strong absorbances while dipoles completely parallel should approach zero. In the ATR mode incident internally reflected light can be polarized perpendicular or parallel to the surface.

Contact angle measurements provide a convenient measurement or indicator of the surface properties of a monolayer [41,50]. Although the wettability of a monolayer is determined by its surface properties and not the properties of the interior medium, the medium would be expected to indirectly affect surface properties by affecting the order, packing, and tilt of surface functional groups. We therefore used contact angles in conjunction with ellipsometry to monitor monolayer-forming reactions and assess surface quality. Contact angle measurements can also provide a route for obtaining the surface tension of the solid by determining the critical surface tension γ_c of a liquid (that surface tension for a liquid that completely wets the surface).

Finally, ellipsometry provides a convenient method for monitoring the monomolecular nature of the films and for monitoring the buildup of multilayers [50].

The results of characterization of several of these measurements for films of OTS (VI), VII, and IX on silicon are shown in Table 4. Our results for OTS are in good agreement with the literature values for this system [57] and provide confidence for comparison with other films. Contact angles of 111° for water and 45° for hexadecane are consistent with close-packed methyl groups at the surface. The low values of critical surface tension (γ_c) for OTS and VII that we obtained are indicative of excellent order at the surface. These data clearly indicate that close-packed monolayer formation while incorporating an aromatic ring is achievable, which is an essential first step to making bulk films with good linear and nonlinear optical properties. The somewhat lower values of wetting angles and higher critical surface tension for IX clearly indicate the essential role of the alkyl chain and importance of its length.

TABLE 4. Physical characteristics of SAM monolayer films.

Property	OTS (VI)	VII	IX
Thickness Å	25	31	24
Wetting angle (H_2O/hexadecane)	111/45	111/45	107/27
γ_c (dynes/cm)	20	20	23.7

FTIR spectra obtained for VII under various conditions are shown in Fig. 10. Several features of the spectra provide insight into orientation and packing in these systems. The peaks in the range from about 2850 cm^{-1} to 2970 cm^{-1} are associated with $-CH_2-$ and $-CH_3$ asymmetric and symmetric stretching modes. In Fig. 11a ν_a ($-CH_2$) and ν_s ($-CH_2$) occur at 2929 cm^{-1} and 2856 cm^{-1}, respectively. The CH_3 stretch is completely obscured by the oscillator strength of the methylene stretch. Under grazing angle conditions (Fig. 11b), the intensity of the methylenes is considerably diminished, consistent with a transition moment in the plane of the film and an electric vector nearly perpendicular. The CH_3 stretches are now clearly visible with ν_a (CH_3) and ν_s (CH_3) at 2965 cm^{-1} and 2880 cm^{-1}, respectively. Several other features are worth noting. The C–O–C stretch occurs at $\nu_{C-O-C} = 1244$ cm^{-1} in solution and 1255 cm^{-1} in the film under grazing angle conditions, and the ring C=C stretch $\nu_{C=C}$ occurs at 1512 cm^{-1} and 1516 cm^{-1} under the two sets of conditions. The intensity in the peaks is consistent with orientation parallel to the electric vector of the incident light. In Fig. 11c the p-polarized ATR spectrum of VII for 45° incident angle is shown. Spectra taken with both s- and p-polarized light show excellent signal-to-noise ratios and can be used to obtain dichroic ratios for the various peaks, thus allowing a determination of orientation within the film.

Grazing angle FTIR of the alkyl regions of monolayers OTS and VII are shown in Fig. 11. It is clear that the peaks associated with the CH_2 stretch at 2919 cm^{-1} and 2851 cm^{-1} are suppressed, indicating nearly perpendicular orientation in film. These peaks are relatively stronger in VII, indicating a higher average tilt angle in these films.

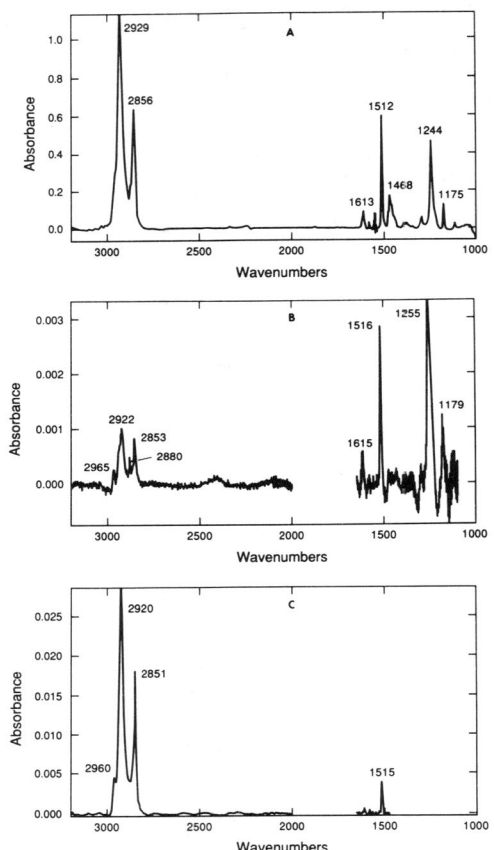

FIGURE 10. IR spectrum of compound 1 as (a) bulk spectrum (CCl_4 solution), (b) grazing angle (76°) external specular reflection spectrum on aluminized silicon, and (c) p-polarized ATR spectrum on silicon prism (45° incidence). Baselines have been adjusted to zero absorbance, and artifacts eliminated where necessary. The opacity of silicon prevents useful detection of IR bands below 1500 cm^{-1} in the ATR.

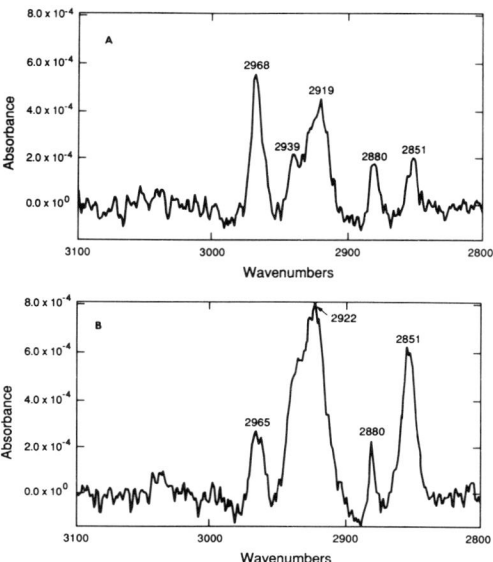

FIGURE 11. C—H stretch region of the grazing angle external specular reflection spectrum of a monolayer of (a) 1 and (b) OTS on aluminized silicon wafers. Spectra were recorded at 76° incidence, 1000 scans, and 2 cm^{-1} resolution, and the baselines have been adjusted to zero absorbance.

Analysis of dichroic ratios by ATR for the various modes leads to a picture of orientation in these two films (Fig. 12). In OTS the chains are tilted on the order of 10 to 15° from perpendicular, whereas in VII the alkyl chains have a 25° tilt and the axis of the phenyl group is 20° in from perpendicular but in the opposite direction. Packing in the monolayer is undoubtedly the result of a number of attractive and repulsive

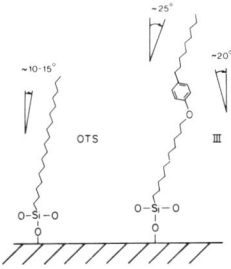

FIGURE 12. Model of orientation for OTS and III on Si surface based on FTIR dichroic ratios.

forces. However, packing cannot be any closer than the spacing
of the largest moiety. A detailed analysis of the dependence
of cross-sectional areas of chains and rings indicates an
increase in Van der Waals attraction in the alkyl chains and a
decrease in the mismatch of the cross-sectional areas of the
phenyl rings and alkyl chains relative to the situation where
chains are perpendicular and the axis of the phenyl is a much
larger angle of 35°.

Finally, we address the problem of multilayer formation.
For this purpose we chose to study mono- and multilayer films
of methyl 23-trichlorosilyltricosanoate XI. Multilayers can
be formed by deposition of a monolayer of XI onto silicon
followed by reduction of the ester group with LiAlH$_4$/THF
solutions [43,53]. Pomerantz et al. [53] found that films
prepared by this method exhibited serious deterioration of
order and packing in the second and third layers.

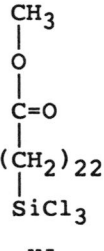

$$
\begin{array}{c}
\text{CH}_3 \\
| \\
\text{O} \\
| \\
\text{C=O} \\
| \\
(\text{CH}_2)_{22} \\
| \\
\text{SiCl}_3
\end{array}
$$

XI

As the data shown demonstrate, we have been able to obtain
25 monolayers without apparent significant deterioration in
film quality. We attribute this success to differences in
handling the films between depositions of the various layers.
In particular, we found that the use of a low-residue aqueous
alkaline wash of the ester surface following deposition
removed excess material that was not bound directly to the
surface. In contrast, the procedures previous workers had
employed involved cleaning the ester surfaces with concentrated HCl and water, followed by Soxhlet extraction with
chloroform [53]. Either residual material or surface
reorganization accompanying these treatments may have been
responsible for the deterioration of the surface.

A plot of film thickness as determined by ellipsometry,
versus layer number is shown in Fig. 13. A slope of 35 Å
layer is obtained which is in very reasonable agreement with
the ~32 Å value obtained for monolayers. Contact angles for
the ester surface after the subsequent deposition of each
monolayer are shown in Fig. 14. The values increased slightly
after the first three or four layers and then stabilized at
69 ± 2° for water and 27 ± 2° for hexadecane. Considerably
more scatter is evident in the contact angle data points for
the hydroxylated surface (Fig. 15). We believe that there is
some pattern to this variation indicative of disorder in a
particular layer followed by self-healing in subsequent
layers. This is evident in the sample represented by the
triangles, which gives a 31° value for the first layer, jumps

215

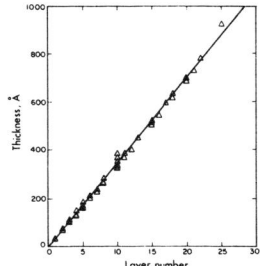

FIGURE 13. Film thickness, determined by ellipsometry, versus layer number, measured on eight different multilayer samples.

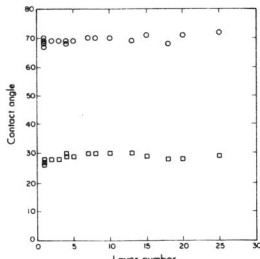

FIGURE 14. Water (circles) and n-hexadecane (squares) advancing contact angles on $MeO_2CC_{22}Si/Si$ monolayers versus layer number, measured on five different multilayer samples.

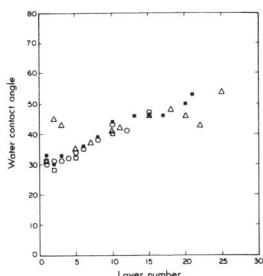

FIGURE 15. Advancing water contact angles on $HOC_{23}Si/Si$ monolayers versus layer number. Measurements were taken using four different multilayer samples, with the values for each plotted using different point types.

to 45° and 43° for the second and third layers, settles to 35°
by the fifth and seems to undergo similar cycles during subsequent depositions.

The experiments described in this section explore some of
the fundamental issues associated with building stable films
with high degrees of order and quality that will be needed for
nonlinear optical applications. In the future we plan to
introduce more complex chromophores that could alter the
balance of forces in the monolayers. This could create
problems or offer opportunities with regard to stability and
packing in the layers. Our initial results on fabrication and
characterization techniques would appear to justify this next
step.

REFERENCES

1. Williams DJ: Angewante Chemie 23, 690 (1984).
2. Chemla DS and Zyss J(eds): Nonlinear Optical Properties
   of Organic Molecules and Crystals Vol. 1. Orlando:
   Academic Press, 1987.
3. Levine BF: Chem. Phys. Lett. 37, 516 (1976).
4. Oudar JL and Chemla DS: J. Chem. Phys. 66, 2664 (1977).
5. Dulcic A and Flytzanis C: Opt. Commun. 25, 402 (1978).
6. Ward J: Rev. Mod. Phys. 37, 1 (1965).
7. Oudar JL: J. Chem. Phys. 67, 446 (1977).
8. Morrell JA and Albrecht AC: Chem. Phys. Lett. 64, 46
   (1979).
9. Lalama SJ and Garito AF: Phys. Rev. A20, 1179 (1979).
10. Docherty VJ, Pugh D, and Morley JO: J. Chem. Soc.
    Faraday Trans. 2 81, 1179 (1985).
11. Oudar JL and Hierle R: J. Appl. Phys. 48, 2664 (1977).
12. Zyss J, Chemla DS, and Nicoud JF: J. Chem. Phys. 74,
    4800 (1981).
13. Panunto TW, Urbanczyk-Lipkowska Z, Johnson R, and Etter
    MC: J. Am. Chem. Soc.
14. Dulcic A and Flytzanis C: Opt. Commun. 25, 402 (1978).
15. Levine BF, Wasserman E, and Leenders L: J. Chem. Phys.
    68, 5042 (1978).
16. Ziolo RF, Gunther WHH, Meredith GR, and Williams DJ:
    Acta Cryst. B 38, 341 (1982).
17. Oudar JL and Zyss J: Phys. Rev. A 26, 2076 (1982).
18. Zyss J and Oudar JL: Phys. Rev. A 26, 2028 (1982).
19. Nicoud JF and Twieg RJ: Chemla DS and Zyss J(eds):
    Nonlinear Optical Properties of Organic Molecules and
    Crystals, Vol. 1. Orlando: Academic Press, p. 193, 1987.
20. Zyss J, Ledoux I, Hierle R, Roj R, and Oudar JL: IEEE J.
    Quantum Electron. QE21, 1286 (1985).
21. Nayar BK: Williams DJ(ed): Nonlinear Optical Properties
    of Organic and Polymeric Materials. ACS Symp. Ser. 233,
    153 (1983).
22. Meredith GR, VanDusen JG, and Williams DJ: Macromolecules
    15, 1385 (1982).
23. Meredith GR, VanDusen JG, and Williams DJ: Williams
    DJ(ed): Nonlinear Optical Properties of Organic and
    Polymeric Materials. ACS Symp. Ser. 233, 110 (1983).

24. Williams DJ: Chemla DS and Zyss J(eds): Nonlinear Optical Properties of Organic Molecules and Crystals, Vol 1. Orlando: Academic Press, p. 405, 1987.
25. Singer KD, Kurzyk MG, and Sohn JE: J. Opt. Soc. Am. B $\underline{4}$, 968 (1987).
26. Thackara GF, Lipscomb GF, Stiller MA, Ticknor AJ, and Lytch R: Appl. Phys. Lett. $\underline{52}$, 1031 (1988).
27. Willand CS, Feth SE, Scozzafava M, Williams DJ, Green D, Weinschenk JS, Hall HR, and Mulvaney JE: Plenum, in press.
28. Willand CS and Williams DJ: Ber. Bunsenges. Phys. Chem. $\underline{91}$, 1304 (1987).
29. Swalen JD: J. Molec. Elec. $\underline{2}$, 155 (1986).
30. Girling IR, Cade NA, Kolinsky PV, and Montgomery CM: Electron. Lett. $\underline{21}$, 169 (1985).
31. Girling IR, Cade NA, Kolinsky, PV, Jones RJ, Peterson IR, Ahmad MM, Neal DB, Petty MC, Roberts GR, and Feast WJ: J. Opt. Soc. Am. B $\underline{4}$, 950 (1987).
32. Cross GH, Girling IR, Peterson IR, Cade NA, and Earls JD: J. Opt. Soc. Am. B $\underline{4}$, 962 (1987).
33. Ledoux S, Josse D, Vidakovic PV, Zyss J, Hahn RA, Gorden BD, Bothwell BD, Gupta SK, Allen S, Robin P, Chastaing E, and Dubois JC: Europhys. Lett. $\underline{3}$, 803 (1987).
34. Lupo D, Pross W, Scheunemann, Laschewsky A, Ringsdorf H, Ledoux S: J. Opt. Soc. Am. B $\underline{5}$, 300 (1988).
35. Girling, IR, Cade NA, Kolinsky PV, Earls JD, Cross GH, and Peterson IR: Thin Solid Films $\underline{132}$, 101 (1985).
36. Girling IR, Kolinsky PV, Cade NA, Earls JD, and Peterson, IR: Opt. Commun. $\underline{55}$, 289 (1985).
37. Neal DB, Petty MC, Roberts GG, Ahmad MM, Feast WJ, Girling, IR, Cade NA, Kolinsky PV, and Peterson IR: Electron. Lett. $\underline{22}$, 460 (1966).
38. Hayden LM, Kowel ST, and Srinivasen MP: Opt. Commun. $\underline{61}$, 351 (1987).
39. Tredgold RH, Young, MCJ, Hodge P, and Khoshdel E: Thin Solid Films $\underline{151}$, 441 (1987).
40. Carr N and Goodwin MJ: Makromol. Chem. Rapid Commun. $\underline{8}$, 487 (1987).
41. Moy R, Sagiv J: J. Colloid Interface Sci. $\underline{100}$, 465 (1984).
42. Tillman NA, Ulman A, Schildkraudt JS, and Penner TL: J. Am. Chem. Soc., accepted for publication.
43. Tillman NA, Ulman A, Schildkraudt JS, and Penner TL: J. Am. Chem. Soc., submitted for publication.
44. Lalama SJ, Sohn JE, and Singer KD: SPIE Proc. $\underline{168}$, 578 (1985).
45. Levine BF and Gethen CG: J. Chem. Phys. $\underline{63}$, 2666 (1975).
46. Onsager L: J. Am. Chem. Soc. $\underline{58}$, 1486 (1936).
47. Penner TL and Ponticello IS: Abstract. Northeast Regional ACS Meeting, p. 153, 1987.
48. Penner TL: private communication.
49. Gun J, Iscovici R, Sagiv J: J. Colloid Interface Sci. $\underline{101}$, 201 (1984).
50. Gun J and Sagiv J: J. Colloid Interface Sci. $\underline{112}$, 457 (1986).

218

51. Cohen SR, Nooman R, and Sagiv J: J. Phys. Chem. <u>90</u>, 3054
     (1986).
52. Finklen HO, Robinson LR, Blackburn A, Richter B, Allara
     D, and Bright T: Langmuir <u>2</u>, 239 (1986).
53. Pomerantz, M, Segmuller A, Netzer L, and Sagiv J: Thin
     Solid Films <u>132</u>, 153 (1985).

SPECTRAL PROPERTIES AND SECOND-HARMONIC GENERATION OF HEMI-
CYANINE DYE IN LANGMUIR-BLODGETT FILMS

P. WINANT, A. SCHEELEN and A. PERSOONS

Department of Chemistry, University of Leuven,
3030 Leuven, Belgium

   Organic molecules with large second-order polarizabilities offer an important potential for applications requiring nonlinear
optical properties, e.g. optical signal processing (1,2).   Unfortunately most molecules crystallize in centrosymmetric structures preventing their use for quadratic nonlinear optics.   The
Langmuir-Blodgett technique (3), which allows the fabrication
of thin films from these molecules in a non-centrosymmetric
lattice, offers good possibilities to circumvent this restric-
tion (4-8).   A requirement for the application of the LB-tech-
nique is the incorporation of an hydrophobic moiety to render
the molecules amphiphilic.

   In this communication we report second-harmonic generation in
LB-films either of the pure dye MO, shown in Fig.1, or of MO-
arachidic acid (AA) mixtures.   As previously observed (6,8),
SHG-efficiency per dye molecule increases upon dilution.   For
multilayer films the expected quadratic increase with number of
layers was not observed.   To investigate these effects we measured absorption and reflectance spectra of the Langmuir-mono-
layers and LB-films.   Interlayer-interactions were investigated
by separating active layers with an increasing number of pure
arachidic layers.

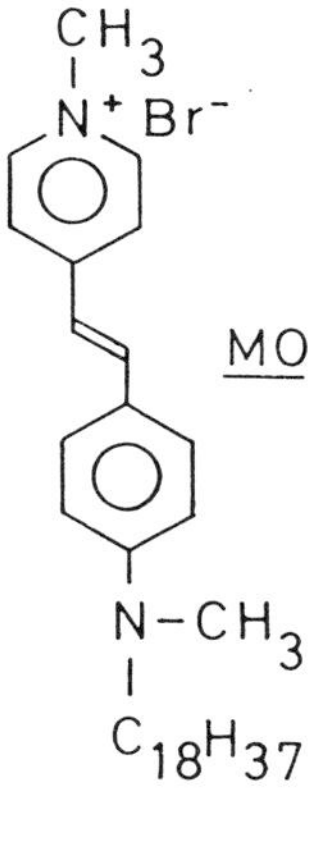

FIGURE 1. Structure of hemicyanine dye (MO).

J. Messier et al. (eds.), Nonlinear Optical Effects in Organic Polymers, 219–224.
© *1989 by Kluwer Academic Publishers.*

LB-films were prepared according to standard practice (3) :
pure dye or a dye-arachidic acid mixture was spread from a chloroform solution onto an aqueous subphase (10 mM NaClO$_4$, pH ~5.5),
the film compressed to a surface pressure of 30 mN/m (Fromherz
through (9)) and, after an appropriate stabilisation time,
transferred on a suitably cleaned glass-substrate (microscope
slide 12 x 38 mm) at a rate of 1.5 mm/min.   In multilayer films
the dye containing layers were separated by an odd number of
inactive arachidic acid layers to preserve the noncentrosymmetry of the active lattice.

Measurements of surface pressure vs. area (π-A) show the formation of mixed films from MO and arachidic acid (Fig.2).   From
the π-A curves the area of the dye-molecule in the Langmuir-
films is determined as 0.38 nm^2, independent of dilution, except at the highest dilution where a slight increase in area
is observed.

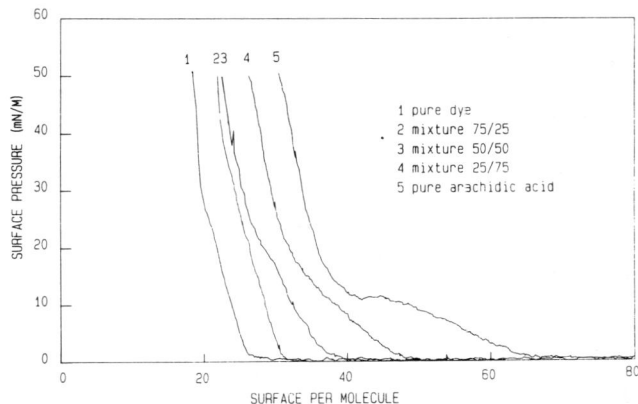

FIGURE 2. π-A curves of pure MO and MO-arachidic acid (AA) mixtures (room temperature)

The absorption spectra -Fig.3- of the LB-monolayers on glass
show a blue shift vs. the spectrum in chloroform. This may be
due to H-aggregation of the dye in the film. However no change
in absorption maximum is observed upon dilution of the dye in
mixed films. Since the dye is in contact with the polar glass
surface some care should be taken in the interpretation of these
absorption spectra. The O.D. of multilayer films increases linearly with the number of dye-layers.

To assess more precisely molecular orientation and structure
of the films reflectance spectra at 45° incidence angle of dye-
arachidic acid Langmuir layers were obtained (10). The layers
were spread on the same aqueous subphase used to deposit the
LB-films and also compressed at 30 mN/m. The reflectance spectra, both for s- and p-polarization, show a pronounced dependence upon dye-concentration in the layer (Fig.4). The blue-

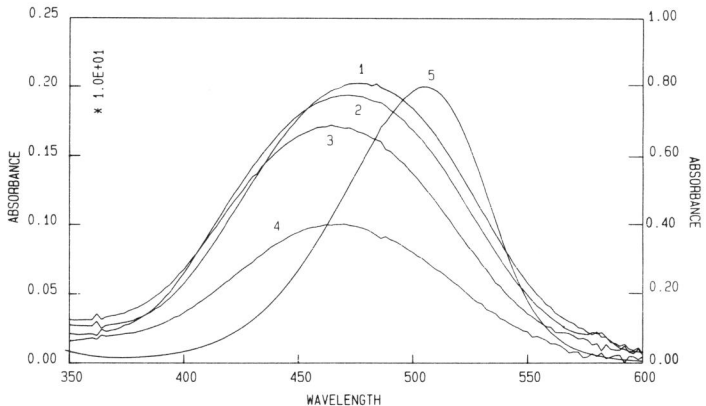

FIGURE 3. Absorption spectra of MO in solution (CHCl$_3$; 0.022 mM) and in monolayer LB-films; (1) pure MO; (2) MO-AA 3:1; (3) MO-AA 1:1 and (4) MO-AA 1:3.

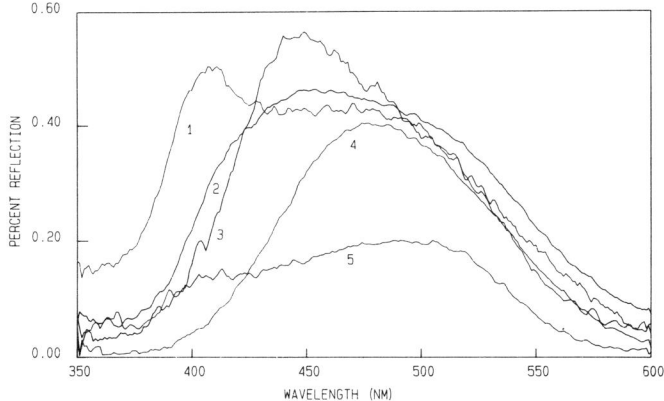

FIGURE 4. Reflectance spectra of Langmuir layers, s-polarization 45° incidence; (1) pure MO; (2) MO-AA 3:1; (3) MO-AA 1:1; (4) MO-AA 1:3; (5) p-polarization reflectance for pure MO (x5).

shift observed with increasing dye-concentration is a clear in-dication of H-aggregation of the dye in the layer. Formally the broad reflection signals can be assigned to reflection from dye-monomers, dimers and higher aggregates ("oligomers"). In the pure dye film the reflection shows a maximum at 420 nm, much more pronounced for s-polarization. From these data an average angle of about 30° between the transition moment of the oligomer and the aqueous surface is derived. The dimer shows a maximum at 450 nm and makes an angle of 20° with the aqueous surface. In the most diluted layers (25 mol% dye) the reflection has a maxi-mum at 500 nm -around the absorption maximum observed in chloro-

form solutions. This reflection is assigned to the monomer
which also makes an angle of 20° with the surface. It should
be noted that the dye-chromophore is always almost parallel to
the surface, a fact which has far-reaching consequences if this
orientation is preserved upon deposition on a glass-substrate.
The reflection intensities are largely independent of dye-concentration indicating a decrease in absorption cross-section
upon aggregation. This may explain why the absorption is largely independent of the dye-concentration in the LB-films (Fig.3).

Transmission SHG from the LB-films, mono- as well as multilayers,
was measured with a p-polarized beam from a 1064 nm Nd-YAG at
power levels 2-10 mJ/pulse (8 nsec pulse-width). The sample
was set on a rotation stage with the rotation axis vertical and
perpendicular to the incident 1064 nm beam. On varying the incidence angle the harmonic signal was detected, after suitable
filtering, by a photomultiplier. SH-signals from glass substrates with arachidic acid LB-films were always below the detection
limit.

The signal measured is identified as SH-radiation by its narrow spectral width and the quadratic dependence upon incident
energy. A typical SH-signal, measured with a monolayer film,
is shown as a function of the incidence angle in Fig.5. The
characteristic fringe pattern is due to the interference from
front and back-layers of the substrate. The fringe spacing depends upon substrate thickness (5,7). The SH-intensity measured with monolayer-films is, to a good approximation, independent of dye-concentration in the films. If the SH-signal is
scaled with the square of the dye-concentration, as may be expected for independent emitters, a marked increase in SHG-efficiency per dye molecule is noted upon dilution, Fig.6. This
is consistent with the results of the reflectance spectra and
indicates a decrease in SHG-efficiency due to dye-aggregation.
This efficiency decrease is more likely due to a diminution of
the hyperpolarizability upon aggregation rather than to a reso-

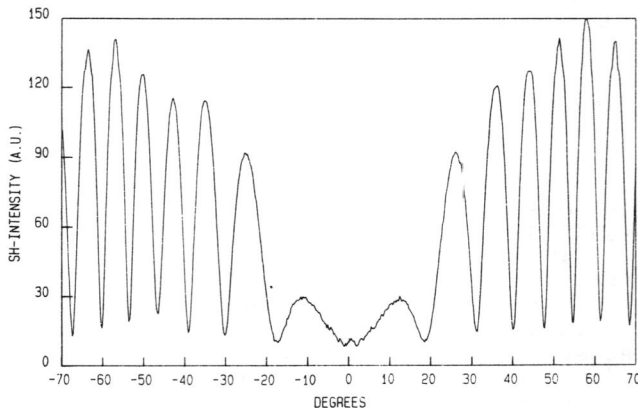

FIGURE 5. SH-intensity as a function of incidence angle for a
double-side LB-monolayer-coated substrate; LB-film of pure MO-
layer.

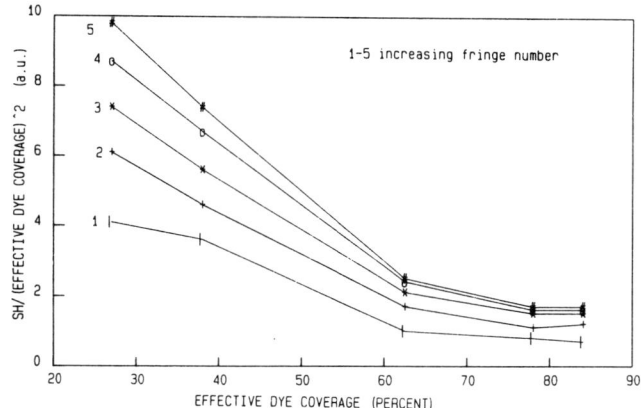

FIGURE 6. SH-signal, normalized to the active dye area vs. effective dye coverage in mixed (MO-AA) LB-monolayer films.

nant enhancement, since absorption spectra of the monolayer-
films do not change upon dilution of the dye.

The SH-intensity measured for films consisting of 1, 2 and 4
pure dye layers, spaced by inactive arachidic acid layers, is
markedly below the expected quadratic increase. LB-films consisting of two pure dye layers separated by 1, 3 or 5 arachidic
acid layers show almost identical SH-intensities. These effects
are generally observed by several research groups. The explanation of this puzzling behaviour may be based upon our results
of the reflectance measurements. Since the dye-molecules have
their chromophores parallel to the LB-films -and taking a random orientation of the chromophores in the film-plane- a dye
molecule in one layer may be antiparallel aligned (incidently
this minimizes electrostatic interaction) with a dye-molecule
in a subsequent layer resulting in a (partial) cancellation of
the second-order susceptibility reducing the efficiency for SHG
substantially in multilayer films.

CONCLUSION
SHG from LB-monolayers of stilbazonium-dyes is a rather efficient process although intra-layer aggregation phenomena strongly decrease SHG-efficiency presumably by diminishing the hyper-
poarlizability by electronic coupling of the molecules in the
aggregate.

The efficiency increase in SHG expected from constructing
multilayer-films with proper symmetry remains yet to be realized. This lack in efficiency-increase is due to antiparallel
molecular alignments between subsequent layers, which is the
more pronounced the lower the angle between dye-molecule and
film-plane. We therefore advance the hypothesis that the qua-
ratic dependence of SHG upon number of layers will be observed
in films where the chromophores are perpendicular to the film-

plane.

This research was supported by Geconcerteerde Acties, grant GOA 87/91-109 and Nationale Loterij, Belgium. P. Winant is aspirant of the N.F.W.O. and A. Scheelen is bursar of IWONL.

REFERENCES

1. D.J. Williams, ed., "Nonlinear Optical Properties of Organic and Polymeric Materials", ACS symposium 233, Washington (1983)
2. D.S. Chemla and J. Zyss, ed., "Nonlinear Optical Properties of Organic Molecules and Crystals", vol. 1 and 2, Academic Press (1987)
3. H. Kuhn, D. Möbius and H. Bücher in "Techniques of Chemistry, vol. I, Part IIIB", A. Weissberger and B.W. Rossiter, eds., Wiley, New York (1972)
4. I.R. Girling, P.V. Kolinsky, N.A. Cade, J.D. Earls and I.R. Peterson; Optics Commun. 55 289 (1985)
5. I. Ledoux, D. Josse, P. Vidakovic, J. Zyss, R.A. Hann, P.F. Gordon, B.D. Bothwell, S.K. Gupta, S. Allen, P. Robin, E. Chastaing and J.C. Dubois; Eurphys. Lett. 3 803 (1987)
6. I.R. Girling, N.A. Cade, P.V. Kolinsky, R.J. Jones, I.R. Peterson, M.M. Ahmed, D.B. Neal, M.C. Petty, G.G. Roberts and W.J. Feast, J. Opt. Soc. Am. B4 950 (1987)
7. D. Lupo, W. Prass, J. Scheunemann, A. Laschewsky, J. Ringsdorf and I. Ledoux; J. Opt. Soc. Am. B5 300 (1988)
8. J.S. Schildkraut, T.L. Penner, C.S. Willand and A. Ulman; Optics Letters 13 134 (1988)
9. P. Fromherz, Rev. Sci. Instr. 46 1380 (1975)
10. M. Oritt, D. Möbius, U. Lehmann and H. Meyer, J. Chem. Phys. 85 4966 (1986)

CUBIC SUSCEPTIBILITY OF ORGANIC MOLECULES IN SOLUTION

F. KAJZAR
Centre d'Etudes Nucléaires de Saclay - IRDI/D.LETI/DEIN/LPEM
91191 GIF SUR YVETTE CEDEX - FRANCE

1. INTRODUCTION

In last years one observes an increasing interest in the search for new highly efficient organic molecules with large optical hyperpolarizabilities. The efforts are going mainly into two directions: synthesis of new molecules and characterization of their nonlinear optical properties. It is well known that the noncentrosymmetric charge transfer complexes with hydrogen bond are suitable for the second order nonlinear optical effects [1] whereas the conjugated one dimensional π electron systems are caracterized by large third order susceptibilities [2-5].

Molecular engineering [1-6] offers a large choice of different molecules with predictable nonlinear optical response. The progress in the synthesis of new materials requires also rapid and relatively precise nonlinear optical properties characterization methods. From the practical point of view such methods have to be simple and assuring at the same time a good precision in the nonlinear susceptibilities. One of the simplest methods giving the electronic part of corresponding nonlinear susceptibilities is harmonic generation technique. This can be done on powders, bulk single crystals, thin films and liquids (solutions). The harmonic generation measurements on powders [7] are approximate only. The results depend on the grain size, absorption of fundamental and/or harmonic wave. The growth of single crystals is costly, time consuming and uncertain. The same is also true for thin films. Especially if one wants to get oriented thin films for the second order effects. Thus the most adapted technique for a rapid and precise screening of microscopic nonlinear optical properties of molecules is harmonic generation in solutions (in the case of soluble molecules). In fact by coupling third harmonic generation (THG) with electric field induced second harmonic generation (EFISHG) one can obtain a complete characterization of second (in the case of noncentrosymmetric molecules) and third order hyperpolarizabilities of molecules in solution. In fact THG gives directly the third order molecular hyperpolarizability

$$\gamma(-3\omega;\omega,\omega,\omega) = \gamma^e(-3\omega;\omega,\omega,\omega) + \gamma^v(-3\omega;\omega,\omega,\omega) \tag{1}$$

where γ^e is the purely electronic contribution and γ^v is the vibrational nonlinearity, whereas the EFISHG technique yields the sum

$$\gamma(-2\omega;\omega,\omega,o) = \gamma^e(-2\omega;\omega,\omega,o) + \gamma^v(-2\omega;\omega,\omega,o) + \frac{\mu\beta(-2\omega;\omega,\omega)}{5\ kT} \tag{2}$$

where μ is the permanent dipolar moment of the molecule and $\beta(-2\omega;\omega,\omega)$ the second order molecular susceptibility.

225

J. Messier et al. (eds.), Nonlinear Optical Effects in Organic Polymers, 225–245.
© *1989 by Kluwer Academic Publishers.*

In the molecular coordinate frame with the dipole moment parallel to the molecular x-axis

$$\beta = \beta_{xxx} + \beta_{xyy} + \beta_{xzz} \tag{3}$$

and

$$\gamma = \frac{1}{5} \left(\gamma_{xxxx} + \gamma_{yyyy} + \gamma_{zzzz} + 2\gamma_{xxyy} + 2\gamma_{xxzz} + 2\gamma_{yyzz} \right) \tag{4}$$

both β and γ being third and fourth rank tensors, respectively.

Although the molecular susceptibilities $\gamma(-3\omega; \omega, \omega, \omega)$ and $\gamma(-2\omega; \omega, \omega, o)$ are different, however far of resonances they are close. In the intermediate region it is possible to get a relationship between these two quantities using a two or three level model [8].

In this paper we review different THG techniques leading to the molecular susceptibility $\gamma(-3\omega; \omega, \omega, \omega)$ determination. We discuss more extensively the case of a complex γ having important practical and fundamental repercussions.

2. THIRD HARMONIC GENERATION TECHNIQUES IN VACUUM

Third harmonic generation is observed in every material medium. Any liquid (solution) has to be kept in a container with transparent windows for harmonic generation purposes. Thus in the general case the laser beam will generate harmonic light in the input and output windows, liquid itself and in the surrounding liquid cell medium (e.g. air). These harmonic fields interfere giving rise to a complicated dependence of harmonic intensity on the interaction length. Moreover, the harmonic intensity dependence on interaction length is also a function of the shape of liquid cell and its environment. Several techniques have been proposed dealing differently with these problems. In the following, for the sake of simplicity we consider the liquid cell placed in vacuum. Influence of surrounding liquid cell medium will be discussed later. We note also that THG measurements are to be performed with focused laser beams and we assume the cell located exactly at the focal point.

2.1. Wedge technique

The wedge technique, originally proposed by Chemla and Kupecek [9] for bulk materials was first used by Meredith et al [10] for THG in liquids. The principle of operation of such a liquid cell is shown in Fig. 1. the interaction length is varied by translating the cell perpendicularly to the laser beam propagation direction. In that case the output harmonic field is a sum of harmonic fields generated in input (G1) and output (G2) windows as well as in liquid compartment (L):

$$E_R^{3\omega}(x) = T_{G1} E_{G1}^{3\omega}(x) + T_{G2} E_{G2}^{3\omega}(x) + T_L E_L^{3\omega}(x) \tag{5}$$

where T's are the corresponding total transmission factors and x is the translation of liquid cell.

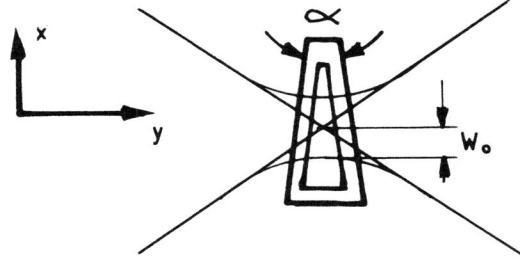

FIGURE 1. Schematic representation of a wedged liquid cell. The light beam propagates along the y direction and the cell is translated along the x-axis.

In the plane wave approximation for incident beam the harmonic field generated in a nonlinear slab with thickness ℓ is given by [11]

$$E_{3\omega}(\ell) = \frac{4\pi\, P_{NL}\, e^{i\psi_{3\omega}}}{\Delta\varepsilon}\, A(e^{i(\psi_\omega - \psi_{3\omega})} - 1) \qquad (6)$$

where we have neglected the multiple reflection effects (what is justified for a wedge).

In Equ. (6) ψ_ω and $\psi_{3\omega}$ are phase mismatches of fundamental and harmonic wave, respectively

$$\psi_{\omega(3\omega)} = \frac{6\,\pi\,\ell}{\lambda_{\omega(3\omega)}}\, n_{\omega(3\omega)}\, \cos\,\Theta_{\omega(3\omega)} \qquad (7)$$

and

$$A = \frac{n_{3\omega}\,\cos\,\Theta_{3\omega} + n_\omega\,\cos\,\Theta_\omega}{n_{3\omega}^o\,\cos\,\Theta_{3\omega} + n_\omega\,\cos\,\Theta_\omega} \qquad (8)$$

where $\Theta_{\omega(3\omega)}$ and $n_{\omega(3\omega)}$ are respectively propagation angles and refractive indices in the medium ($n^o_{3\omega}$ corresponds to the medium after that under consideration), $\Delta\varepsilon = n_\omega^2 - n_{3\omega}^2$ is dielectric constant dispersion and

$$P_{NL} = \frac{1}{4}\,\chi^{(3)}\,(-3\omega;\omega,\omega,\omega)\,E_\omega^3 \qquad (9)$$

is nonlinear polarization created in the medium. $\chi^{(3)}\,(-3\omega;\omega,\omega,\omega)$ is third order susceptibility responsible for the THG process and E_ω is the

fundamental electric field. By translating the wedge the interaction
length varies with x as follows

$$l(x) = l_0 + 2x \; tg \; \frac{\alpha}{2} \qquad\qquad (10)$$

where α is the wedge angle (cf. Fig.1) and l_0 is its initial thickness.
For perfectly plane parallel input and output windows the corresponding
harmonic fields E_{G1} and E_{G2} do not depend on translation x. However, it is
difficult to determine their magnitude which depends on the wavelength of
operation. Practically, it is also difficult to get perfectly plane
parallel windows. In order to overcome this difficulty Meredith et al [10]
used wedged liquid cell windows, with different wedge angles. In this case
both E_{G1} and E_{G2} depend on translation x, but in a different way. This
leads to complicated harmonic light interference spectra which can be
resolved numerically. There is a large number of free parameters :
refractive index dispersions in window materials and in liquid, input and
output windows wedge angles and corresponding initial thickness l_0's (cf.
Equ. (10)), $\chi^{(3)}$ of liquid, its phase and l_0 for liquid compartment.
Although some of these parameters can be measured independently (wedge
angles, l_0's), the precision however is not sufficient and the corrrespon-
ding quantities can be used only as input parameters in a fitting procedure.

2.2. Wedge technique with zero emission windows

In order to avoid contributions to harmonic field from liquid cell
windows Thalhammer and Penzkofer [12] manufactured the windows in such a
way that their thickness is exactly equal to the even number of coherence
lengths ($l_c = \lambda_\omega/6\Delta n$). In this case both harmonic fields E_{G1} and E_{G2} are
equal to zero (complete cancelling between free and bound waves) and the
resultant field is that generated in the liquid compartment. Translating
the cell one obtains regular Maker fringes with nearly $sin^2 \Delta\phi$ dependence
of harmonic intensity on phase mismath $\Delta\phi(\Delta\phi = \psi_\omega - \psi_{3\omega})$. In this case one
determines easily the coherence length (or Δn) and χ^3 of liquid.

The principal drawback of this technique is that the liquid cell works
at one wavelength only (coherence length of window material depending on
wavelength). It requires also a very high precision in the manufacturing
of such liquid cell and the window flatness and paralleism. Typically the
coherence length of e.g. silica varies from 2.9 μm at 0.8 μm fundamental
wavelength to 18.1 μm at 1.9 μm. It means that the precision should be
better than 1000 Å in first case to get an error less than 7 % for comparable $\chi^{(3)}/\Delta\varepsilon$ values with those of window material (this requirement will
be higher in the case of lower $\chi^{(3)}/\Delta\varepsilon$ values for liquid).

As in general on has to perform the THG measurements in function of
concentration the error will depend on it.

One can, of course, imagine a liquid cell which can be taken into pieces
and the input and output windows set up in such a way (e.g. by rotating
them) to get a zero emission from each one independently. Aside a practical difficulty the principal drawbak is precision. Let $E_{G1}^M = 0.1 \; E_L^M$,
where superscript M refers to the maximum value. The amplitude of harmonic
field generated in window will be about 100 times smaller than that in
liquid thus difficult to detect, whereas for a filled cell $I_{3\omega} \sim (E_G \pm E_L)^2$
giving an error of about 20%. With a similar reasoning for the output
window one gets a larger cumulative error.

229

2.3. Plane parallel liquid cell

Kajzar and Messier [11] have used a plane parallel liquid cell for THG measurements in liquids. The interaction length variation is obtained by rotation of the cell around an axis perpendicular to the beam propagation direction (cf. Fig. 2). In this case Equ. (5) holds with dependence of all fields (E_{G1}, E_L and E_{G2}) as well as of transmission factors on the rotation (incidence) angle Θ. One obtains quite complicated harmonic intensity dependence on Θ examplified in Fig. 3. This can be resolved numerically giving all parameters of interest. The present technique is very similar to the wedged liquid cell technique with wedge shaped windows and is characterized by the same number of free parameters. In the present case these are: input and output window thickness ℓ_{G1}, ℓ_{G2}, liquid compartment thickness ℓ_L, refractive index dispersions in window material and in liquid (Δn_G and Δn_L respectively) $\chi^{(3)}$ of liquid and its phase. The number of free parameters can be diminished by doing measurements on empty cell. It allows to determine independently ℓ_{G1}, ℓ_{G2} and Δn_G. Filling up the cell in situ with studied liquid these parameters can be kept constant (cf. Kajzar and Messier [13]. The technique is much less sensitive to the quality of window polishing and their flatness as it is the case of wedged liquid cell. This is essentially due to the fact that only a limited volume of the cell is experimented by the laser beam in contrast to other techniques translating the cell. It gives not only the modulus of $\chi^{(3)}$ but also its phase. At the same time the fitting procedure yields precisely the coherence length of liquid.

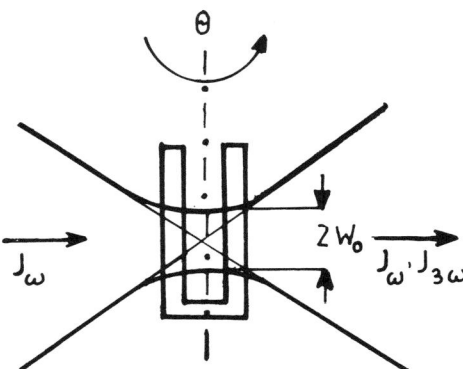

FIGURE 2. Principle of third harmonic generation in a plane parallel liquid cell. The cell is rotated along an axis perpendicular to the beam propagation direction (y).

The sum in Equ. (5) can be represented graphically in a complex plane (cf. Fig. 4) and the resultant field X and Y components can be obtained by summing corresponding projections of E_{G1}, E_{G2} and E_L on X, Y axes. As it is seen from Equ. (6) the harmonic fields generated by a plane wave in any

230

nonlinear medium is a subtense of a circle whose radius depends on $\chi^{(3)}/\Delta\epsilon$ value. When translating the wedge (or rotating the cell) all vectors change giving rise to complicated harmonic intensity dependence on x (or Θ). Of course, in the summation of the harmonic fields projections one has to take account of transmission factors (cf. Eq. (5)) which in the case of rotated cell give rise to the envelope function (cf. Fig. 3).

The above considerations are true if the liquid cell is placed under vacuum. In air one has to take account of extra harmonic fields generated before and after liquid cell by focused laser beam.

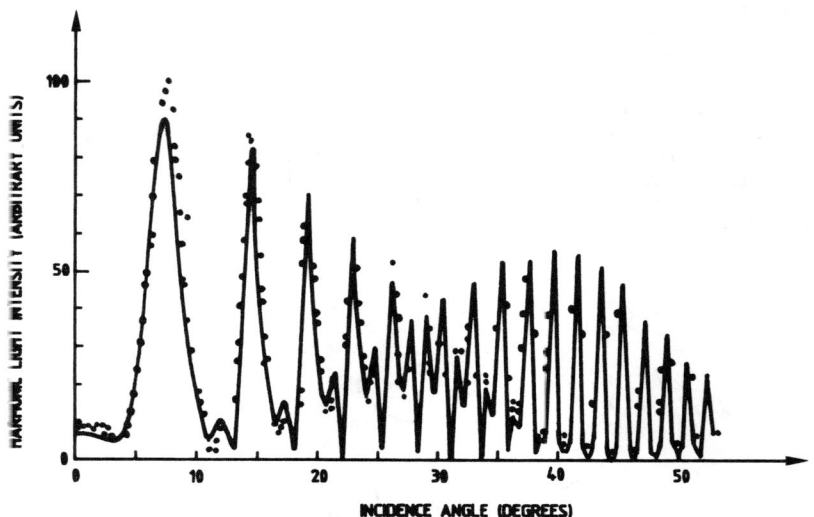

FIGURE 3. Harmonic intensity as a function of incidence angle from a rotated liquid cell. Points show experimental values and solid line the calculated ones.

3. ENVIRONMENTAL EFFECTS

In order to understand the mechanism of harmonic generation in air before and after the liquid cell as well the principle of operation of other THG in liquid techniques we will discuss shortly the formalism of harmonic generation in an infinite medium by focused laser beam.

According to Ward and New [14] the resultant harmonic field at a point (x,y,z) for the Gaussian shape fundamental beam propagating in the y-direction

$$E_{3\omega}(x,y,z) = \frac{4\pi P_{NL}\ I(y)}{\Delta\epsilon}\ e^{ik_{3\omega}y}\ e^{-\frac{3(x^2 + z^2)}{W_o^2\ (1 + i\tau)}} \qquad (11)$$

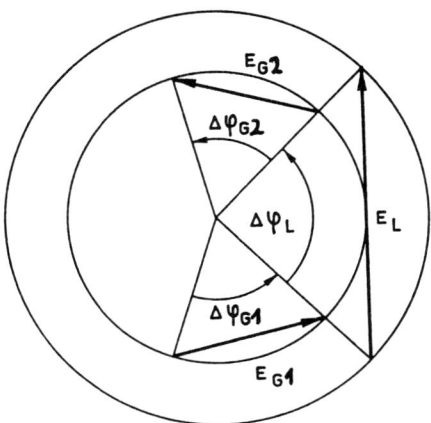

FIGURE 4. Graphical representation in complex plane of the sum of harmonic fields generated in input (G1), output (G2) windows and in liquid compartment (L). The resultant harmonic field X and Y components are obtained by summation of projections of corresponding harmonic fields on X and Y axes with corresponding transmission factors.

where P_{NL} is nonlinear polarization in the medium (cf. Equ. (9)),

$$\tau = \frac{2y}{w_0^2 \, k_\omega} \tag{12}$$

with $2W_0$ the beam waist (cf. Fig. 5)

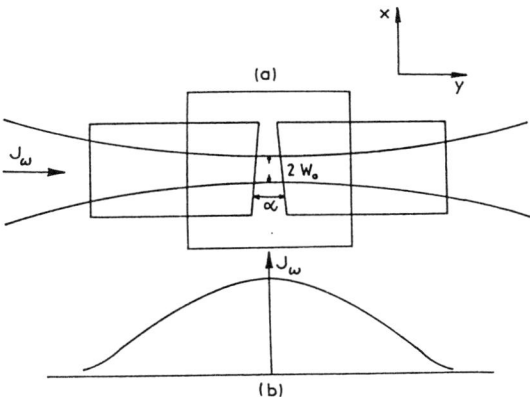

FIGURE 5. Principle of operation of a wedged liquid cell with thick windows in a focused laser beam (a) and schematic representation of light beam intensity distribution along the cell (b) (ref. [18]).

The function I(y) in Equ. (11) is a sum of all contributions from the points between the source (y = −∞) and the point y :

$$I(y) = \frac{3\omega}{c} \, \Delta n \int_{-\infty}^{y} dy' \, \frac{e^{i\Phi}}{(1 + i\tau')^2} \tag{13}$$

where the phase mismatch

$$\Phi = \frac{3\omega}{c} \, (y' - y) \, \Delta n \tag{14}$$

The function I(y) is antisymmetric with respect to the focus position (y = 0). It means that there is no resultant harmonic field at the output of an infinite medium:

$$\int_{-\infty}^{0} dy' \frac{e^{i\Phi}}{(1+i\tau')^2} = - \int_{0}^{\infty} dy' \, \frac{e^{i\Phi}}{(1+i\tau')^2} \tag{15}$$

The resultant harmonic field at the point y can be thus considered as a sum of small contributions, increasing in amplitude when approaching the focal point. Such sum can be presented graphically in the complex plane as it is shown in Fig. 6; each elementary vector representing an element of the above sum. Because of the y dependence of the length of these elementary vectors, they will describe a developping spiral when approaching the focal point and a closing spiral when leaving it, as it is shown in Fig. 6. The corresponding resultant fields are the sums of these elementary vectors (vectors OA and OA', respectively). It is seen also that the corresponding fields are bound waves (cf. Oudar [15]). Thus, as we mentionned before, at the output of an infinite medium the resultant harmonic field is equal to zero. Breaking the symmetry (cf. Equ. (15)) by putting at the focal point a sample this is no more true and both: the harmonic fields generated before and after the sample will interfere with that generated in the liquid cell, modifying significantly the harmonic intensities.

|a| |b|

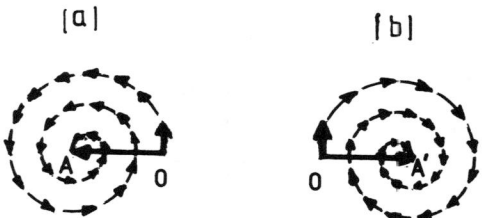

FIGURE 6. Graphical representation of the sum of "elementary" harmonic fields generated in an infinite medium by a focused laser beam before (a) and after (b) the focal point (0). Resultant fields are respectively OA and OA' vectors.

The influence of air in THG intensities have been first observed by Meredith et al [10] and studied in details by Kajzar and Messier [11]. The air contribution depends on the focal length, liquid cell thickness and the operation wavelength. Kajzar and Messier [11] have given an experimental correction based on the above formalism, which allows to take account of air contribution. In Fig. 7 the modification of Maker fringes from a rotated nonlinear plane parallel slab is shown at three extreme cases: air contribution smaller, equal and greater than the harmonic field generated in the slab itself. Besides change in the harmonic intensity one observes also an important modification of the envelope function and in the second case (Fig. 7b) even a complete cancellation (no harmonic signal at all !).

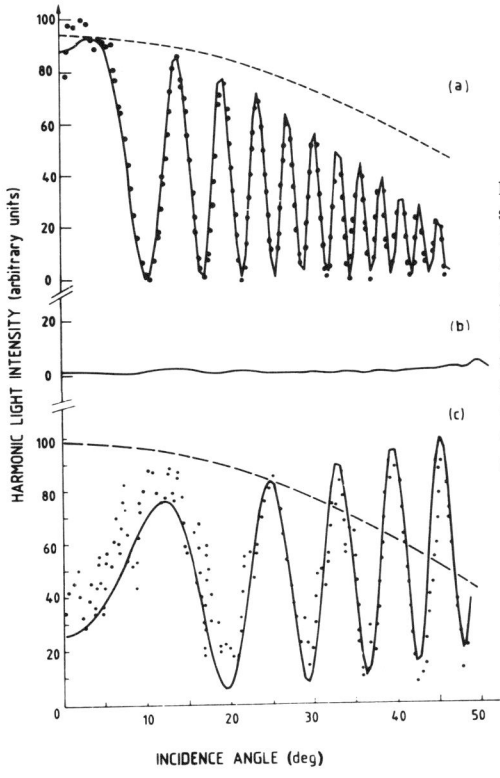

FIGURE 7. Harmonic intensity from a silica plate in air in function of incidence angle for (a) small and (c) large air contribution compared to the field generated in plate. Dashed line shows enveloppe function in vacuum. Points are experimental values and solid line shows the calculated ones. Figure (b) shows a (simulated) complete cancellation of harmonic fields generated in air and in plate (ref. [11]).

4. THG TECHNIQUES USING THE PRINCIPLE OF HARMONIC GENERATION IN AN INFINITE MEDIUM

4.1. Long chamber liquid cell

Meredith et al [16] (see also Stevenson and Meredith [17]) have proposed an originally designed liquid cell whose principle of operation is based on harmonic generation in an infinite medium by focused laser beams. The liquid compartment is long enough (~ 6 cm) that this principle applies to THG inside. The input window is wedge shaped in first version [16]. The harmonic field generated in output window is negligible compared to those created in liquid compartment and in the input window. By translating the liquid cell perpendicular to the beam propagation direction only harmonic field generated in the input window varies. The resultant harmonic field is the sum of harmonic fields generated in input window and in the liquid compartment. In this version the cell has to be kept in vacuum. In the latter version [17] the authors used thick input window. In this case the harmonic field generated in such a window (see discussion below) does not depend on translation and the measurements can be done in air.

At the same time the authors divide the cell into two parts. One compartment is filled with a reference liquid and the second one with the measured one. The technique is very simple and give precise results ($\sim$ 1% precision in $\chi^{(3)}_{xxxx}$ value [17]. The only drawbacks of the method are:

1. Because of large optical interaction in liquid compartment ($\sim$ 6 cm) it applies only for nonabsorbing liquids, at both fundamental and harmonic frequencies.

2. One needs an independent measurement of the coherence length l_c. In fact the precision in $\chi^{(3)}$ depends also on the precision in l_c determination.

4.2.Modified wedge technique

Kajzar and Messier [18] – [19] used a liquid cell with thick (2 ÷ 3 cm) input and output windows and a wedge shaped thin (100 ÷ 200 µm) liquid compartment (cf. Fig. 5)). The wedge angle is very small, typically 0.3 ÷ 0.7 degrees. Both windows can be considered as infinite media and the formalism described before (Eqs. (11)-(14)) fully applies. For a not very tightly focused incident light the laser beam in liquid compartment can be considered as a plane wave. Consequently the harmonic field generated in liquid is given by Equ. (6).

Let us denote by E_{G1} the amplitude of harmonic field generated in the input window:

$$E_{G1} = \pi \left\{ \frac{\chi^{(3)}}{\Delta\varepsilon} \right\}_G \left\{ E_\omega t_\omega \right\}^3 \tag{16}$$

where

$$t_\omega = \frac{2}{1 + n_\omega^G} \tag{17}$$

is transmission factor on air/glass interface. The harmonic field generated in the output window is given by

$$E_{G2} = - E_{G1} e^{i \phi_\omega^L} (t_\omega^{GL} t_\omega^{LG})^3 \tag{18}$$

where similarly as before ϕ_ω^L is phase mismatch in liquid compartment

$$\phi_\omega^L = \frac{6\pi\ell(x)}{\lambda_\omega} n_\omega^L \cos \Theta_\omega^L \tag{19}$$

and $\ell(x)$ is given by Equ. (10).

Thus at the output of liquid cell the resultant harmonic field will be given by (we take in factor the phase factors in both windows)

$$E_{3\omega}^R = E_{G1}\, e^{i\,\psi_{3\omega}^L}\, t_{3\omega}^{GL}\, t_{3\omega}^{LG}\, t_{3\omega}^{GO} + E_L\, \{t_\omega\, t_\omega^{GL}\}^3\, t_{3\omega}^{GO} - E_{G2}\, e^{i\,\psi_\omega^L}$$

$$\{\, t_\omega\, t_\omega^{GL}\, t_\omega^{LG}\, \}^3\, t_{3\omega}^{GO} \tag{20}$$

where

$$t_{\omega(3\omega)}^{ij} = \frac{2n_\omega^j(3\omega)}{n_\omega^i(3\omega) + n_\omega^j(3\omega)} \tag{21}$$

are corresponding transmission factors between i and j media at fundamental (ω) or harmonic (3ω) frequency (0 refers to air). Thus the harmonic intensity at output of the liquid cell is given by

$$J_{3\omega} = \frac{64\pi^4}{c^2}\, |\, \{ \frac{\chi^{(3)}}{\Delta\varepsilon} \}_G\, e^{i\psi_{3\omega}}\, A_G\, |^2\, |\, t_{3\omega}^{GL}\, t_{3\omega}^{LG} - e^{i\Delta\psi}\, \{ t_\omega^{LG}\, t_\omega^{GL}\, \}^3\, +$$

$$\rho\, \{ e^{i\Delta\psi} - 1 \}\, \{ t_\omega^{GL}\, \}^3\, |^2\, \{ t_{3\omega}^{GO}\, \}^2\, (t_\omega)^6\, J_\omega^3 \tag{22}$$

where

$$\rho = \{ \frac{\chi^{(3)}}{\Delta\varepsilon} \}_L\, /\, \{ \frac{\chi^{(3)}}{\Delta\varepsilon} \}_G\, |\, \frac{A_L}{A_G}\, |^2 \tag{23}$$

where A_L and A_G arise from boundary conditions and practically $(A_L/A_G) = 1$ and

$$\Delta\psi = \psi_\omega^L - \psi_{3\omega}^L \tag{24}$$

is phase mismatch in liquid compartment. By translating the liquid cell perpendicular to the beam propagation direction only the harmonic field generated in the liquid compartment varies with x. For close liquid and window material refractive indices the harmonic intensity is approximated by a very simple formula:

$$J_{3\omega} \sim (1 - \rho)^2\, J_\omega^3\, \mathrm{SIN}^2\, \frac{\Delta\psi}{2} \tag{25}$$

giving rise to Maker fringes with constant amplitude (in a transparent medium) in the function of translation x (cf. Fig. 8a).

236

Ir a medium absorbing at fundamental and/or harmonic frequency the
corresponding indices are complex and the phase mismathes ψ^2_ω, $\psi^2_{3\omega}$ contain
a non-zero imaginary part giving rise to a damping term. The amplitude of
harmonic intensity is no more constant in function of translation x with
non-zero minima (cf. Fig. 8b). Complex are also corresponding transmission
factors t_ω, $t_{3\omega}$. However in practice the complex part of refractive index
is very small (but not the damping terms). Typically for a liquid cell
with optical pathlength of 100 μm and absorbance at harmonic frequency of 2
being the maximum acceptable value, the imaginary part of refractive index
at 1.064 μm operation wavelength is equal to 1.3 x 10^{-3} which is negligible
compared to the real part.

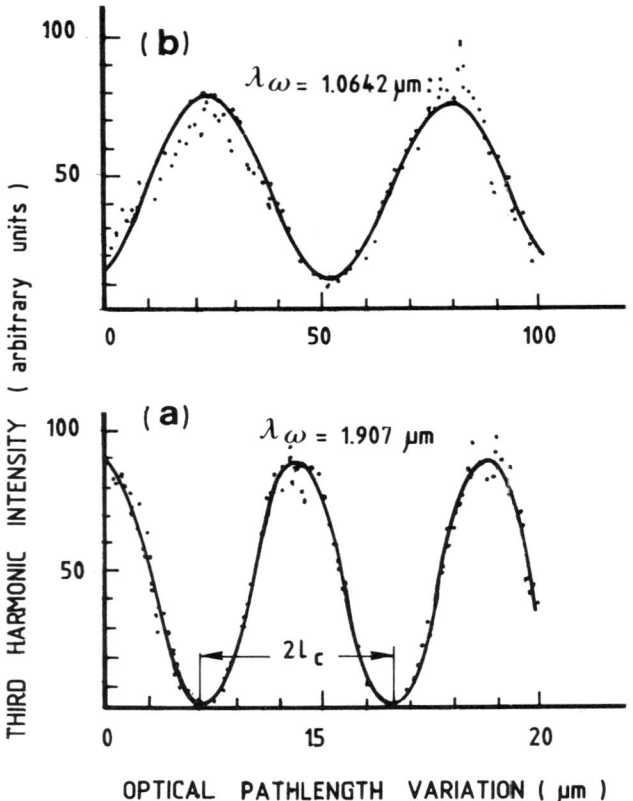

FIGURE 8. Harmonic light intensity in function of translation (optical
pathlength variation) from the wedged liquid cell with thick windows for
the case of liquid transparent (a) and absorbing at harmonic frequency
(b).

For $\chi^{(3)}$ determination of liquids we have to use a solvent as standard solvent. In this case the harmonic intensities can be directly normalised to those of a standard (5) giving

$$J_{3\omega}^N = J_{3\omega}^L/J_{3\omega}^S = |e^{i(\psi_{3\omega}^L - \psi_{3\omega}^S)}|^2 \ |t_{3\omega}^{GL} \ t_{3\omega}^{LG} - e^{i\Delta\psi} (t_\omega^{LG} \ t_\omega^{GL})^3 + \rho(e^{i\Delta\psi}-1)$$

$$(t_{GL}^\omega)^3 \ |_L^2 \ / \ | \ (t_{3\omega}^{GL} \ t_{3\omega}^{LG} - e^{i\Delta\psi} (t_\omega^{LG} \ t_\omega^{GL})^3 + \rho \ (e^{i\Delta\psi} - 1) \ (t_{GL}^\omega)^3 \ |_S^2 \quad (26)$$

5. THIRD HARMONIC GENERATION IN SOLUTIONS

For a solution of two species: solvent (s) and solute (p) with the relative mass concentration $C = m_p/m_s$, where m_p and m_s are corresponding masses we can write

$$\chi^{(3)} = N_p \ f_p \ \gamma_p + N_s \ f_s \ \gamma_s \quad (27)$$

where N_p and N_s is respectively number of solute and solvent molecules, γ_p, γ_s are corresponding molecular hyperpolarizabilities and f_p, f_s are local field factors.

In Lorenz-Lorentz approximation

$$f_{p(s)} = \{ \frac{n_\omega^2 + 2}{3} \}_{p(s)}^3 \ \{ \frac{n_{3\omega}^2 + 2}{3} \}_{p(s)} \quad (28)$$

The quantities N_p and N_s can be expressed in terms of mass concentration C yielding

$$\chi^{(3)} = \frac{f(C) \ d(C) \ N_A}{1 + C} \ \{ \frac{C\gamma_p}{M_p} + \frac{\gamma_s}{M_s} \} \quad (29)$$

where $d(C)$ is density, $f(C)$ local field factor; both depending on concentration. M_p and M_s are corresponding molecular masses and N_A is the Avogadro number.

In deriving Equ. (29) we have assumed the same local field factors for solute and solvent molecules (cf. discussion by Oudar [15] and Kajzar et al [21]).

In the dilute limit $1/(1+C) = 1-C$ and Equ. (29) can be rewritten in a more convenient form

$$\chi^{(3)} (C) = C \ \chi_p^{(3)} + (1-C) \ \chi_s^{(3)} \quad (30)$$

where we have kept only linear terms in C. Equation (30) is valid only in dilute limit. At higher concentrations one has to use Equ. (29).

6. CASE OF COMPLEX HYPERPOLARISABILITY

The quantum mechanical formula for cubic susceptibility responsible for THG process and obtained by the time dependent perturbation method in

238

dipolar approximation (22) - (23) reads :

$$\chi^{(3)} \ (-3\omega;\omega,\omega,\omega) \ \alpha \ \sum_g \sum_{nfn'} \ q(g) \ \Omega_{gn}\Omega_{nf}\Omega_{fn'}\Omega_{n'g}$$

$$x \ \frac{1}{(E_{ng} - 3\omega) \ (E_{fg} - 2\omega) \ (E_{n'g} - \omega)} + \frac{1}{(E_{ng} + \omega) \ (E_{fg} - 2\omega) \ (E_{n'g} - \omega)}$$

$$+ \frac{1}{(E_{ng} + \omega) \ (E_{fg} + 2\omega) \ (E_{n'g} - \omega)} + \frac{1}{(E_{ng} + \omega) \ (E_{fg} + 2\omega) \ (E_{n'g} + 3\omega)} \Bigg]$$

(31)

where $q(g)$ is the fundamental state density function, $\Omega_{ij} = \langle i|e\vec{r}|j\rangle$ are dipolar moment transition matrix elements, $\omega_{ij} = (E_i - E_j)/\hbar$ is the energy difference between i and j states in $\hbar$ units. (n, n' are allowed and f forbidden for one photon transition levels, cf. Fig. 10).

Equation (31) shows that in the case when the photon energy, its double or triple matches one of the excited level of unperturbed system with a corresponding symmetry (selection rules) a resonance enhancement will occur in $\chi^{(3)}$ $(-3\omega;\omega,\omega,\omega)$.

Consequently the cubic susceptibility will be complex. The one and three photon resonances will occur when the fundamental or respectively harmonic wavelength falls in the absorption band and it can be easily foreseen. This is no more true for the two photon resonance which can occur for both fundamental and harmonic wavelength lying in the transparency range. Therefore we can not foreseen apriori whether or not $\chi^{(3)}$ is real and even if it is, the sign of $\chi^{(3)}$. Thus a more detailed analysis of $\chi^{(3)}$ is required. Because of interference between different fields generated in a liquid cell the resultant harmonic intensity will depend not only on the modulus of liquid $\chi^{(3)}$ but also on its phase (we assume $\chi^{(3)}$ of window to be real and positive what can be assured by a subsequent choice of window material). In Fig. 9 we have displayed the TH intensity as a function of $\chi^{(3)}$ phase ϕ keeping constant its modulus. It shows that the measurements at one concentration only can be erronous if we are not sure of the phase of $\chi^{(3)}$. In order to check it one has to do the THG measurements in function of solute concentration.

From Equ. (25) it is seen also that for real and positive $\chi^{(3)}$ of liquid there exist two solutions; the correct one can be obtained from concentration dependence of normalized harmonic intensity.

Introducing a complex $\chi^{(3)}$ in Equ. (22) one obtains following dependence of the normalized harmonic intensity as a function of mass concentration C.

$$J_N(C) = J_{3\omega}(C)/J_{3\omega}(o) = \frac{1}{\beta^2} \ (\ | \ (A+B(C) \ \{ \ \frac{C\gamma'_p}{M_p} + \frac{\gamma_s}{M_s} \ \} \ | ^2$$

$$+ \ C^2 \ \{ \ \frac{\gamma''}{M_p} \ B(C) \ \}^2 \ ) \ / \ \{ \ A + B(o) \ \frac{\gamma_s}{M_s} \ \}^2$$

(32)

where

$$A = t_{3\omega}^{GL} t_{3\omega}^{LG} - \alpha\beta \ e^{i\Delta\psi} \ \{ \ t_{\omega}^{GL} t_{\omega}^{LG} \ \}^3$$

(33)

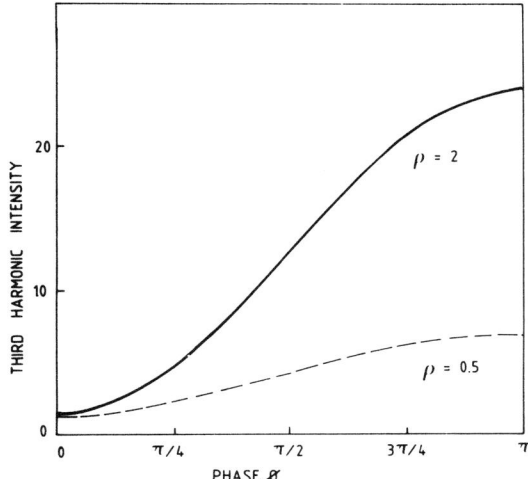

FIGURE 9. Calculated harmonic intensity dependence on $\chi^{(3)}_L$ phase of liquid for two different values of $\rho = |\chi^{(3)}_L|/\chi^{(3)}_G$. For the sake of simplicity the refractive indices of fundamental and harmonic frequencies are assumed to be the same as in window material (silica).

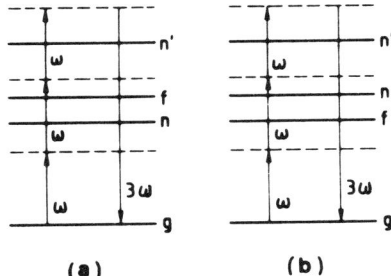

FIGURE 10. Schematic representation of third harmonic generation process with two photon level above (a) and below (b) one photon state. Solid lines show fundamental (g) and excited unperturbed states whereas dashed lines represent virtual levels.

240

and

$$B(C) = \frac{f(C)\ d(C)\ N_A}{(1 + C)_G \chi^{(3)}} \quad (\alpha\beta e^{i\Delta\psi} - 1)\ (t_\omega^{GL})^3\ t_\omega^{LG}\ \frac{(\Delta\epsilon)_G}{(\Delta\epsilon)_L} \tag{34}$$

Depending on relative ratios of real and imaginary parts of solution with respect ot those of window material this dependence will be different.

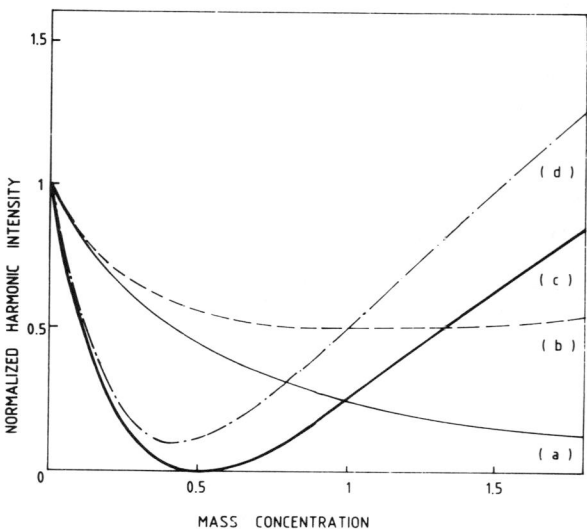

MASS CONCENTRATION

FIGURE 11. Mass concentration dependence of harmonic intensity for different values of real and imaginary parts of $\rho_p = \chi^{(3)}_p / \chi^{(3)}_G$ (p-solute, G-window) and $\rho_s = 2(\rho_s = \chi^{(3)}_s / \chi^{(3)}_G$, s-solvent). a - $\rho'_p = 1$, $\rho''_p = 0$; b - $\rho'_p = -1$, $\rho''_p = 1$, c - $\rho'_p = -1$, $\rho''_p = 0$; d - $\rho'_p = -1$, $\rho''_p = 1$. $\chi^{(3)}_G$ and $\chi^{(3)}_s$ are assumed to be real and positive.

The coefficients α and β in Equ. (32) - (34) take account of absorption at fundamental and harmonic frequency, respectively and read

$$\alpha = e^{3\omega K_\omega \ell/c} \tag{35}$$

$$\beta = e^{3\omega K_{3\omega} \ell/c} \tag{36}$$

where K_ω, $K_{3\omega}$ are imaginary parts of refractive indices at ω, 3ω frequency, respectively.

In Figs 11 and 12 we have shown theoritical $J_N(C)$ dependence for two different cases $\rho = 0.5$ ($\rho = \chi^{(3)}_L / \chi^{(3)}_G$) and $\rho = 2$, respectively. For the

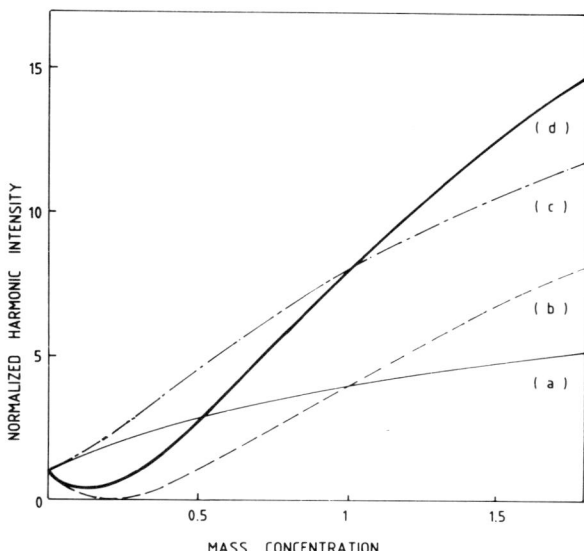

FIGURE 12. Mass concentration dependence of harmonic intensity for larger values of real and imaginary parts of ρ_p : a – $p'_f = 4$, $\rho''_p = 0$; b – $\rho'_p = -4$, $\rho''_p = 0$; c – $p'_p = 4$, $p''_p = 4$, d – $\rho'_p = -4$, $\rho''_p = 4$. Other details as in fig. 11.

sake of simplicity we have assumed the same refractive indices for liquid and window material. Different curves show the calculated normalized intensity variation in function of mass concentration for several values and signs of real and imaginary parts of solute $\chi^{(3)}$ (with respect to that of solvent). It is seen that for a negative real part of solute $\chi^{(3)}$ one expects a strong decrease of $J_N(C)$. The non zero imaginary part of $\chi^{(3)}_L$ gives rise to a strong dependence of $\chi^{(3)}$ on mass concentration. Thus from the concentration dependence of normalized harmonic intensity one gets the modulus and the phase of solute $\chi^{(3)}$.

In figs. 13 and 14 we have shown as an illustration the concentration dependence of the normalized harmonic intensity for merocyanine solutions in ethanol at two different wavelengths: 1.064 and 1.907 μm [24], respectively. At 1.064 μm fundamental wavelength the normalized harmonic intensity decreases at small concentration and strongly increases at larger concentrations. Such a behaviour is characteristic for a complex $\chi^{(3)}$ with negative real part (cf. Figs. 11-12). At 1.907 μm $J_N(C)$ varies slowly with mass concentration C, and the corresponding susceptibility is real and positive (cf. Table I).

242

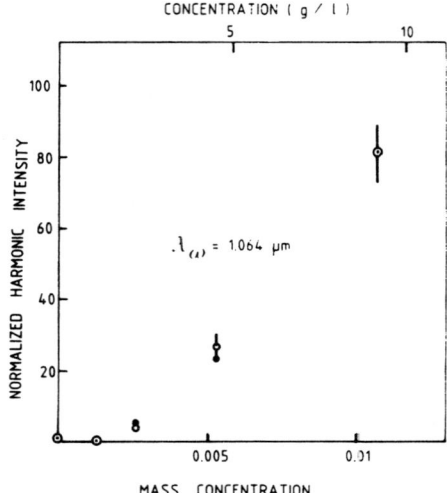

FIGURE 13. Mass concentration variation of normalized harmonic intensity from merocyanine solutions in ethanol at 1.064 μm.

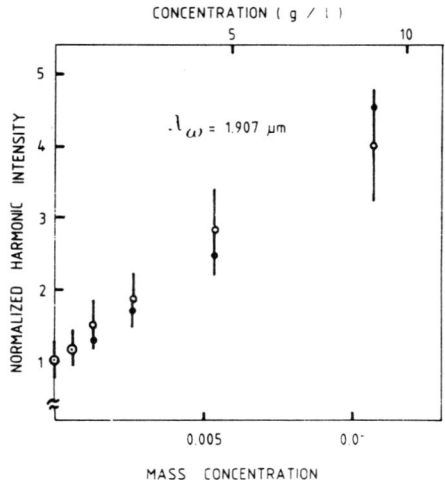

FIGURE 14. Mass concentration variation of normalized harmonic intensity from merocyanine solutions in ethanol at 1.907 μm.

7. SUMMARY

In this paper we have described different THG techniques for liquids. These are very useful for the study of molecular hyperpolarizability and at the actual stage of development they allow a rapid and precise determination of γ for soluble molecules. The most extensively described wedge technique with thick input and output windows gives not only the modulus but also the phase of γ and allows the THG measurements for liquids absorbing at fundamental and/or harmonic frequencies. In Table I we have listed the values of real and imaginary parts of γ for several solvable conjugated molecules using this technique.

The THG technique coupled with EFISHG gives both γ and β [8]. Both THG and EFISHG measurements can be done simultaneously using the same liquid cell.

The knowledge of the phase of γ is of both fundamental and practical importance. In fact, γ is complex in the vicinity of electronic and/or vibronic transition. Thus the knowledge of γ phase allows to position corresponding excited states. Of special interest are the states with the same symmetry as fundamental one where direct transitions are forbidden. These give rise to a two photon resonant enhancement in γ (cf. Equ. (31)) for fundamental wavelength lying in the transparency range. It leads also to an incident ligth intensity dependent complex refractive index

$$n = n_o + n_2 I \tag{37}$$

where

$$n_2 = 12 \ \pi^2 \ \chi^{(3)} \ (-\omega;\omega,\omega,-\omega) \ / \ n_o^2 c \tag{38}$$

Although the Kerr susceptibility $\chi^{(3)} (-\omega;\omega,\omega,\omega)$ is not the same as that responsible for THG process but it is also resonantly enhanced in the vicinity of the two photon transition and is also complex. For special cases of n_2 real and positive and n_2 real and negative the refractive index n will increase and decrease with incident light intensity, respectively. This can lead to different practical applications in e.g. switching systems. On the other hand a complex n_2 may lead to an active optical bistability.

TABLE 1. Molecular hyperpolarizability with real (γ') and imaginary (γ'') parts as determined by different techniques for some conjugated 1D molecules

| Molecule | Molecular hyperpolarizability in 10^{-33} e.s.u. | | | |
| | $\lambda_\omega = 1.0642$ μm | | $\lambda_\omega = 1.907$ μm | |
	γ'	γ''	γ'	γ''
Merocyanine in ethanol[a]	− 2.3	± 1.6	0.8	0
Merocyanine in propanol	− 0.6	± 2.1	0.4	0
Polydiacetylene TS-12[b]	− 1.7	− 1.1	1.2	0
Polydiacetylene TS-12[c] in $CHC\ell_3$	− 1.8	± 0.4		
Polydiacetylene TS-12[c] in DMF	− 2.0	± 2.3		
CS_2			0.0044	

a − Ref. 24
b − Ref. 13
c − Ref. 21
d − Ref. 10

REFERENCES

1. J. Zyss and D.S. Chemla in Nonlinear Optical Properties of Organic Molecules and Crystals, D.S. Chemla and J. Zyss eds, Academic Press, Orlando 1987, pp 23-191.
2. J. Ducing, in Nonlinear Optics, NATO ASI, P.G. Harper and B.S. Wherett eds, Academic Press, London 1975, pp 11-45.
3. P.N. Prasad, Nonlinear Optical Interactions in Polymer Thin Films, SPIE Proceedings, Vol. 682, Molecular and Polymeric Optoelectronic Materials: Fundamentals and Applications, Washington 1987, pp. 120-124.
4. F. Kajzar and J. Messier, in Nonlinear Optical Properties of Organic Molecules and Crystals, D.S. Chemla and J. Zyss eds., Academic Press, Orlando 1987, pp. 51-83.
5. J. Zyss, J. Mol. Electron. 1, 25-45 (1985).

6. J.F. Nicoud and R.J. Twieg, in Nonlinear Optical Properties of Organic Molecules and Crystals, D.S. Chemla and J. Zyss eds., Academic Press, Orlando 1987, pp. 227-296.
7. S.K. Kurtz and T.T. Perry, J. Appl. Phys. $\underline{39}$, 3798 (1968).
8. M. Barzoukas, P. Fremaux, D. Josse, F. Kajzar and J. Zyss, in Nonlinear Optical Properties of Polymers, A.J. Heeger, J. Orenstain and D.R. Ulrich eds., Materials Research Society Symposium Proceedings, vol. 109, Pittsburgh 1988.
9. D.S. Chemla and P. Kupecek, Revue Phys. Appl. $\underline{6}$, 31-50 (1971).
10. G.R. Meredith, B. Buchalter and C. Hanzlik, J. Chem. Phys. $\underline{78}$, 1533-42 (1983).
11. F. Kajzar and J. Messier, Phys. Rev. $\underline{A32}$, 2352-2363 (1985).
12. M. Thalhammer and A. Penzkofer, Appl. Phys. $\underline{B32}$, 137 (1983).
13. F. Kajzar and J. Messier, in Polydiacetylenes Structure and Electronic Properties. NATO ASI series, series E: Applied Sciences, n° 102. D. Bloor and R.R. Chance eds, Martinus Nijhoff Publ., Dordrecht 1985 pp. 325-334.
14. J.F. Ward and G.H.C. New, Phys. Rev., $\underline{185}$, 57-72 (1969).
15. J.L. Oudar, J. Chem. Phys. $\underline{67}$, 446-457 (1977).
16. G.R. Meredith, B. Buchalter and C. Hanzlik, J. Chem. Phys. $\underline{78}$, 1543-51 (1983)
17. S.H. Stevenson and G.R. Meredith, in Molecular and Polymeric Optoelectronic Materials, Proceedings SPIE, Vol. 682, pp. 147-152.
18. F. Kajzar and H. Messier, Rev. Sci. Instrum. $\underline{58}$, 2081-85 (1987).
19. F. Kajzar and J. Messier, J. Opt. Soc. Am. B, $\underline{4}$, 1040-46 (1987).
20. The author is grateful to Dr E. Chauchard for bringing him in mind the error concerning omission of a transmission factor in Equ. (15), ref. (18).
21. F. Kajzar, I. Ledoux and J. Zyss, Phys. Rev. $\underline{A36}$, 2210-19 (1987).
22. J.F. Ward, Rev. Mod. Phys. 37, 1-18 (1965).
23. B.J. Orr and J.F. Ward, Mol. Phys. $\underline{20}$, 513-26 (1971).
24. F. Kajzar, Complex Hyperpolarizability and Solvatochromic Effects in Merocyanine Solutions, Proceedings of Symposium on Nonlinear of Organics and Semiconductors, Tokyo, July 25-26, 1988, Springer Verlag (in print)

MEASUREMENT OF THE NON-LINEAR REFRACTIVE INDEX OF SOME METALLOCENES BY THE OPTICAL POWER LIMITER TECHNIQUE.

C.S.WINTER, S.N.OLIVER AND J.D.RUSH.

British Telecom Research Laboratories,
Martlesham Heath,
Ipswich IP5 7RE,
England.

1.INTRODUCTION.

Although many measurements have been made on the optical non-linearities of organic materials (1), few such studies have been carried out on organo-metallics. Those that have been published have been limited to second order phenomena (2,3). However many of the molecular phenomena exploited to give large non-linearites in organics can be used with organo-metallics. We have thus begun a program to explore a range of organo-metallic systems for their third order non-linearities. In this paper the results obtained on a number of metallocenes are summarised, and the mechanism and size of the response discussed.

2.PROCEDURE.

The materials listed in table 1 were used as-purchased from Aldrich, with the exception of bis(trimethylsilyl) ferrocene (BTMSF in the tables)which was synthesised as described in reference 4. The materials were measured in the molten state; to minimise decomposition during the measurement procedure they were initially degassed and then sealed under helium prior to melting. Some decomposition occurred with the hafnocene and zirconocene dichlorides. Bis(trimethylsilyl) ferrocene is a liquid at room temperature, and was measured at 20°C. A solution of ferrocene in ethanol was also studied for comparison.

The measurement technique was based on that developed by Soileau *et al* (5-7); in this method the onset of optical power limiting is used to characterise the material. Information on the size of the response and, in some cases, the mechanism can be obtained. The basic technique consists of tightly focusing a spatially gaussian beam into the sample and measuring the throughput through the lens/pin-hole system shown in figure 1. When the power of the incident radiation is increased beyond a critical threshold self-focusing occurs; in a pulsed laser system this results in a time-dependent movement of the focus point. The movement of the focus point defocuses the second lens and pins the output at the critical power. Although the formation of the filament arises only from the non-linear refractive index, the observed power limiting is a combination of both non-linear refractive effects and non-linear absorption in the high intensity filament. The latter leads to material breakdown, which is manifested as a bright flash or streamer in the material, upon self-focusing. If measurements are made in the absence of the pin-hole a distinct break in the plot of input-output power is still seen, although in this case it is due only to non-linear absorption in the self-focused region.

Other mechanisms apart from self-focusing (which is associated with a positive n_2 coefficient) can lead to a similar limiting phenomena. However these all result from *intensity*-dependent mechanisms e.g. two photon absorption, thermal effects arising from linear absorption and other mechanisms that give rise to a negative n_2

J. Messier et al. (eds.), Nonlinear Optical Effects in Organic Polymers, 247–251.

coefficient. They are thus distinct from *power*-dependent self-focusing (5-8) associated with a positive n_2 and can be distinguished by varying the focal length of the input lens. The area of the beam at the focal spot scales as the square of the input lens' focal length, and thus the intensity scales similarly. Should an intensity dependent phenomena occur then varying the lens should vary the critical power; this is discussed further in the results section and is a powerful tool in unravelling the mechanism.

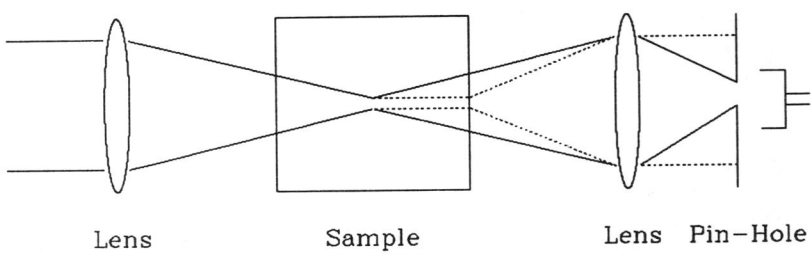

Lens   Sample   Lens   Pin−Hole

FIGURE 1. Schematic of the apparatus showing lenses, pin-hole and sample. The dotted line shows the effect of self-focusing.

If the mechanism arises from a positive n_2 coefficient then it is possible to derive the value of n_2 from the limiting power. This requires numerical solutions of the non-linear wave equations. For a tightly-focused gaussian beam Marburger (8) has derived the relationship between n_2 and the critical power given below -

$$P_{c2} = 3.72\ c\lambda^2/32\pi^2 n_2 \quad \text{(esu)}$$

where P_{c2} is the critical power, n_0 the linear refractive index, λ the wavelength and n_2 the non-linear refractive index defined by -

$$n = n_0 + n_2 <\xi^2>$$

where $<\xi^2>$ is the time-averaged optical electric field. Hence there is a direct relationship between the measured critical power for self-focusing and the non-linear refractive index.

For the experiments described below a 10 ns, 1.06 μm Nd:YAG was used (Quartel 585) operating in the TEM_{00} mode, with a 7mm beam diameter. All the input lenses were of a singlet 'best-form' construction. The beam energy was monitored on a shot-shot basis using pyroelectric energy meters, the temporal characteristics were monitored by fast silicon detectors. The spatial quality of the beam was confirmed using a multiple-element silicon photodiode; this permits single shot monitoring of the spatial profile of the beam. No evidence was seen of higher order modes on the traces taken.

3.RESULTS AND DISCUSSION.

Measurements were made on four molten metallocenes, a liquid derivative of ferrocene and a solution of ferrocene in ethanol. The results are given in Table 1 for all the materials. Results on ferrocene and ruthenocene were carried out at more than one lens focal length, and the results are presented in the same table. Figure 2 shows a typical data plot for ruthenocene. The critical power and optical power limiting can be clearly seen. In general the critical power can be obtained with an accuracy of 15-20%. Results are also given for nitrobenzene as a comparison. The nitrobenzene result is in good agreement with literature values of n_2 (5,9).

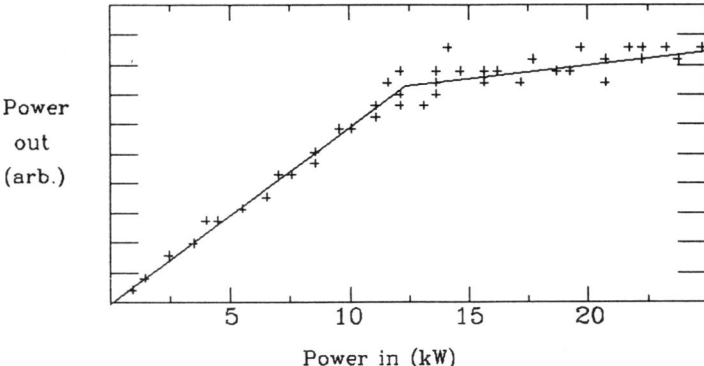

FIGURE 2. Typical Experimental Data for Ruthenocene (75mm lens)

TABLE 1. Measured values of P_c and calculated n_2's.

Material	Lens (mm)	P_c (kW)	n_0	n_2 (esu)
Nitrobenzene	75	25	1.56	1.6×10^{-11}
Ferrocene (soln.)	500	1500	1.36	2.6×10^{-13}
Ferrocene (melt)	75	28		1.48×10^{-11}
BTMSF	50	26	1.55	1.51×10^{-11}
BTMSF	75	28		
BTMSF	150	25		
Ruthenocene	75	12		3.3×10^{-11}
Ruthenocene	150	15		
Hafnocene dc	75	30		1.3×10^{-11}
Zirconocene dc.	75	38		1.0×10^{-11}

The results on molten, solution and liquid forms of ferrocene can be compared by calculating the molecular hyperpolarisability, γ, for each example. The molecular hyperpolarisability is related to n_2 by -

$$n_2 = 4\pi\chi^{(3)}/n_0$$

and

$$\gamma = \chi^{(3)}/L^4N$$

where N is the number density of molecules, L the local field correction and n_0 the linear refractive index. Assuming a simple Lorentz model, we calculate L as being $(n_0^2 + 2)/3$. The refractive index for the molten metallocenes was estimated to be similar to that of bis(trimethylsilyl) ferrocene obtained from reference 9. The results are summarised in Table 2. Note that large errors can occur due to the approximate refractive index used to calculate L.

TABLE 2. Calculated values for γ for the ferrocene samples.

Material	Molecs/cm³	γ (esu)
Ferrocene (soln)	5×10^{19}	1.2×10^{-32}
Ferrocene (melt)	3×10^{21}	3.9×10^{-32}
BTMSF	2×10^{21}	6.4×10^{-32}

The results for ferrocene, in its different forms, and for ruthenocene are consistent with the mechanism arising from a positive n_2; as can be seen by inspecting the lens, and thus intensity, dependence of the critical powers measured. In all cases the phenomena are power dependent, and, as discussed earlier this rules out contributions from thermal effects, two photon absorption and negative n_2.

The most obvious contribution to a positive n_2 in a liquid is molecular re-orientation, as is believed to occur in nitrobenzene and carbon disulphide. Since the laser used in these experiments delivers 10 ns pulses and the molecular relaxation time in a bis(trimethylsilyl)ferrocene can be estimated from the Debye formula at about 500 ps this seems the most likely mechanism. However there are two pieces of evidence that conflict with this view. First, the viscosity of the melt phase for all the metallocenes studied is much greater than the liquid form of ferrocene and the corresponding rotational re-orientation times are similarly lengthened; and secondly, the results presented below on the polarisation dependence of n_2 conflict with previous theory and experimental results on other systems.

The simplest theory predicts that the ratio of the n_2's for linear and circular polarised light in an isotropic medium should be 4, for molecular re-orientation (10). Experimental studies on nitrobenzene and carbon disulphide have measured a ratio in the region 1.9-2.2 (6,7), and all theories agree that the linear value should be greater than the circular one. In solids, where electro-striction and electronic effects dominate, values of 1.1-1.3 have been observed (11). Table 3 shows the results for bis(trimethylsilyl) ferrocene and ruthenocene. In these materials values of 0.6-0.8 have been measured.

TABLE 3.Circular vs Linear Polarisation Data.

Material	P_c (circ)	P_c (lin)	$n_2(\text{lin})/n_2(\text{circ})$
Nitrobezene	58	25	2.1
BTMSF	18	26	0.7
Ruthenocene	7	12	0.6

We suggest here that the most likely explanation for the observed data is that the mechanism is electronic rather than rotational. This needs further study and a picosecond system is currently being set-up to measure the response at other timescales, shorter than the projected molecular rotation.
For device potential it is necessary to consider both the size of the non-linear coefficient and the linear absorption. The only material for which the absorption has been measured is the liquid ferrocene derivative, where α is about 1 cm^{-1}. The melt phases have not been measured; however bis(trimethylsilyl) ferrocene is a dark yellow-brown liquid, whereas ruthenocene is a pale pink colour and thus the absorption should be much less.

4.CONCLUSIONS

The result, if correctly interpreted, is interesting since it would be a large electronic effect for a non-polymeric organic system. However further work is needed to clarify the origin of the anomalous ratio of linear to circular polarised light responses. The observed values of n_2 are useful in suggesting that large effects may be achieved in organo-metallic systems which have all the advantages of organics for molecular engineering and further range of variable properties based on the metallic component. The optical power limiter technique is a simple, useful way of screening materials for reasonable size coefficients and obtaining some information on the mechanism on the non-linear response.

5.ACKNOWLEDGEMENTS.

The Director of Research and Technology British Telecom plc for permission to publish this paper.

6.REFERENCES

1. Chemla DS and Zyss J: Nonlinear Optical Properties of Organic Molecules and Crystals. London: Academic Press, 1987
2. Frasier CC, Harvey MA, Cockerham MP, Hand HM, Chauchard EH and Lee CH: J.Phys.Chem., **90**, 1986, 5703.
3. Calabrese JC and Tam W: Chem.Phys.Lett., **133**, 1987, 244.
4. Marr G and White TM: J.Chem.Soc.(C), 1970, 1789.
5. Soileau MJ, FRanck JB and Veatch TC: NBS Spec., **620**, 1980, 385.
6. Soileau MJ, Wlliams WE and Van Stryland EW: IEEE J.Quant.Elect. **QE-19**, 1983, 731.
7. Wlliams WE, Soileau MJ and Van Stryland EW: Opt.Commun., **50**, 1984, 256.
8. Marburger JH: Prog.Quant. Elect., **4**, 1975, 35.
9. Weast (ed.): Handbook of Chemistry and Physics 60th Ed.. Florida: CRC Press, 1980.
10. Shen YR: Phys.Lett., **20**, 1966, 378.
11. Feldmann A, Horowitz D and Waxler RM: NBS Spec. Publ., **372**, 1972, 92..

POLYMERIZATION AND X-RAY STRUCTURE OF NEW SYMMETRICAL AND UNSYMMETRICAL DIACETYLENES

M. BERTAULT, L. TOUPET

Laboratoire de Physique Cristalline, UA 804 du CNRS, Université de Rennes I, Campus de Beaulieu, 35042 - Rennes Cedex (France)

and A. COLLET, J. CANCEILL

Chimie des Interactions Moléculaires, ER 285 du CNRS, Collège de France, 11, place Marcelin-Berthelot, 75231 - Paris Cedex 05 (France)

1. INTRODUCTION

The topochemical polymerization of certain crystalline diacetylenes R-C≡C-C≡C-R' such as PTS (where R is R' : CH_3-Ph-SO_2-CH_2O) and analogous compounds yields macroscopic crystals of conjugated polymer chains extended along crystallographic directions [1]. Systems of this type are potentially useful for the design of linear and non-linear optical devices [2].

Unsymmetrical diacetylenes are a potential source of non-centrosymmetrical crystals. In connection with previous study in this area [3,4], we report some preliminary solid-state properties of a new series of symmetrical and unsymmetrical polymerisable derivatives of 2,4 - hexadiyne 1,6-diol (1). These compounds R^1OCH_2-C-C-C-C-CH_2OR^2 have in common the presence of a benzyl urethane side group R^1 which has not been tried so far for this purpose (R^1 is $PhCH_2$-NHCO).

2. POLYMERISATION

The preparation and the eventual recrystallisation of the symmetrical (3) named BCDA (where R^2 is R^1) and unsymmetrical corresponding diacetylenes (4)-(13) obtained from the mono-urethane (2) have been already described [5] : these compounds are listed in the table and, with exception of (11), derivatives (2)-(13) were found to polymerize in the crystalline state upon thermal annealing or exposure to light, U.V. or γ-radiation. (13) has incomplete polymerization.

$$R^1\text{-}OCH_2\text{-}C{\equiv}C\text{-}C{\equiv}C\text{-}CH_2O\text{-}R^2$$

Table . Structures and properties of new diacetylenes and related compounds.

	R^1	R^2	M.p./°C	Polym.
(1)	H	H	113	
(2)	$PhCH_2$-NHCO	H	69	(+)
(3)	$PhCH_2$-NHCO	=R^1	140	(+)
(4)	$PhCH_2$-NHCO	Ph-SO_2	71	(+)
(5)	$PhCH_2$-NHCO	p-$MeC_6H_4SO_2$	68	(+)
(6)	$PhCH_2$-NHCO	p-$CF_3C_6H_4SO_2$	99	(+)
(7)	$PhCH_2$-NHCO	p-$FC_6H_4SO_2$	90	(+)
(8)	$PhCH_2$-NHCO	p-$ClC_6H_4SO_2$	87	(+)
(9)	$PhCH_2$-NHCO	p-$MeOC_6H_4SO_2$	75	(+)
(10)	$PhCH_2$-NHCO	p-$NO_2C_6H_4SO_2$	98	(+)
(11)	$PhCH_2$-NHCO	p-$NO_2C_6H_4NHCO$	146	(−)
(12)	$PhCH_2$-NHCO	p-$MeC_6H_4SO_2NHCO$	168	(+)
(13)	$PhCH_2$-NHCO	S-$PhCH^*(Me)NHCO$	73	(+)

253

J. Messier et al. (eds.), Nonlinear Optical Effects in Organic Polymers, 253–256.
© 1989 by Kluwer Academic Publishers.

The isothermal polymerisation of BCDA, as the one under γ-ray irradiation, exhibits kinetics showing the absence of an induction period. The corresponding time conversion curves (polymer content vs annealing time) are shown on figure 1-a) and b). This behaviour is similar to that of MBS (where R^1 is R^2 : $p\text{-MeOC}_6\text{H}_4\text{SO}_2$), but contrasts with that of PTS and on several of its congeners which display polymerization kinetics with a typical S-shaped time conversion curve (shown in figure 1.c)) characterized by a comparatively long induction period ($X\leqslant.10$) followed by a fast autocatalytic range. The unsymmetrical diacetylenes (6) and (8) polymerize readily with similar kinetics than BCDA, whereas the corresponding symmetrical bis (sulphonyl)esters (where R^1 is R^2) are unreactive.

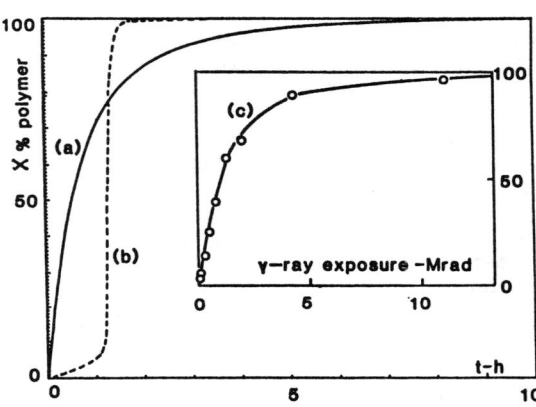

Fig.1 :time conversion curves for BCDA (a) and PTS (c) isothermal polymerization and BCDA (b) γ—ray induced polymerization

There is a "polymerogenic" effect of the benzyl urethane side groupe R^1 which has the ability to promote the solid state polymerization of unsymmetrical diacetylenes even if the analogous symmetrical are not reactive. This high reactivity seems to be due to their large positive activation entropies (+5, +11 and +8 cal mole^{-1}K^{-1} for respectively BCDA, (6) and (8)) as compared with the large negative value for PTS (-11cal mol^{-1}K^{-1}in the autocatalytic range[7]), which almost offset the influence of their activation enthalpies. This fact seems also be demonstrated on the basis of X-ray structural data of BCDA.

3. STRUCTURAL DATA OF BCDA

Small plates of BCDA were studied on an automatic diffractometer at 143K (monomer) and at room temperature (polymer) after annealing of the monomer crystal at 60°C for 48hrs. Both structures were isomorphous[10], triclinic, space group $P\bar{1}$, Z=2. The lattice parameters are :

monomer : a = 5.79Å , b = 7.73Å, c = 22.02Å
          = 101.85°, = 100.2°, = 90.40°, V = 949Å^3
polymer : a = 5.63Å, b = 7.96Å, c = 21.32Å
          = 99.40°, = 92.65°, = 90.30°, V = 941Å^3

Neither the monomer nor the polymer showed any phase transition from room temperature to 130K.

Projections of the monomer and the polymer structures on the plane containing one of the ab diagonals(the direction of polymerization) and the diacetylene rods are shown on figure 2.a) and b) : the packing parameters d = 4.82Å, Φ = 47° and R = 3.77Å fulfill the geometrical criteria for polymerizability.

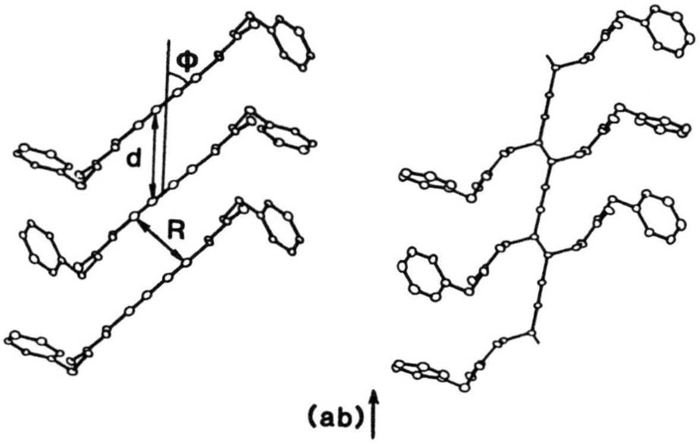

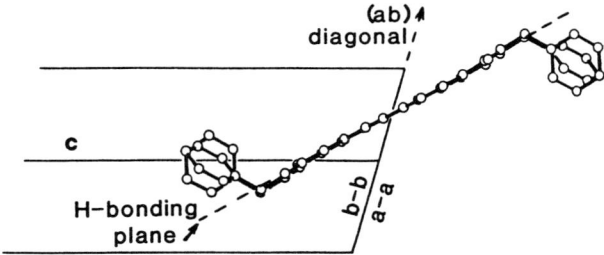

Fig.2:projection of the crystal structure of BCDA on the plane containing (ab)

diagonal and diacetylene rods:(a) monomer and (b) polymer

Each monomer molecule is H-bonded to the neigh-bouring molecules along planes containing the second ab diagonal (perpendicular to the first ab diagonal, the stacking direction) and a direction not far from c axis (figure 3) : the H-bonding planes form flat ribbons that are very roughly perpendicular to the plane of polymerization (plane of figure 2.a)) which is near the plane of figure 3.

Fig.3:projection of the crystal structure of monomer BCDA on the plane

perpendicular to the H-bonding plane

The important feature, which perhaps explains the high reactivity of BCDA, is the fact that the polymerization direction is different from, and independant of the H-bonding network which holds the molecules together and probably governs the packing. As a consequence, the H-bonding pattern is almost completely unaffected on going from the monomer to the polymer, as shown in figures 4.a) and b) : the reaction can proceed smoothly to completion.

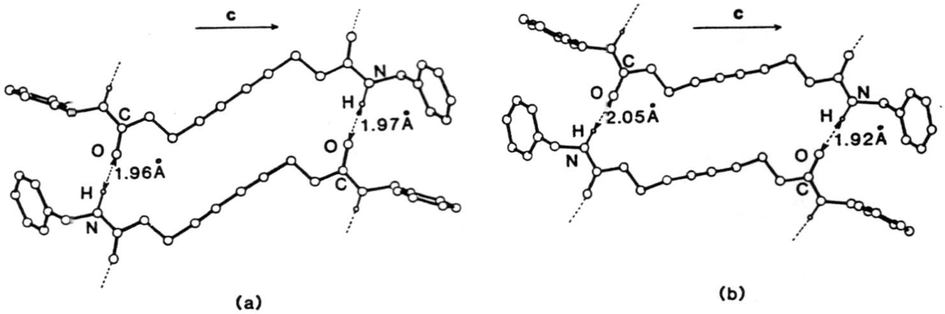

Fig.4:H-bonding planes in monomer (a) and polymer (b) BCDA

This behaviour is in contrast with that of other bis (phenylurethanes) analogous of BCDA, as HDPU[8] (R^1 is $CONHC_6H_5$) or HDCPU[9] (R^1 is $CONHC_6H_4cl$) in which the polymerization direction and the H-bonding network coincide : these compounds polymerize partially, because the structural changes that occur during the reaction take place within the H-bonding planes and the packing of the unreacted monomers progressively deviates from the conditions allowing polymerization.

REFERENCES

1. For recent reviews, see : "polydiacetylenes", edited by Cantow H.J., Adv. Polym. SCi. , 63 (1984) ; "polydiacetylenes", edited by Bloor D. and Chance R.R., NATO ASI Series, Martinus Nijhoff Publishers, 1985.
2. For recent review see : "non-linear properties of organic molecules and crystals", eds. Chemla D.S. and Zyss J. , Academic Press, vol. 1-2, 1987.
3. Bertault M., Toupet L., Canceill J. and Collet A. ,Makromol. chem. , Rapid Commun.8, 443 (1987).
4. Strohriegl P. , Makromol. chem. , Rapid Commun., 8, 437 (1987).
5. Bertault M. , Canceill J. , Collet A. and Toupet L., J. Chem. Soc. , Chem. Comm. , 163 (1988).
6. Ando D.J. , Bloor D. , Hubble C.L. and Williams R. , Makromol. Chem. , 181 , 453 (1980).
7. Bertault M. , Schott M. , Brienne M. - J. and Collet A. , chem. Phys. , 85 , 481 (1984).
8. Whuler A. , Spinat P. and Brouty C. , Acta Cryst. , Sect. C, 40 693 (1983).
9. Brouty C. , Spinat P. and Whuler A. , Acta Cryst. , Sect. C, 39, 594 (1983) ; 40 , 1619 (1984).
10. All the crystallographic data have been deposited at the Cambridge Crystallographic Data Centre.

THIRD–ORDER NONLINEAR GUIDED–WAVE DEVICES

GEORGE I. STEGEMAN, RAY ZANONI, K. ROCHFORD, and COLIN T. SEATON

Optical Sciences Center, University of Arizona
Tucson, Arizona, USA 85721

SUMMARY: The strong beam confinement provided by integrated optics waveguides, combined with the nonresonant third–order nonlinearities available in nonlinear polymeric films, makes them ideal for all–optical switching devices.   The operating characteristics of such devices are discussed, and the material requirements are compared with those available from polymeric materials.   Recent advances in grating devices are summarized.

1. INTRODUCTION

Third–order optical nonlinearities have been under intensive study recently for a number of potential applications to all–optical signal processing.   These applications can be separated into two well–defined areas.   The first is called parallel processing and has been identified with parallel optical computing and image identification [1].   The second, serial processing, has been targeted for ultra–fast all–optical processing of high–speed serial information [2].   The material requirements and competing material systems are quite different for the two cases.

For parallel processing, two–dimensional arrays of all–optical logic functions are implemented with arrays of parallel nonlinear elements, each of which acts as a small etalon [1].   A number of interrelated parameters are required for this application [1], and they are best deduced from a simple example.   First, a minimum nonlinear phase shift of approximately $\pi/10$ is needed to switch each element, at a total power of 10 mW per element.   Assuming an optimum cross–sectional area of 2 μm^2 and the maximum nonlinearity obtained for nonlinear organics to date, namely $n_2 \cong 10^{-14}$ m^2/W near and on resonance [3], then the required etalon thickness is $\cong$ 1 mm.   However, for this value of nonlinearity, the attenuation coefficient is $\cong 10^4$ cm^{-1}, which means that the throughput is negligible.   This should be compared with the case for GaAlAs, in which milliwatt switching powers have been achieved with etalon thicknesses of a few micrometers and throughputs of tens of percent [4].   In addition, the SEED devices also perform at these power levels [5].   There are, however, some guided–wave approaches, coupled with nonresonant nonlinearities, which could substantially improve the case for nonlinear organics.   The only clear advantage that organics have is in terms of speed, which is not

257

J. Messier et al. (eds.), Nonlinear Optical Effects in Organic Polymers, 257–276.

believed to be a critical issue for parallel processing. Nevertheless, at present it does not appear that nonlinear organics are competitive with semiconductors for all-optical parallel processing applications.

The situation is just the opposite for serial processing applications [2]. Here, the nonlinearity recovery time, which limits the processing speed, should be sub-picosecond and currently this is only available from nonlinear organic materials and glasses (whose nonlinearity is small). This will take all-optical processing of serial data streams into a speed regime inaccessible to electronic devices. Here guided-wave approaches are needed because the large nonlinear phase shifts necessary, varying from π to 4π, require long propagation distances [2]. High throughput dictates nonresonant nonlinearities, and no nonlinear material system can compete with organics in this respect [6].

Essentially three types of waveguides can be used. The most common are optical fibers, in which the transverse waveguide dimension traps light to a radial cross-section of typically a few micrometers. Picosecond all-optical switching utilizing a number of different phenomena has recently been implemented in fibers [7-9]. If nonlinear polymers could be introduced into the fiber, switching powers could be reduced from the present kilowatt levels to sub-watt powers. Planar waveguides that offer confinement in one dimension are usually only the first step toward the channel waveguides that are crucial to low-power switching. A number of nonlinear phenomena have been observed in planar waveguides, including the results of the first experiments involving nonlinear organics [2,6,10-15]. Switching devices have also been implemented in channel waveguides of semiconductor-based materials [16-19]. Ultimately, nonlinear polymers appear to be the most promising waveguide systems for sub-picosecond all-optical switching.

Waveguides can be fabricated with any approach that leads to a region of high index surrounded by media of lower index. All of the standard film-deposition techniques, such as MBE, CVD, thermal and e-beam evaporation, and sputtering, can and have been used. When dealing with organic systems, there are additional possibilities for making waveguides since molecular structure can be modified by electromagnetic waves, electron bombardment, and so on [20]. Guided waves can be excited by externally incident radiation fields, using a number of techniques, the most popular being distributed coupling by means of prisms or gratings, or end-fire coupling. These approaches are summarized in Fig. 1.

In this paper we discuss a variety of all-optical guided-wave devices. In particular, the method of analysis leading to device operating parameters is presented, and the corresponding material requirements and figures of merit summarized. Essentially all of the all-optical device responses are shown to belong to just a few generic types. Experimental progress toward implementing such devices in a variety of material systems is discussed, and the potential for nonlinear organics to improve device performance is highlighted.

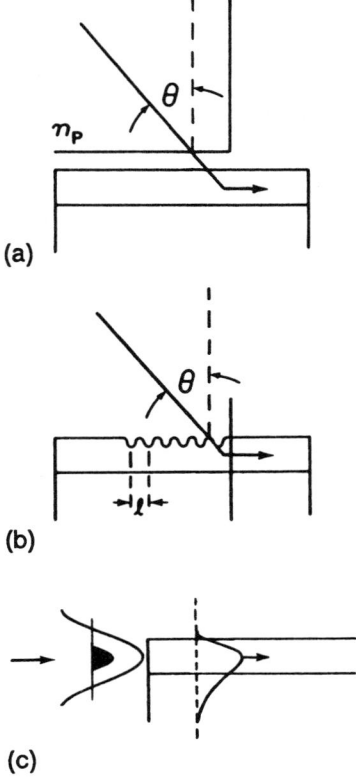

Fig. 1. Techniques for coupling an external radiation field into optical waveguides: (a) prism coupling; (b) grating coupling; (c) end-fire coupling.

2. GENERIC NONLINEAR GUIDED-WAVE INTERACTIONS

Over the last fifteen years a variety of integrated optics devices have been developed for the coupling, modulation, and switching of light. The basic phenomena involved can be understood either in terms of wavevector matching between coupled guided-wave fields, or in terms of interference between two guided waves whose relative phase has been modulated externally (typically by means of the electro-optic effect). All of these devices can be operated in an all-optical mode by utilizing nonlinear materials in the waveguiding regions. To date, grating and prism couplers [11-15,21-27], grating reflectors [6,28], directional couplers [16-19,29-33], Mach-Zehnder interferometers [34,35], and mode sorters [36,37] have all been considered. The device which has drawn the most theoretic attention is the directional coupler, because it is a four-port device with two input and output channels [16-19,29-33]. A single input signal can be routed to either of the two outputs, depending on the intensity of the input signal; or the second input channel can be used to control the choice of output channel. These are very useful characteristics for switching, optical logic, and multiplexing. However, most of the experimental work has centered on the nonlinear coupling between a guided wave and an externally incident radiation field using a prism or a grating [11-15,21-27]. This phenomenon has proven very useful for measuring the magnitude and sign of the film nonlinearity, as well as determining its physical origin [15,38,39]. Here we discuss the operational characteristics of a variety of all-optical guided-wave devices, concentrating on these two cases.

The key point in all of these devices, when operated in an all-optical mode, is that the guided-wave wavevector, βk_0 ($k_0 = \omega/c$) depends on the guided-wave power; that is, $\beta \rightarrow \beta(P)$. This is a consequence of at least one of the waveguiding media having an intensity-dependent refractive index. Nonlinear organics are believed to obey the classic Kerr-law form for the nonlinearity, namely

$$n = n_i + n_{2i}I ,\tag{1}$$

where I is the local intensity and the subscript i identifies the i-th medium. The net effect of such media on a guided wave requires appropriate averaging of the nonlinearities over the field distributions. First we write the guided-wave field distribution for the simplest case of TE modes as

$$E(r,t) = \frac{1}{2}\, \hat{e}\, f(x,y)\, a_{gw}(z)\, \exp[j(\omega t - \beta k_0 z)] + c.c. ,\tag{2}$$

where the transverse field distribution $f(x,y)$ is normalized so that $|a_{gw}(z)|^2 \equiv P$ is the guided-wave power. This leads, by means of coupled-mode theory, directly to an effective index for the guided wave, given by [40]

$$\beta = \beta_0 + \Delta\beta_0 P ,\tag{3a}$$

where P is the guided-wave power, either in watts for a channel waveguide, or in watts/meter for a planar guide. A simple application of coupled-mode theory gives [40]

$$\Delta\beta_0 = \frac{c^2\epsilon_0^2}{4}\; \frac{\displaystyle\int_{-\infty}^{\infty}dx \int_{-\infty}^{\infty}dy\; n^2(x,y)n_2(x,y)\, f^4(x,y)}{\left|\displaystyle\int_{-\infty}^{\infty}dx \int_{-\infty}^{\infty}dy\; f^2(x,y)\right|^2} .\tag{3b}$$

All of the nonlinear materials in the waveguide are sampled in the integrations over the transverse dimensions. Therefore, the parameter $\Delta\beta_0$ is effectively the nonlinearity $[n_2(x,y)]$ averaged over the transverse intensity distribution. Implicit in Eq. (3) is the assumption that the field distributions are not a function of guided-wave power, which is not always the case, especially for weakly guiding systems. Corresponding results for TM modes can be found in Ref. 40.

A cumulative nonlinear phase shift $\Delta\phi^{NL} = \Delta\beta_0 k_0 LP$ occurs after a propagation distance L at power P. Therefore, if the high-power beam which undergoes a nonlinear phase shift is mixed with a reference beam whose phase does not change with power, the resulting interference effect becomes power-dependent. Also, since the wavevector is now power-dependent, wavevector matching (or phase matching) also becomes power-dependent. The resulting nonlinear response, based on an ideal Kerr-law medium such as a nonlinear polymer, of a variety of integrated optics devices is shown in Fig 2. Note that in each case the response of the device is tuned by varying the optical power. The sharpest switching characteristics are obtained for the nonlinear directional coupler and the nonlinear Bragg reflector. For other materials with a complicated nonlinear response, such as semiconductors, the all-optical response characteristics are much more complex because of loss and saturation of the achievable index change [41,42].

In almost every case shown in Fig. 2, there are two normal modes involved which are nonlinearly coupled, and which interfere with one another. Their coupling is described by a generalized version of the coupled-mode equations first introduced by Jensen [29]:

$$\mp j \frac{d}{dz} a_{gw1}(z) = \kappa e^{j\Delta\beta kz} a_{gw2}(z) + k_0[\Delta\beta_{11}|a_{gw1}(z)|^2 + G\Delta\beta_{12}|a_{gw2}(z)|^2]a_{gw1}(z) , \qquad (4a)$$

$$\pm j \frac{d}{dz} a_{gw2}(z) = \kappa e^{-j\Delta\beta kz} a_{gw1}(z) + k_0[\Delta\beta_{22}|a_{gw2}(z)|^2 + G\Delta\beta_{21}|a_{gw1}(z)|^2]a_{gw2}(z) , \qquad (4b)$$

where the $\pm$ signs refer to contradirectional and codirectional wave interactions respectively. κ is the coefficient that describes the coupling between the two guided waves, and it leads to a beat length of $2\pi/\kappa$ for codirectional interactions. $\Delta\beta k$ is the initial wavevector mismatch between the two modes identified with the linear device. The $\Delta\beta_{11}$ and $\Delta\beta_{22}$ parameters are the $\Delta\beta_0$ defined for the individual modes 1 and 2 through Eq. (3), and the nonlinear wavevector change induced in mode 1 resulting from field overlap with mode 2 (and vice versa) is given by

262

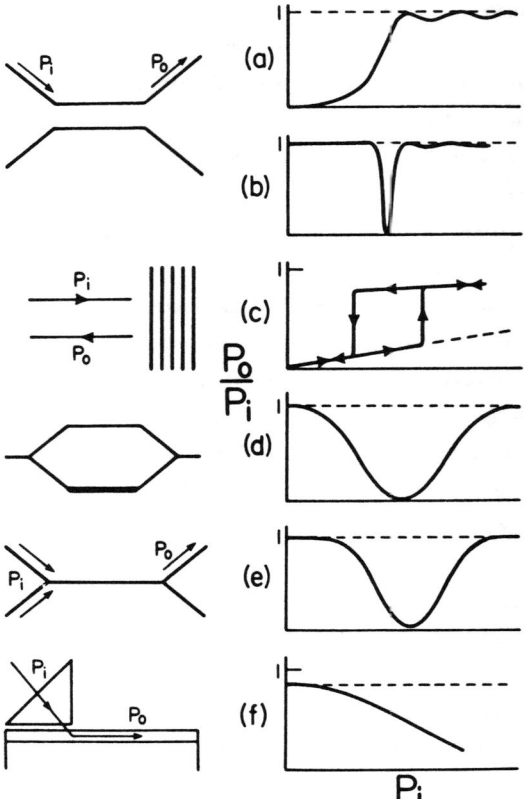

Fig. 2. Standard integrated optics devices and their response to optical power with and without nonlinearities: (a) 1/2 beat length directional coupler; (b) 1 beat length directional coupler; (c) distributed-feedback grating; (d) Mach-Zehnder interferometer; (e) mode sorter; and (f) prism coupler.

$$\Delta\beta_{12} = \Delta\beta_{21} = \frac{c^2\epsilon_0^2}{4} \frac{\int_{-\infty}^{\infty} dx \int_{-\infty}^{\infty} dy \; n^2(x,y)n_2(x,y) \; f_1^2(x,y)f_2^2(x,y)}{\left| \int_{-\infty}^{\infty} dx \int_{-\infty}^{\infty} dy \; f_1^2(x,y) \right| \left| \int_{-\infty}^{\infty} dx \int_{-\infty}^{\infty} dy \; f_2^2(x,y) \right|}. \tag{5}$$

Finally, G is a constant which depends on the nature of the nonlinearity and the polarization of the guided modes. For example, for two TE polarized waves and a Kerr-law nonlinearity, G = 2, and for two orthogonally polarized waves G = 2/3.

2.1 Nonlinear Distributed Coupler

Prism and grating couplers are routinely used to couple external plane-wave fields into guided modes, and vice versa. In both cases, it is necessary to match the incident wavevector parallel to the surface to βk_0. For a grating coupler, the additional wavevector component parallel to the surface is provided by the grating wavevector. In a prism coupler, the incident wavevector is enlarged by a factor n_p by virtue of the light being incident through the prism. A typical incident-beam/prism-waveguide geometry is shown in Fig. 3. Light is incident onto the base of the prism such that the wavevector component parallel to the base of the prism matches that of the guided wave to which coupling is desired. This is the "synchronous coupling condition." Light "tunnels" across the gap between the prism and the film, and the guided wave grows with propagation distance. Hence the term "distributed coupler;" that is, the coupling typically occurs over 100 μm to 1 cm distances along the waveguide surface. This technique requires pulsed lasers for studying the nonlinear properties of organic film waveguides, and hence the following analysis refers to temporal pulses. We write the incident field (for example, inside a prism of refractive index n_p) as [38]

$$E(r,t) = \frac{1}{2}\,\hat{e}_{in}a_{in}(z,t)\,\exp[j(\omega t - n_p k_0[\sin\theta z - \cos\theta x])] + c.c. \ . \tag{6}$$

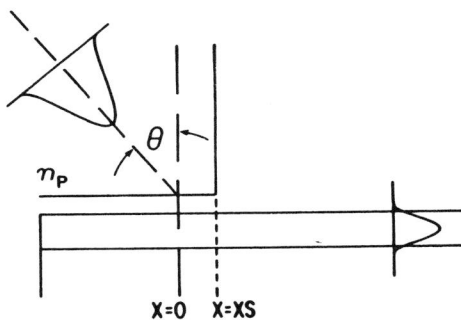

Fig. 3. The coupling geometry. A beam, Gaussian along the x-axis, is incident at an angle θ onto the prism base whose 90° corner is located at xs. A guided wave is generated and propagates for x > xs in the free waveguide.

Here $a_{in}(z,t)$ describes the temporal pulse envelope and spatial field distribution of the incident field amplitude along the guided-wave propagation direction at the base of the prism. (Henceforth everything will be normalized to unit distance along the wavefront y axis.) The transverse field distribution is given by $f(x)$, and $\beta(z,t)$ is the guided-wave wavevector whose dynamics in space and time will be determined by the nature of the nonlinearity. Coupled-mode theory leads to a modified version of Eq. (4a), namely

$$\frac{d}{dz} \, a_{gw}(z,t) = \hat{t} \, a_{in}(z,t) \, \exp[j\phi(z,t)] - \left[\frac{\alpha_{gw}(z,t)}{2} + \frac{1}{\ell}\right] a_{gw}(z,t) \, , \tag{7}$$

$$\phi(z,t) = [\beta(z,-\infty)k_0 - n_p k_0 \sin\theta]z + \phi^{NL}(z,t) \, , \tag{8}$$

where $\hat{t}$ is the transfer coefficient that takes into account all of the details of the coupling geometry; for example, gap thickness and the refractive indices of the various waveguide and prism materials. Furthermore, α_{gw} is the guided-wave power attentuation coefficient and ℓ is the characteristic distance for re-radiation into the prism from the guided wave. That is, the guided wave can couple out of the waveguide into the prism, corresponding to a loss mechanism for the guided wave.

The details of the evolution of $\phi^{NL}(z,t)$ in both space and time are determined by the nature of the nonlinearity and its time response. In a typical organic material, one would expect that both electronic and thermal nonlinearities would be present. For the Kerr-law response, we assume an n_{2e} whose "turn-on" and "turn-off" times are assumed to be instantaneous on the time scale of the pulses used. For the thermal nonlinearity, absorption of radiation leads to local heating and an index change through dn/dT. The thermally induced index change is written as

$$\delta n_t(x,z,t) = \frac{\partial n}{\partial t} \, \delta T(x,z,t) \, , \tag{9}$$

where δT is the local temperature change. Assuming, for simplicity, no spatial diffusion of the temperature change, which requires pulses shorter than 100 ns,

$$\frac{d}{dt} \, \delta T(x,z,t) = \frac{\alpha_0}{\rho C_p} \, I(x,z,t) - \frac{\delta T(x,z,t)}{\tau_t} \, , \tag{10}$$

where τ_t is the relaxation time for thermal effects (which will depend on the details of the waveguide geometry), and α_0, ρ, C_p are the attenuation coefficient, density, and heat capacity, respectively. For $\Delta t \gg \tau_t$, where Δt is the pulse width, the equilibrium value of the nonlinearity, in the absence of diffusion effects, can be defined as

$$n_{2t} = \frac{\partial n}{\partial T} \, \frac{\alpha_0 \tau_t}{\rho C_p} \, . \tag{11}$$

Note that although this expression neglects spatial diffusion effects, it does provide a useful order-of-magnitude estimate, especially when dealing with pulses short enough that diffusion essentially does not occur. In the other limit, $\Delta t \ll \tau_t$,

$$\delta n_t(x,z,t) = \frac{\partial n}{\partial T} \, \frac{1}{\rho C_p} \int_{-\infty}^{t} dt' \alpha_0 I(x,z,t') \, , \tag{12}$$

which is called an "integrating nonlinearity." The local index change accumulates over the duration of the pulse since the nonlinearity relaxation time is much longer than the pulse width. Substituting Eq. (10) into Eq. (12) gives

$$\delta n_t(x,z,t) = \frac{n_2}{\tau_t} \int_{-\infty}^{t} dt' I(x,z,t') \ . \tag{13}$$

Approximating the integral by $\Delta t \overline{I(x,z,t)}$ where $\overline{I(x,z,t)}$ denotes the time-averaged intensity,

$$n_{2eff} \cong n_{2t} \frac{\Delta t}{\tau_t} \ , \tag{14}$$

where clearly the effective nonlinearity encountered by a pulse of width Δt for the integrating case is reduced by the ratio of the pulse width to the relaxation time. Therefore, the effect of the thermal nonlinearity can be reduced relative to the electronic effect by decreasing the pulse width. The last case of interest here corresponds to $\tau_t \gg \Delta t \gg \tau_e$:

$$\delta n(x,z,t) = n_{2e} I(x,z,t) + \frac{n_{2t}}{\tau_t} \int_{-\infty}^{t} dt' I(x,z,t') \ . \tag{15}$$

The above equations, which describe the material dynamics, refer to local material properties. To obtain the changes in the waveguide properties, for example, the guided-wave effective index $\beta(z,t)/k_0$, it is necessary to average over the transverse field distributions $f(x)$[23], as discussed previously. Thus the guided-wave nonlinear phase shift is finally given by

$$\frac{\partial}{\partial x} \phi^{NL}(z,t) = A_e k_0 |a_{gw}(z,t)|^2 + A_t k_0 \frac{1}{\tau} \int_{-\infty}^{t} dt' |a_{gw}(z,t')|^2 \ , \tag{16}$$

with

$$A_{e,t} = \frac{\displaystyle\int_{-\infty}^{\infty} dx \ n_{2e,t}(x) |f(x)|^4}{\left| \displaystyle\int_{-\infty}^{\infty} dx \ f(x)^2 \right|^2} \ . \tag{17}$$

The nonlinear prism or grating coupler can be used to measure the sign and magnitude of the waveguide nonlinearity, as well as to differentiate between integrating (thermal) and electronic nonlinearities. A number of specific measurements can be made to obtain this information. Measuring the coupling efficiency as a function of incidence angle gives the magnitude and sign of the nonlinearity. An example of calculations of nonlinear grating coupling used to analyze polyamic-acid thin-film waveguides [15] is shown in Fig. 4. Note that there is a shift in the optimum coupling angle with increasing pulse energy, and that the direction of the shift gives the sign of the nonlinearity. This shift is linear with peak power as long as the shift is small compared to the angular width of the coupling efficiency curve. However, the shift obtained per unit power decreases rapidly when the peak shift becomes comparable to or greater than the angular width of the low-power coupling curve. There is also a decrease in the optimum coupling efficiency with increasing power. Note that although the magnitude and sign of n_{2e} or n_{2eff} can be evaluated, this measurement does not identify the nature of the nonlinearity (electronic versus thermal). If the incidence is maintained at the optimum coupling angle at low powers, Fig. 4 shows that the coupling efficiency decreases with increasing power, exhibiting limiter action. However, this measurement also is not capable of distinguishing between thermal and electronic nonlinearities. In fact, for experiments in which only one pulse width is available, the only way of differentiating between electronic and thermal effects is to measure the distortion in the in-coupled pulse shape. If the distortion remains symmetric in time about the peak power, the nonlinearity is electronic. If the distortion leads to an asymmetric pulse profile, the nonlinearity is thermal, with a relaxation time longer than the pulse width. Examples of asymmetric distortion are shown in Fig. 5.

Another approach to differentiating between nonlinear mechanisms is to perform the angular scan coupling experiments with two different pulse widths. The key idea is that the effective thermal nonlinearity decreases with decreasing pulse width. Based on typical polymer parameters, we estimate that $n_2/\tau \cong 2.5 \times 10^{-9}$ m^2/J, based on an absorption coefficient $\alpha_0 = 0.25$ cm^{-1}. Therefore, for 10 ns pulses, $n_{2eff} \cong 2.5 \times 10^{-17}$ m^2/W. Taking the nonlinearity for CS$_2$ as the reference ($n_2 \cong 4.5 \times 10^{-18}$ m^2/W), thermal effects would dominate the nonlinear response. However, for 100 ps pulses, $n_{2eff} \cong 2.5 \times 10^{-19}$ m^2/W and the electronic nonlinearity dominates. This is exactly the approach used recently to study polyamic-acid film waveguides, and to measure separately the thermal and electronic components in the waveguide.

2.2 Dual Directional Coupler

A dual directional coupler, as indicated in Figs. 2(a) and (b), consists of two parallel, identical channel waveguides. In a linear device, the light is usually injected into one channel and the light is coupled back and forth between the channels with propagation distance. The minimum distance required for all of the light to reappear in the incidence channel is called the beat length and is defined as $2\pi/\kappa$. For the nonlinear case [29], the

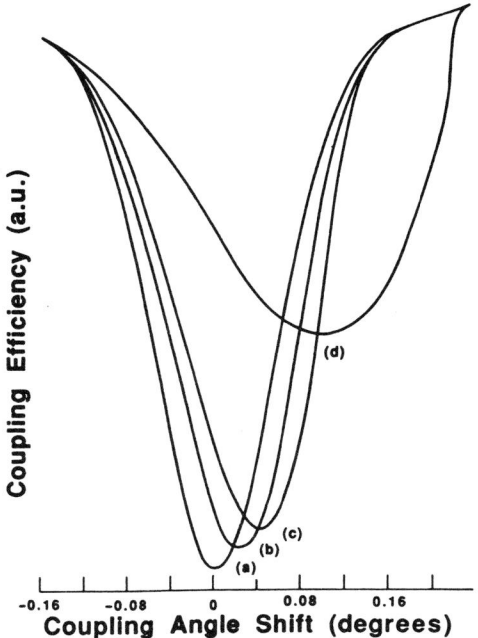

Fig. 4. Coupling efficiency versus coupling angle for grating coupling to the TE_0 mode of 0.6-μm-thick polyamic acid film. Here $n_2 = 4.5 \times 10^{-18}$ m^2/W, and the pulse energies are (a) 10^{-7} J/mm (very low powers); (b) 10^{-5} J/mm; (c) 2×10^{-5} J/mm; (d) 10^{-4} J/mm.

two channels are wavevector matched only when there is an equal amount of light in the two channels. Otherwise, the optical power produces a wavevector mismatch $k_0 \Delta\beta_0 (P_1 - P_2)$, $[\Delta\beta_{11} = \Delta\beta_{22}]$, and hence a change in the beat length. Note that usually the cross terms $\Delta\beta_{ij}$ $i \neq j$, are negligible and the $+$ sign in Eq. (4b) is appropriate for this case. The equations can be solved analytically to give

$$a^2_{gw1}(z) = a^2_{gw1}(0) \frac{1 + cn(2z,m)}{2} \, , \tag{18}$$

where m $= a^2_{gw1}(0)/P_c$ and cn( ) is the appropriate Jacobi elliptic integral. The critical power associated with the solutions is given by [29]

$$P_c = \frac{8\kappa}{k_0(\Delta\beta_{11} + \Delta\beta_{22} + 2G\Delta\beta_{12})} \, . \tag{19}$$

For a one-half beat length coupler, defined as the minimum length for which a low-power wave injected into one channel emerges completely from the parallel cross channel, the power-dependent response is shown in Fig. 2. The key point is that the transmission changes by 100% from its low-power to its high-power value in both channels, corresponding to an all-optical switch. The required switching power $P_s = 1.25 P_c$.

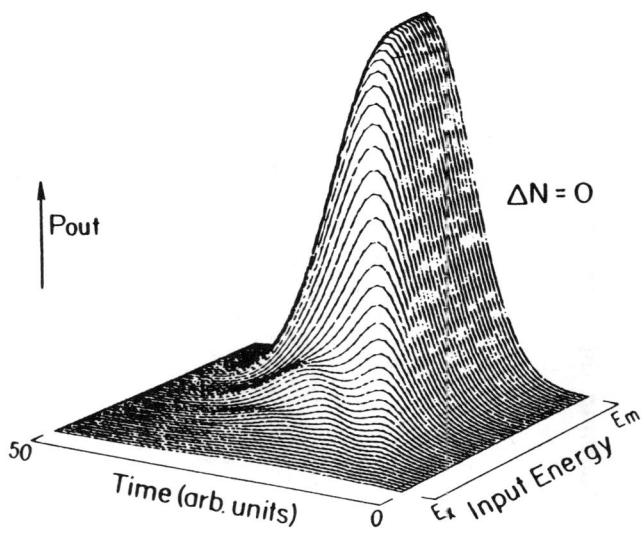

Fig. 5. Variation in pulse profile with incident pulse energy which varies logarithmically from $10^{-7}(E_m)$ J/mm to $10^{-4}(E_x)$ J/mm for an integrating nonlinearity with $n_{2eff}(thermal) = 10^{-14}$ m^2/W.

This type of all-optical switching device has not yet been implemented with nonlinear organic materials. All-optical switching in this device geometry has been demonstrated in MQW GaAlAs waveguides with cw lasers, but the question remains whether the switching was attributable to thermal or electronic nonlinearities [16-18]. Picosecond switching, clearly the result of electronic nonlinearities, has been observed in semiconductor-doped glass waveguides [19], and in dual core fibers [9].

Equation (19) can be rewritten in terms of the minimum phase shift, $\Delta\phi^{NL}$, required in any one channel of length L (equal to the half-beat length, π/κ) for switching to occur. Approximating $\Delta\beta_0 \cong n_2/A$, where A is the effective waveguide area, this gives

$$P_c \cong \frac{\Delta\phi^{NL} A}{k_0 L n_2} . \qquad (20)$$

Detailed analysis has shown that $\Delta\phi^{NL} > 4\pi$ for a nonlinear directional coupler (summarized in Table 1) [41,42]. Assuming that $A \cong 1$ μm^2, L = 1 cm, $\lambda = 1$ μm, and $n_2 \cong 10^{-16}$ m^2/W, then $P_c \cong 2$ W peak power. Furthermore, for 80% throughput, the waveguide loss must be less than 0.1 cm^{-1}. All of these numbers look very promising.

Table 1. The minimum nonlinear phase shift, $\Delta\theta^{NL}$, and the minimum dimensionless material parameter W (> 80% transmission) required for various nonlinear guided-wave devices.

Nonlinear Device	$\Delta\theta^{NL}$	W
Directional Coupler 1/2 beat length	4π	10
Directional Coupler 1 beat length	$\cong 3.3\pi$	8
Mach-Zehnder Interferometer	2π	5
Distributed-Feedback Grating	π	2.5

2.3 Nonlinear Gratings

Gratings have long been used for reflecting guided waves in integrated optics, having found, for example, wide application as distributed reflectors for semiconductor lasers. When a guided wave is incident onto a grating imbedded in a waveguide, the wave is reflected, provided that the Bragg condition $\beta_r k_0 = \beta_i k_0 + K$ is satisfied, where $\beta_i k_0$ and $\beta_r k_0$ are the incident and reflected guided-wave wavevectors, respectively, and K (which equals $2\pi/\Lambda$, where Λ is the grating period) is the grating wavevector. When the waveguide contains a nonlinear material, $\beta \rightarrow \beta(P)$, and the reflection properties of the grating become power dependent [28,43,44]. For this case, the negative sign is appropriate in Eq. (4b) and κ is the distributed reflection coefficient. For example, for the simplest case of a sinuosoidal grating of peak-to-peak amplitude $2u_0$ at the $x = x'$ interface of a waveguide, [44]

$$\kappa = \frac{\omega\epsilon_0}{8} u_0 \int [n^2(x = x_+') - n^2(x = x_-')] [E_i(x = x_+') \cdot E_r(x = x_-')] \, dy \ . \tag{21}$$

Since the most efficient reflection occurs between two modes of the same polarization and mode number, we assume $\beta_i = \beta_r$ and define the detuning from the Bragg condition for contradirectional guided waves as as $\Delta\beta = 2\beta - K$. Although there are analytical solutions for this case, they are too complicated to reproduce here in detail. Again there is a critical power [43] given by

$$P_c = \frac{2\kappa}{3\Delta\beta_0 k_0 L} \, , \qquad (22)$$

at which interesting switching effects occur. As shown in Fig. 2, bistability is predicted for sufficient detuning $\Delta\beta$. This type of device has not been demonstrated in any nonlinear material system to date.

What can be expected from nonlinear distributed-feedback grating structures when implemented with nonlinear organic materials? It is possible to recast Eq. (22) into the form of Eq. (20). The problem is that it is not possible to specify the magnitude of κL required for some useful switching without evaluating the solutions in detail. Such an analysis, including saturation effects, has shown that a minimum $\Delta\phi^{NL}$ of π is required for switching. Therefore the power required for switching a nonlinear distributed-feedback grating is only 1/4 of that required for the nonlinear directional coupler discussed previously.

There are variants on all-optical switching of gratings, however, that have been demonstrated using thermal nonlinearities in a polymeric film. Consider the case where a strong (control) optical beam is incident from above onto the the guided-wave grating, and tunes the Bragg condition of the grating for weak-incident and Bragg-reflected guided-wave beams. In that case, the quantity in the square brackets in Eqs. (4a) and (4b) is replaced by $\Delta\beta_0 I$, where I is the externally incident intensity. Assuming uniform illumination from above, the wavelength of the reflection maximum of the grating is shifted with increasing incident power. Therefore, for a given guided-wave wavelength, the fraction of the incident guided wave reflected by the grating varies with the external beam power.

This particular interaction has been implemented recently using a highly reflecting distributed-feedback (DFB) grating fabricated into a glass substrate overcoated with a guiding film of PMMA. Two additional gratings are also used to couple radiation into and out of the waveguide on either side of the DFB grating. A tunable dye laser operating in the 0.60 to 0.61 μm range was used to investigate the wavelength response of the reflectivity of the DFB grating whose center, low-power wavelength was 0.6055 μm. The externally incident control beam consisted of radiation from a CO_2 laser. The PMMA film has a strong absorption at the operating wavelength, resulting in a thermally induced change in the effective index of the waveguide modes.

The transmission of the DFB structure with the control CO_2 beam on and off is shown in Fig. 6. The shift in the center wavelength of the grating by about one-half of the grating filter bandwidth is clear. To demonstrate all-optical modulation of the guided-wave beam, the guided-wave wavelength was chosen so that the grating transmission is minimized in the off state (control beam off). The transmission can be made to increase to about 50% by turning on the CO_2 laser power. When the CO_2 beam is chopped, the transmission of the grating is correspondingly modulated on and off very efficiently.

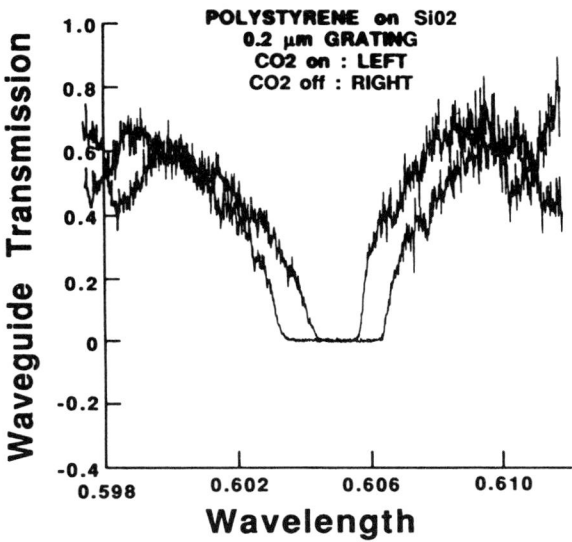

Fig. 6. *The guided-wave power transmitted through a distributed-feedback grating for a polystyrene waveguide as a function of wavelength when the CO_2 laser is off, and when the CO_2 laser is on. The grating is illuminated from above by the CO_2 beam.*

2.4 Other Nonlinear Guided-Wave Devices

As mentioned previously, virtually any standard integrated optics device can be made into an all-optical device when one or more of the waveguiding media has an intensity-dependent refractive index. Three of the more promising have been discussed in the preceding sections. Those that rely on the coupling between two guided modes with a coupling coefficient κ (at low powers) independent of propagation distance have responses similar to that of a nonlinear directional coupler. The situation is more complicated when the coupling coefficient varies with propagation distance; for example, when two channel waveguides actually cross. There the operating parameters depend on the details of the device design and the typical values required for $\Delta\phi^{NL}$ are still not known.

There are also devices in which the cumulative nonlinear phase shift in decoupled channels leads to all-optical device characteristics. The simplest example is the nonlinear, asymmetric Mach-Zehnder interferometer [35], as shown in Fig. 2. The incident guided light, which is split into two separate branches, undergoes different phase shifts in the two branches. The amount of light in the recombined channel depends on the interference condition between the two interferometer arms. If the phase difference is $2m\pi$, for $m = 0, 1, 2$, etc., then maximum throughput is obtained, and $(2m+1)\pi$ minimizes the throughput. Thermal stability requires that the two arms be of approximately equal length, and hence unequal phase shifts can be implemented either by having the nonlinearity in only one arm, or by using unequal channel cross-sectional areas (since the intensity, or power/area, determines the phase shift). A detailed analysis of typical waveguides, including the effects of saturation on the index change, has shown that a minimum $\Delta\phi^{NL}$ of approximately 2π is required.

There is another very interesting class of devices that has not been discussed here, primarily because their potential implementation is not as imminent as those summarized in Fig. 2. They rely not on cumulative phase shifts, but on a minimum optically induced change in the refractive index. For the previously discussed devices, it was assumed that the optically induced index changes did not perturb the field distribution significantly and only induced small changes in the guided-wave wavevector, linear with guided-wave power. This is essentially a first-order perturbation approach. If the optically induced refractive index change, Δn_o, becomes comparable to, or larger than, the index difference between the high-index guiding region and the bounding media, then new, all-optical waveguide behavior becomes possible. The potential device characteristics take on a thresholding character, leading to optical limiters and lower-threshold devices. In addition, devices based on the exchange of solitons between waveguides should also be possible in the future [45].

A specific example of such a device is shown in Figs. 2(a) and (b). There are two parallel waveguides, separated by a nonlinear slab sufficiently thick that no coupling occurs between the waveguides from field overlap. When light is coupled into one waveguide, its transmission as a function of input power exhibits power limiting and discontinuities. The power in the second waveguide is zero up to a critical input power in the first waveguide, and then jumps up to about 80% of the input power. This is achieved by the emission of a spatial soliton from the first waveguide and its trapping by the second waveguide. The beam propagation simulation of this effect is also shown in Fig. 8.

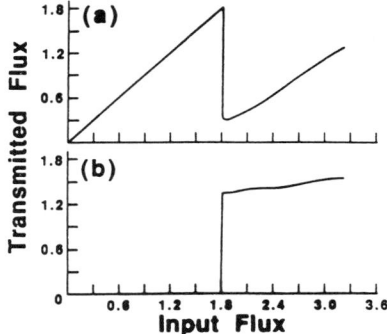

Fig. 7. Transmitted flux versus input flux for (a) the input waveguide, and (b) the parallel waveguide.

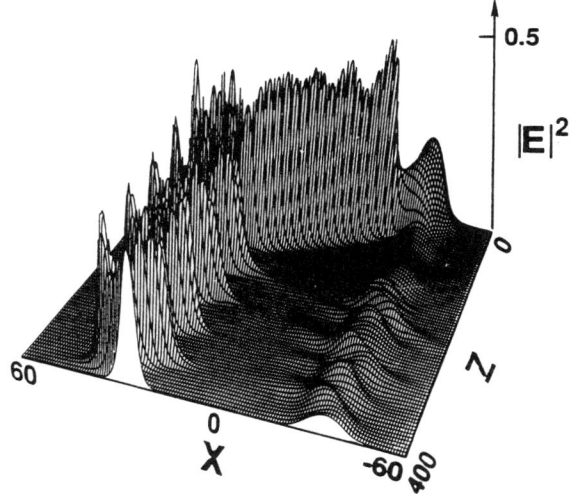

Fig. 8. Evolution of the intensity profile in a soliton coupler with propagation distance z in free-space wavelengths. A large fraction of the light transfers through the nonlinear medium to the parallel channel, leaving the incidence channel weakly excited.

3. SUMMARY

A number of nonlinear integrated optics devices have been discussed and their operating parameters summarized. Such devices can be used for all-optical switching and logic operations, and a limited number of devices already have been demonstrated in a variety of material systems. There are many different figures of merit that can be used in differentiating the suitability of various material systems. These include device viability, throughput, speed, and operating power. The figure of merit which simultaneously includes device viability and throughput is $W = \Delta n_{sat}/\alpha\lambda$. Values on the order of 10 to 100 for this parameter are projected for nonlinear polymers. Glass is the only other material with superior properties, in terms of this parameter. Currently these two are the only material classes with sub-picosecond response times as well. However, the small values of n_2 available in glasses, as compared to those of nonlinear organic materials, result in kilowatt switching powers in glasses, versus the watts predicted for nonlinear polymers. (It is important to note, however, that the only picosecond switching devices demonstrated to date were performed in fibers.) Based on all of these considerations, nonlinear organic materials appear to be a preferred system for the implementation of nonlinear guided-wave devices.

This research was supported by Hoechst-Celanese, and the Joint Services Optics Program of the Air Force Office of Scientific Research and the Army Research Office.

REFERENCES

1. Caulfield HJ, Horovitz S, Tricoles GP, Von Winkle WA (eds.): Special Issue on Optical Computing, Proc. IEEE 72, 755, 1985.

2. Stegeman GI, Zanoni R, Finlayson N, Wright EM, Seaton CT: J. of Lightwave Tech. 6, 953, 1988.

3. Carter GM, Thakur MK, Chen YJ, Hryniewicz JV: Appl. Phys. Lett. 47, 457, 1986.

4. Gibbs HM, Jewell JL, Moloney JV, Rushford MC, Tai K, Tarng S, Weinberg DA, Gossard AC, McCall SL, Passner A, Wiegmann W: Appl. Phys. Lett. 29, 171, 1982.

5. Miller DAB, Chemla DS, Damen TC, Gossard AC, Wiegmann W, Wood TH, Burrus CA: Appl. Phys. Lett. 45, 13, 1984.

6. Stegeman GI, Zanoni R, Seaton CT: Mat. Res. Soc. Symp. Proc. 109, 53, 1988.

7. Trillo S, Wabnitz S, Stolen RH, Assanto G, Seaton CT, Stegeman GI: Appl. Phys. Lett. 49, 1224, 1986.

8. Gusovskii DD, Dianov EM, Maier AA, Neustreuev VB, Shklovsii EI, Shcherbakov IA: Sov. J. Quant. Electron. 15, 1523, 1985; Friberg SR, Silberberg Y, Oliver MK, Andrejco MJ, Saifi MA, Smith PW: Appl. Phys. Lett. 51, 1135, 1987.

9. Trillo S, Wabnitz S, Banyai WC, Finlayson N, Seaton CT, Stegeman GI, Stolen RH: Appl. Phys. Lett., in press.

10. Stegeman GI, Seaton CT: Appl. Phys. Rev. (J. Appl. Physics) 58, R57, 1985.

11. Carter GM, Chen YJ, Tripathy SK: Appl. Phys. Lett. 43, 891, 1983.

12. Goodwin MJ, Glenn R, Bennion I: Electron. Lett. 22, 789, 1987.

13. Singh BP, Prasad P: J. Opt. Soc. Am. B 5, 453, 1988.

14. Sasaki K, Fujii K, Tomioka T, Kinoshita T: J. Opt. Soc. Am. 5, 457, 1988.

15. Burzynski R, Banhu P, Prasad P, Zanoni R, Stegeman GI: Appl. Phys. Lett., submitted.

16. Li Kam Wa P, Stich JE, Mason NJ, Roberts JS, Robson PN: Electron. Lett. 21, 26, 1985.

17. Cada M, Keyworth BP, Glinski JM, SpringThorpe AJ, Mandeville P: J. Opt. Soc. Am. B 5, 462, 1988.

18. Berger PR, Chen Y, Bhattacharya P, Pamulapati J, Vezzoli GC: Appl. Phys. Lett. 52, 1125, 1988.

19. Finlayson N, Banyai WC, Wright EM, Seaton CT, Stegeman GI, Cullen TJ, Ironside CN: Appl. Phys. Lett., in press.

20. Lalama SJ, Sohn JE, Singer KD: SPIE Proceedings of Symposium on Integrated Optical Engineering II 578, 168, 1985.

21. Fortenberry RM, Moshrefzadeh R, Assanto G, Mai X, Wright EM, Seaton CT, Stegeman GI: Appl. Phys. Lett. 49, 687, 1986.

22. Pardo F, Chelli H, Koster A, Paraire N, Laval S: IEEE J. Quantum Electron. QE-23, 545, 1987.

23. Lukosz W, Pirani P, Briguet V: in *Optical Bistability III*, Springer-Verlag, Berlin, 1986, p109.

24. Lukosz W: Opt. Lett. 10, 143, 1985; Lukosz W, Briguet V: Thin Solid Films 126, 197, 1985.

25. Assanto G, Svensson B, Kuchibhatla D, Gibson UJ, Seaton CT, Stegeman GI: Opt. Lett. 11, 644, 1986.

26. Chen YJ, Carter GM, Sonek GJ, Ballantyne JM: Appl. Phys. Lett. 48, 272, 1986.

27. Valera JD, Seaton CT, Stegeman GI, Shoemaker RL, Mai X, Liao C: Appl. Phys. Lett. 45, 1013, 1984.

28. Seaton CT, Stegeman GI, Winful HG: Opt. Eng. 24, 593, 1985; Winful HG, Stegeman GI: Proc. SPIE 517, 214, 1984.

29. Jensen SM: IEEE J. Quantum Electron. QE-18, 1580, 1982.

30. Daino B, Gregori G, Wabnitz S: J. Appl. Phys. 58, 4512, 1985.

31. Wabnitz S, Wright EM, Seaton CT, Stegeman GI: Appl. Phys. Lett. 49, 838, 1986.

32. Kitayama K, Wang S: Appl. Phys. Lett. 43, 17, 1983.

33. Thylen L, Wright EM, Stegeman GI, Seaton CT, Moloney JV: Opt. Lett. 11, 739, 1986.

34. Lattes A, Haus HA, Leonberger FJ, Ippen EP: IEEE J. Quantum Electron. QE-19, 1718, 1983.

35. Thylen L, Finlayson N, Seaton CT, Stegeman GI: Appl. Phys. Lett. **51**, 1304, 1987.

36. Silberberg Y, Stegeman GI: Appl. Phys. Lett. **50**, 801, 1987.

37. Finlayson N, Seaton CT, Stegeman GI, Silberberg Y: Appl. Phys. Lett. **22**, 1562, 1987.

38. Assanto G, Fortenberry RM, Seaton CT, Stegeman GI: J. Opt. Soc. Am. B **5**, 432, 1988.

39. Fortenberry RM, Assanto G, Moshrefzadeh R, Seaton CT, Stegeman GI: J. Opt. Soc. Am. B **5**, 425, 1988.

40. Stegeman GI: J. Quant. Electron. **QE-18**, 1610, 1982.

41. Caglioti E, Trillo S, Wabnitz S, Daino B, Stegeman GI: Appl. Phys. Lett. **51**, 293, 1987.

42. Trillo S, Wabnitz S, Caglioti E, Stegeman GI: Opt. Commun. **63**, 281, 1987.

43. Winful HG, Marburger JH, Garmire E: Appl. Phys. Lett. **35**, 379, 1979.

44. Stegeman GI, Liao C, Winful HG: in *Optical Bistability II*, Plenum Press, New York, 1984, p389.

45. Heatley DR, Wright EM, Stegeman GI: Soliton coupler, Appl. Phys. Lett., in press.

ORGANIC INTEGRATED OPTICAL DEVICES

R. LYTEL, G.F. LIPSCOMB, M. STILLER, J.I. THACKARA, AND A.J. TICKNOR

Research and Development Division
LOCKHEED MISSILES AND SPACE COMPANY, INC.
D-9720, B-202, 3251 Hanover St., Palo Alto, CA 94304

1. INTRODUCTION

Organic and polymeric materials have emerged in recent years as promising candidates for advanced device and system applications[1-4]. This interest has arisen from the promise of extraordinary optical, structural, and mechanical properties of certain organic materials[5-12], and from the fundamental success of molecular design[13-15] performed to create new kinds of materials. From an optical standpoint, organics offer temporal responses ranging over fifteen orders of magnitude, including large nonresonant electronic nonlinearities (fsec-psec), thermal and motional nonlinearities (nsec-msec), configurational and orientational nonlinearities (msec-sec), and photochemical nonlinearities (psec-sec). Additionally, organic and polymeric materials can exhibit high optical damage thresholds, broad transparency ranges, and can be polished or formed to high-optical quality surfaces. Structurally, materials can be made as thin or thick films, bulk crystals, or liquid and solid solutions, and can be formed into layered film structures, with molecular engineering providing different optical properties from layer to layer. Mechanically, the materials can be strong and resistant to radiation, shock, and heat. When coupled with low refractive indices and low D.C. and microwave dielectric constants, the collective properties of these extraordinary materials show great promise towards improving the performance of existing electro-optic and nonlinear optical devices, as well as allowing new kinds of device architectures to be envisioned.

This paper reviews our current research toward fabricating electro-optic, organic integrated optical devices. We begin in section 2 with a brief review of the important properties of glassy, electro-optic polymer materials for integrated optics, and show how their unique properties make them ideally suited to integrated optic devices. In section 3, we describe our methods for fabricating and poling slab waveguide modulators for the characterization of the glassy polymer films and the demonstration of new fabrication techniques[16]. In section 4, we describe a new method[17,18], called the selective poling procedure (SPP), for producing active, buried-channel waveguide devices. The SPP is illustrated by the fabrication of several integrated optical devices, including a Y-branch interferometer, a directional coupler, and a GHz traveling-wave modulator. We conclude with a summary of our work and point to new directions for future research.

2. ORGANIC MATERIALS IN INTEGRATED OPTICS

The synthesis of glassy polymer films containing molecular units with large nonlinear polarizabilities has led to the rapid implementation of organic integrated optics. These films, either spun, cast, or dipped, are amorphous as produced, and can be processed to achieve a macroscopic alignment for the generation of second-order nonlinear optical effects by electric-field poling. Typically, the films are poled[15,19] by forming an electrode-polymer-electrode sandwich, and applying an electric field normal to the film surface. This produces films with their nonlinear molecular units oriented normal to the film. In this state, the electric field of an optical beam propagating through the film can be maximally modified when the field is parallel to the oriented molecular units, that is, when the propagation vector lies in the film plane. As such, the films are ideally suited for guided wave applications. The successful development of guest-host and side-chain polymer systems incorporating molecular units with large nonlinear

J. Messier et al. (eds.), Nonlinear Optical Effects in Organic Polymers, 277–289.

polarizabilities in this manner has thus led to the availability of organic thin film materials for integrated optics.

Organic electro-optic materials offer a variety of potential advantages over conventional materials for integrated optical device applications. Table 1 provides a comparison of the potential of organic materials with the current commercial technology, Ti-indiffused LiNbO3 in three major areas of importance: materials parameters, processing technology, and fabrication technology. Some of these advantages have already been realized in our work on E-O modulators using poled polymer films, described below.

- CURRENT TECHNOLOGY: Ti:LiNbO$_3$
  - Materials Dev. Began in 1960s

- r = 32 pm/V
  - Larger Modulating Voltage
  - Little Improvement Expected

- LIMITED FABRICABILITY
  - 1000°C Processing
  - Depth Limited to 5 μm
  - Low Index Change Δn
  - Loss > 0.1 dB/cm
  - Optical Damage (Photorefractor)

- LARGE DIELECTRIC CONSTANT (28)
  - Longer Time Constants = RC
  - Large Velocity Mismatch in Traveling Wave Modulator

- MASS PRODUCTION DIFFICULT

- PROPERTIES OF POLYMERIC ORGANIC E-O MATERIALS
  - Materials Dev. Began 1975

- r = 14-53 pm/V (poled films [1])
  - Lower Modulating Voltage
  - Potentially Much Larger r

- FLEXIBLE FABRICATION
  - Low Temperature Processing
  - Flexible Dimensions
  - Controllable Index Change Δn
  - Loss < 0.8 dB/cm [1]
  - High Optical Damage Threshold

- LOW DIELECTRIC CONSTANT (4)
  - Shorter Time Constants = RC
  - Smaller Velocity Mismatch

- POTENTIAL FOR MASS PRODUCTION

TABLE 1. Comparison of Integrated Optics Technologies: Current Ti-LiNbO3 and Projected Organics Technologies

The major advantages are due to the intrinsic differences in E-O mechanisms in inorganic and organic materials. Organics should provide flat E-O response well beyond a GHz, and, indeed, measurements of the E-O coefficients and SHG coefficients of certain poled polymer films show little or no dispersion. Second, the E-O coefficients of poled polymer films can be made nearly as large as LiNbO3. Pure MNA crystals already exhibit larger E-O coefficients than LiNbO3, but it is our sense that practical, organic integrated optical devices will be made with films, not crystals. It is also true that the dielectric constant of poled polymer films is substantially lower than that of LiNbO3, implying smaller RC time constants and wider frequency bandwidths. In this connection, the bandwidth-length product for a typical poled polymer is of order 120 GHz-cm, compared to 10 GHz-cm for LiNbO3. Finally, the processing technology for integrated optical devices based on poled polymer films is relatively straightforward and fast, requiring only moderate temperatures (100-200 degrees C) for poling, and standard semiconductor fabrication equipment for fabrication of layered waveguide structures.

There are some potential drawbacks to polymers for integrated optics, as well. The microwave loss tangents in glassy poled polymers are not yet known, but are expected to be small. However, the materials have low thermal conductivity, which may lead to power dissipation problems in practical devices, as well as produce instabilities of the poled states.

Further, the stability of poled states at nominal temperatures is not yet well known. These issues are of a practical nature, and are currently being addressed by us.

In light of the potential benefits of poled polymer films for integrated optical devices discussed above, many research groups in the United States, Europe, and Japan have embarked upon dedicated materials synthesis and device fabrication programs to bring this field to fruition in commercial and military products. In the remainder of the paper, we review our progress toward developing an understanding of the use of the materials in device architectures, and toward achieving high levels of performance in prototype devices based upon poled polymer films.

Our work is a cooperative effort with the Hoechst Celanese Research Division (HCRD), Hoechst Celanese Corporation, and the specific materials discussed in this paper were supplied to us by HCRD[5].

3. SLAB WAVEGUIDE DEVICES AND MATERIALS PROPERTIES

The Lockheed organic integrated optical devices effort initially focused on the fabrication of simple slab guided wave structures made from polymer films. These structures allow the determination of fabrication techniques for organic devices, as well as direct measurement of Kerr and electro-optic (E-O) coefficients in the waveguide configuration. Figure 1 illustrates a typical slab waveguide device and some of the components used as substrates, electrodes, buffer layers, and the polable, glassy polymers. The polymers used include: 1) guest-host MNA/PMMA films, both poled and unpoled, 2) PC6S, a yellow, pendant side-chain polymer, 3) C-22, a red, pendant side-chain polymer, and 4) HCC-1237, a more active version of C-22. Waveguide structures such as that in the figure were built up by spin-coating the various layers. The polymers were then poled to produce a non-centrosymmetric structure exhibiting a nonzero electro-optic effect.

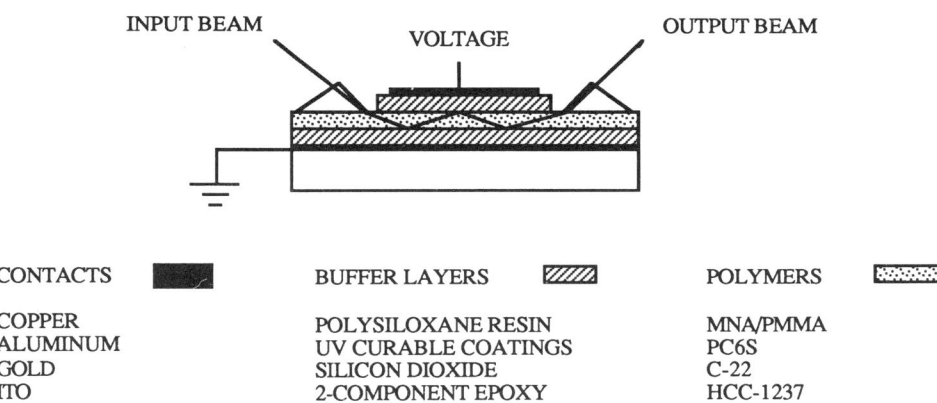

CONTACTS	BUFFER LAYERS	POLYMERS
COPPER	POLYSILOXANE RESIN	MNA/PMMA
ALUMINUM	UV CURABLE COATINGS	PC6S
GOLD	SILICON DIOXIDE	C-22
ITO	2-COMPONENT EPOXY	HCC-1237

FIGURE 1. Typical Slab Waveguide Modulator and Components

The poling procedure typically consists of first spin-coating the electrode-coated substrate with the bottom buffer layer and then the active polymer, applying a top electrode, and applying a voltage of order 1 MV/cm to the structure. The entire procedure is monitored in real time to optimize the poling. The top electrode is then removed, and a top buffer layer and new electrode can be applied to the poled structure. Guiding of 830 nm light from a semiconductor laser over a 1-3 cm dimension with minimal loss is then achieved by locating the prism couplers directly over the ends of the poled region.

Inital work was performed with PC6S. The optical response was measured by placing the slab modulator in one arm of an external interferometer, and is illustrated in figure 2. The device half-wave voltage was about 48 volts. Frequency response was flat out to 400 kHz. The electro-optic coefficient of the poled PC6S film was measured to be 2.8 pm/V. Similar slab modulators were fabricated with the C-22, although different buffer layers were used with the C-22 devices. The poling fields were comparable to those for PC6S, but the C-22 glass transition temperature was slightly higher. We observed flat E-O modulation out to 85 MHz with the C-22 modulator, and measured $r_{33}=16\pm2$ pm/V, about half that of LiNbO3.

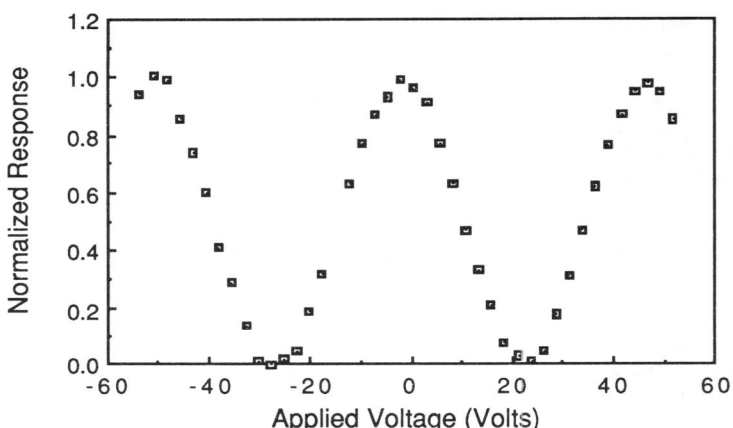

FIGURE 2. Interferometric Measurement of PC6S Modulator Response

Table 2 summarizes the measured physical parameters for PC6S and C-22 slab waveguides. It is most significant that the linear losses in the C-22 guides are below a dB/cm. This allows path lengths of order several cm, and, consequently, half-wave voltages approaching TTL levels. Both materials exhibit even better performance at a wavelength of 1.3 μm.

The slab waveguide experiments proved that poled polymers could be made into sandwich waveguide devices with standard spin-coating fabrication equipment and existing electrodes and buffer compounds. Further, fabrication is cheap, easy, and fast. However, the major application of organics to integrated optics will be in the form of complex channel waveguide devices on a single substrate, with full integration with sources and detectors. Therefore, most of our effort is devoted to the development of new methods for the fabrication of channel waveguides, and extension of modulation frequencies to beyond a GHz. This work is reported next.

(wavelength = 0.83 μm)	PC6S	C-22
E-O COEFFICIENT (pm/V)	2.8	16.0
TM REFRACTIVE INDEX (POLED)	1.7	1.58
TM INDEX DIFFERENCE (POLED-UNPOLED)	0.06	0.005
WG LENGTH (cm)	1.8	2.5
MEASURED LOSS (dB/cm)	2-3	0.8
HALF-WAVE VOLTAGE (volts)	48	7

TABLE 2. Physical Parameters Measured for PC6S and C-22 Slab Waveguides

4. CHANNEL WAVEGUIDE DEVICES

As discussed above, polymer waveguides can be made by spinning thin films onto high quality optical substrates. As spun, the films are isotropic and exhibit no second-order nonlinear susceptibility. In addition to inducing a non-centrosymmetric structure to achieve a macroscopic electro-optic effect, a second major transformation must be engineered in the material to enable the fabrication of integrated optic circuits. Channel waveguides must be formed to confine and guide the light from one active element of the integrated optic circuit to another. The formation of channel waveguides and the poling of the material to produce an active, E-O channel must both be accomplished for device prototypes, and would usually be performed in two distinct steps: fabrication and poling.

Figure 3 illustrates a new, powerful method developed by us, called the selective poling procedure (SPP), by which active, poled channel waveguides can be fabricated in a single fabrication step. An electrode pattern defining the channel waveguides is first deposited onto a substrate using standard photolithographic techniques. A planar buffer layer is then applied to optically isolate the active waveguide layer from the metal electrodes. The buffer material must be chosen to have an index lower than the guiding layer and to be compatible with the required processing. Thus, different buffer layers must often be used with different nonlinear polymers. A planar electrode is evaporated directly onto the nonlinear polymer for poling. The nonlinear layer is then poled by applying an electric field above the polymer glass transition and cooling the sample to room temperature under the influence of the field. The degree of alignment induced and the resultant electro-optic coefficient can be calculated based on a statistical average of the molecular susceptibilities. In this case only those regions of the material defined by the electrode pattern on the substrate are poled.

Since most organic nonlinear optical molecules also possess an anisotropic microscopic linear polarizability, the poled region becomes birefringent. The poled regions are uniaxial, with n_e oriented along the direction of the poling field. Consequently, TM and TE waves propagating in vertical and transverse device structures respectively will experience a greater refractive index in the poled regions than in the unpoled regions, and so can be confined in the lateral dimension. Thus, by applying the poling fields using electrodes patterned to define the waveguide network, including both active and passive sections, no further patterning of the organic NLO layer is required to form the channel E-O waveguide structures. The devices are then completed by etching off the planar poling electrode, applying an upper buffer layer and depositing the patterned switching electrodes, as shown in Figure 3. If electrode removal is not possible, the device may be finished first, and then be selectively poled through the buffer layer.

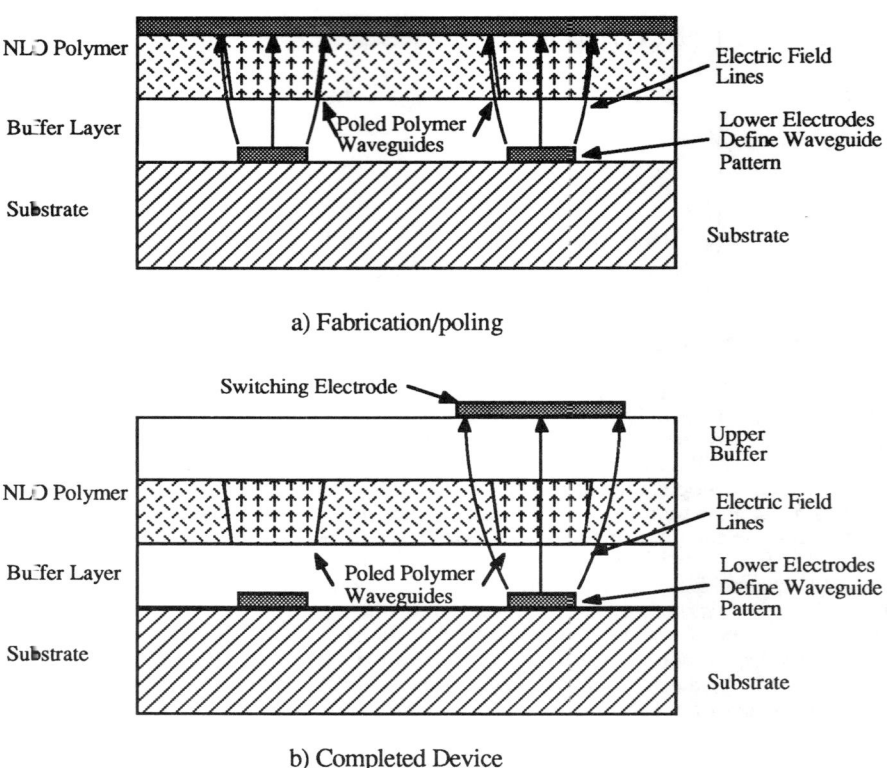

a) Fabrication/poling

b) Completed Device

FIGURE 3. Selective Poling Procedure for Active Channel Waveguides

The SPP permits both vertical and transverse poling of strips, as illustrated in figure 4. The vertical method locates the electrodes directly over the guides, implying that parallel channels can be located closer together without having an electrode in the middle of the channels. Vertically poled channels should guide TM radiation, while transverse poled channels should guide TE radiation.

Using the SPP, channel waveguides were constructed with both the PC6S and the C-22 materials, and light guiding was observed in both cases. The index change was measured by determining the prism coupling angles in poled and unpoled guides and by measuring the phase change from a double pass reflection through poled and unpoled regions. PC6S has a large index change of $\Delta n = 0.06$ for TM waves, while the C-22 material exhibited a $\Delta n = 0.005$ for TM waves. These index differences can be fine tuned by adjusting the poling field or by alterations in the nonlinear optical material. Figure 5 illustrates the confinement of 830

283

nm light in a poled PC6S channel, as compared with the unconfined light in a PC6S slab. In addition to lateral confinement, no birefringence was observed outside the electrode region with a scanned, focused laser beam, indicating that the fringing fields do not cause the guiding region to spread much beyond the electrodes.

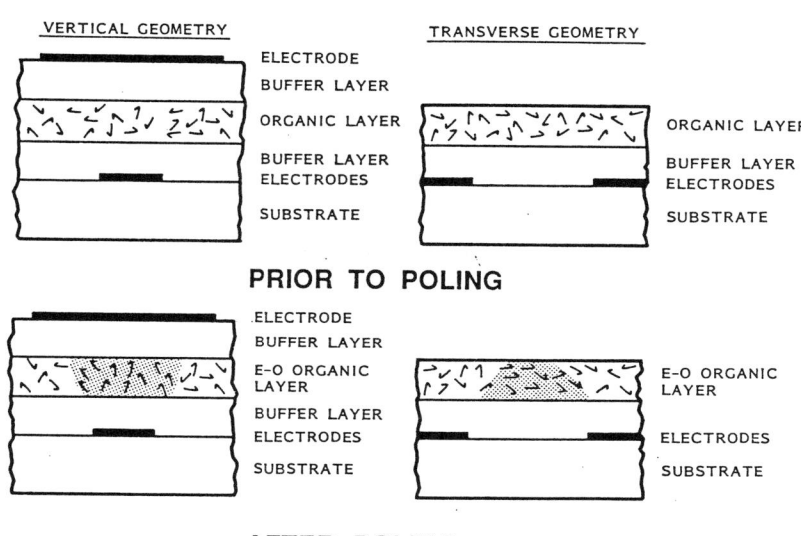

PRIOR TO POLING

AFTER POLING

FIGURE 4. Vertical and Transverse Electric Field Strip Poling using SPP

In addition to the obvious simplification of the device fabrication process, this technique has several other advantages over other methods of producing channel waveguides in poled polymer films. Fringing of the poling fields in the buffer and E-O layers acts to smooth out the edges of the guiding regions. If the edge roughness of the electrodes is small compared to the buffer layer thickness (of order 2 μm), the roughness of the waveguide boundaries should be independent of the resolution of the photolithographic process used, and should result in lower scattering losses in the waveguides. This has been confirmed by us in PC6S channel guides. Figure 6 displays the loss measurements in both a slab and channel guide. The losses are equal, within experimental error. Scattering losses from surface roughness are a major problem in channel guides defined by etching the nonlinear optical material or by channel filling in the substrate. Another advantage of the SPP is that the waveguides defined by the poling process have significantly different index changes for the TE and TM waves, making possible polarization selective elements.

Potential drawbacks to the SPP include partial lateral confinement of TE waves near the guide boundaries due to nonuniformly poled polymer and the fact that the guide cladding is intrinsically the unpoled polymer and cannot be arbitrarily chosen for a particular application.

In order to model the performance that can be expected for integrated optic devices, we developed a wave propagation design tool. Figure 7 shows the results of a model of the input of a Y-branch interferometer. The normalized guide width is 5 λ/n_{eff} and an induced index difference of $\Delta n=0.005$ is assumed, corresponding to the C-22 polymer. Clear confinement in the lateral dimension is predicted for a C-22 Y-branch fabricated by the SPP (Fig. 7a), while no confinement is predicted for an unpoled guide (Fig. 7b).

284

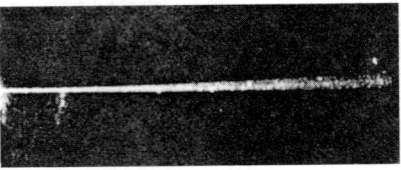

a) Slab Waveguide (beam fanning)

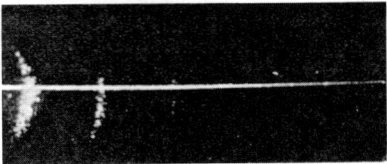

b) Channel Waveguide (lateral confinement)

FIGURE 5. Lateral Confinement of Light in a Poled PC6S Channel Waveguide

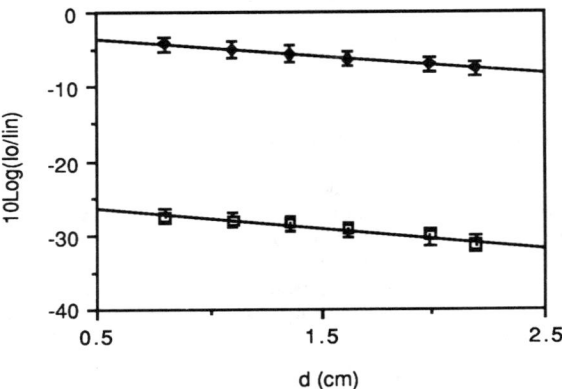

FIGURE 6. Loss Measurements in PC6S Slab Guides (circles) and 150 μm Channel Guides (squares)

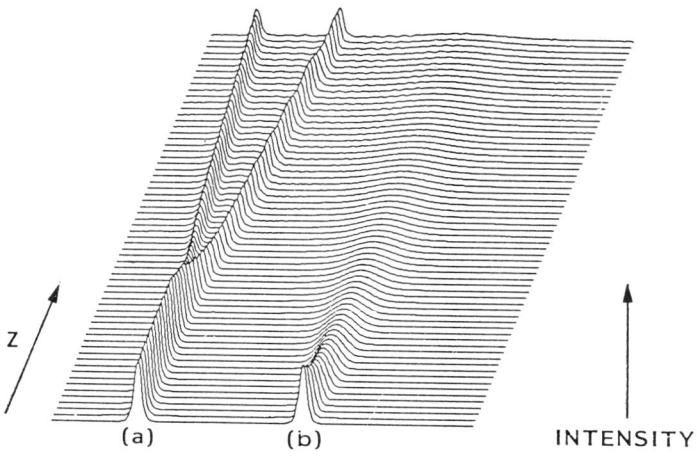

FIGURE 7. Predicted Lateral Confinement in a C-22 Y-Branch

The SPP has been used to fabricate three classes of integrated optical device structures: a Y-branch interferometer, a directional coupler, and a traveling wave phase modulator. These device experiments were aimed at developing the necessary processing techniques for poled polymer devices, which differ significantly form those of Ti:LiNbO₃, and to determine the effects of secondary, as well as primary, materials parameters on device performance. Device optimization was not carried out for the first prototypes, but is now a routine part of our work.

The prototype Y-branch interferometer was fabricated by first defining the waveguide pattern in an aluminium electrode on a glass substrate. The guides were 7 μm wide. The dimensions of these particular electrodes were chosen to facilitate tests of the poling waveguide formation process and were not scaled to optimize the completed devices made form different materials. A 3 μm lower buffer layer was deposited on the substrate using UV curing epoxy, and then a 2 μm layer of the PC6S was spun onto the substrate. A gold poling electrode was deposited directly onto the the the PC6S and the material was poled at 90°C for 5 minutes with an electric field of 100V/μm. The poling electrode was etched off with dilute Aqua Regia and a glass slide containing the upper electrode over one arm of the interferometer was glued on with a 3 μm thick layer of optical epoxy. The epoxy also served as the upper buffer layer. To facilitate construction the upper electrode was much larger than necessary resulting in increased device capacitance and a reduced maximum modulation rate Figure 8 illustrates the structure of the completed device and its dimensions. Prism coupling was used to inject 780 nm light into the device and guiding was observed in each arm. Modulation to a few kHz was detected in the output beam, indicating successful confinement and poling of the Y-branch.

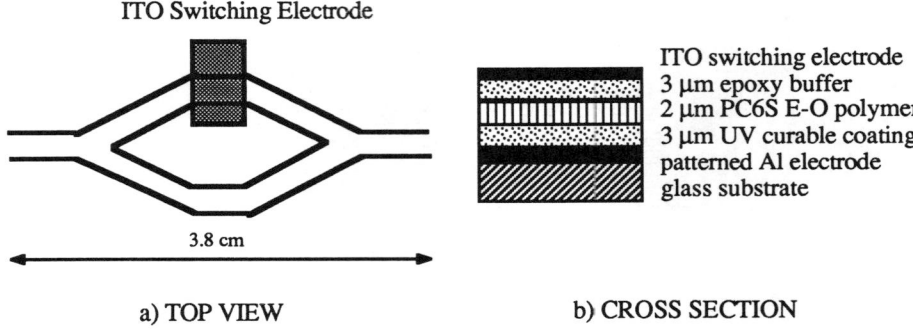

a) TOP VIEW   b) CROSS SECTION

FIGURE 8. Structure of the PC6S Y-Branch Interferometer Made By SPP

A second prototype device, a directional coupler, was fabricated using the C-22 material. The electrode defining the waveguide sections consisted of two 7 μm wide sections joining together to form a 13 μm wide, 1 cm long common section at a crossing angle of 3°. Two more symmetric 7 μm sections then diverged at the same angle, resulting in a total device length of 3.8 cm. The structure consisted of the lower aluminium electrode on a glass substrate covered with a 3 μm UV curing epoxy buffer layer. A 2 μm layer of the C-22 material was then spun on and covered with a 3 μm layer of polysiloxane. A planar gold electrode was then deposited for poling and the device was poled at 105°C for 15 minutes with an electric field of 90 V/μm in the C-22 layer. The upper gold electrode was then patterned to form the switching electrode over the central section of the coupler. Figure 9 illustrates the structure of the completed device and its dimensions. Prism coupling was used to inject 830 nm light into one input arm. With an applied voltage of 125 V the output was concentrated in the arm on the same side as the input channel, producing a bar state. With an applied voltage of 65 V a significant fraction of the light was switched to the opposite output arm producing a partial crossed state, as illustrated in figure 10. Complete switching was not observed due to the large crossing angle and the non-optimum waveguide dimensions.

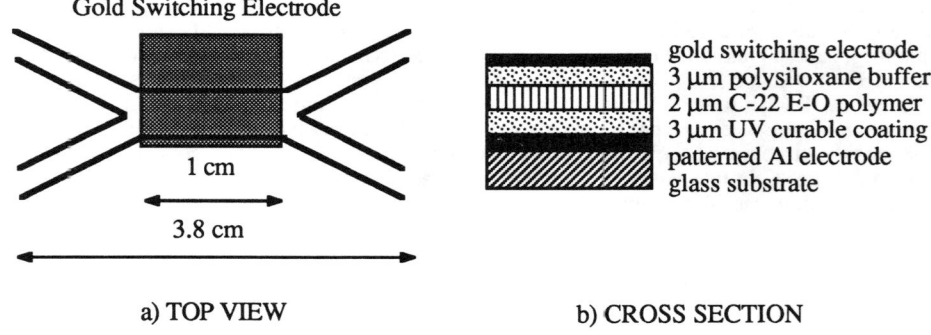

a) TOP VIEW   b) CROSS SECTION

FIGURE 9. Structure of the C-22 Directional Coupler Made by SPP

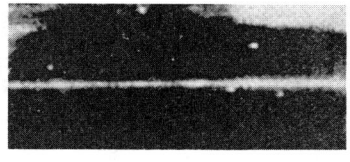

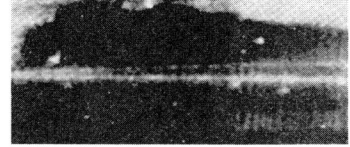

a) Bar State   b) Cross State

FIGURE 10. Bar and Cross States of the C-22 Directional Coupler

A third device prototype, a traveling-wave phase modulator based on C-22, was fabricated using methods similar to those described above. This device, illustrated in figure 11, was designed to modulate light efficiently at 270 MHz. The device impedence was 9.5 Ω, and a quarter-wave transformer was used to drive the modulator with a 0.5 W electrical input. The total active length was 3 cm. Figure 12 illustrates the response of the device when placed in one arm of an external interferometer. The modulation achieved at 270 MHz (Fig. 12a) was over 60%, almost exactly what we calculate it should have been by using the value of r_{33} measured at low frequencies, indicating little dispersion in the E-O coefficient from DC to 270 MHz. Figure 12b illustrates the response at 1.0 GHz, which was small due to the impedance mismatch. However, this represents the first reported measurement of an E-O effect in a poled-polymer channel waveguide above a GHz. The combination of the unique features of the SPP and the properties of the C-22 and newer polymers shows some of the promise of organic materials for integrated optics.

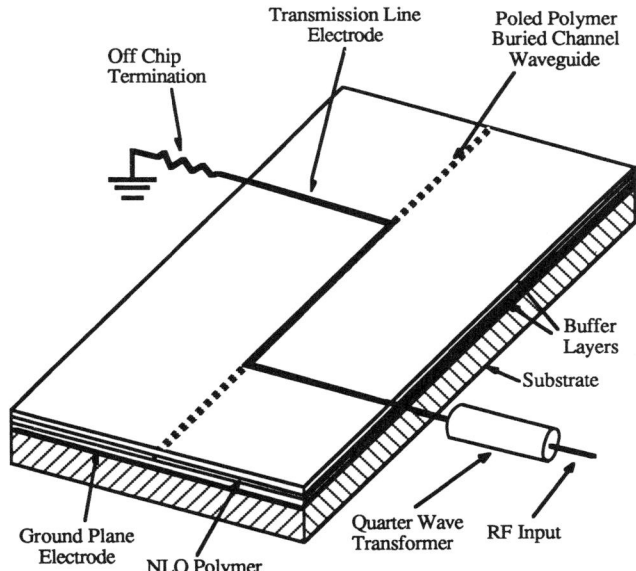

FIGURE 11. Traveling Wave Phase Modulator Made by SPP

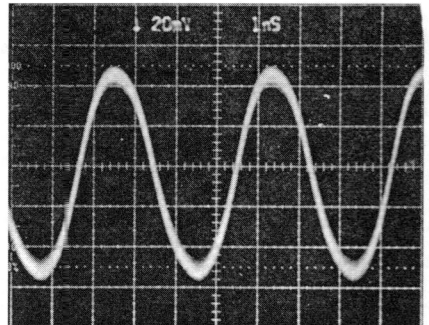 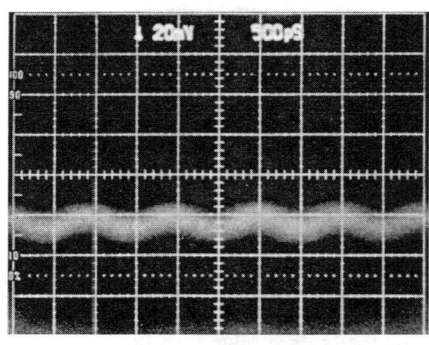

a) 270 MHz   b) 1 GHz

FIGURE 12. Frequency Response of the Traveling Wave Modulator

5. CONCLUSIONS

We have demonstrated the application of a new fabrication technique, the selective poling procedure (SPP), for producing active, buried-channel waveguide devices in poled polymer films. The SPP combines the fabrication and alignment steps required to make an active device. The method has been demonstrated in several device formats showing promise for real applications, and is under current intensive investigation at Lockheed for more advanced integrated optical structures. We have reported for the first time the achievement of a GHz response in a poled polymer channel waveguide, and have observed little or no dispersion in the E-O coefficient in our materials. The combination of unique fabrication methods and good material properties implies that the field of organic integrated optics, now in its infancy, is well on its way toward achieving much of the promise and expectations of organic electro-optic materials.

Future research on electro-optic polymer devices must address several research topics in poled polymer waveguides. From a materials standpoint, larger electro-optic coefficients and lower linear absorption are required, and higher glass transition temperatures are desirable. From a device standpoint, microwave losses must be measured and evaluated, the stability of the poled states must be determined within the operating range of the devices, and numerous other device fabrication issues, such as buffer layer selection, end-surface preparation, and fiber pigtailing must be solved. Finally, devices must eventually be integrated with semiconductor sources and detectors. Such advances are expected to occur within the next several years, and should produce a new class of high-speed, cost-effective devices for integrated optics.

6. REFERENCES

1. R. Lytel, G.F. Lipscomb, and J.I. Thackara, "Recent Developments in Organic Electro-optic Devices", in Nonlinear Optical Properties of Polymers, A.J. Heeger, J. Orenstein, and D.R. Ulrich , ed., Proc. Materials Research Society Vol. 109, 19 (1988).
2. R. Lytel and G.F. Lipscomb, "Nonlinear and Electro-optic Organic Devices", in Nonlinear Optical and Electro-active Polymers, P.N. Prasad and D.R. Ulrich, ed., Plenum Press, New York (1988), p. 415.

3. R. Lytel, G.F. Lipscomb, and J.I. Thackara, "Advances in Organic Electro-optic Devices", Proc. SPIE Vol. 824, pp.152-161 (1987).
4. R. Lytel, G.F. Lipscomb, P. Elizondo, B. Sullivan, and J. Thackara, "Optical Nonlinearities in Organic Materials: Fundamentals and Device Applications", Proc. SPIE 682, 125 (1986).
5. R.N. Demartino, E.W. Choe, G. Khanarian, D. Haas, T. Leslie, G. Nelson, J. Stamatoff, D. Stuetz, C.C. Teng, and H. Yoon, "Development of Polymeric Nonlinear Optical Materials", in Nonlinear Optical and Electro-active Polymers, P.N. Prasad and D.R. Ulrich, ed., Plenum Press, New York (1988), p. 169.
6. D.J. Williams, "Nonlinear Optical Properties of Guest-Host Polymer Structures", in Nonlinear Optical Properties of Organic Molecules and Crystals, Vol. 1, D. Chemla and J. Zyss, ed., Academic Press, FLA (1986), p. 405.
7. G.F. Lipscomb, A.F. Garito, and R.S. Narang, "An Exceptionally Large Linear Electro-optic Effect in the Organic Solid MNA", J. Chem. Phys. 75, 1509 (1981).
8. D. Pugh and J.O. Morley, "Molecular Hyperpolarizabilities of Organic Materials", in Nonlinear Optical Properties of Organic Molecules and Crystals, Vol. 1, D. Chemla and J. Zyss, ed., Academic Press, FLA (1986), p. 193.
9. Nonlinear Optical Properties of Organic Molecules and Crystals, Vol. 1 and 2, D. Chemla and J. Zyss, ed. (Academic Press, FLA) 1986.
10. G.J. Bjorklund et. al, "Organic and Polymeric Materials", in Research on Nonlinear Optical Materials: An Assessment, Appl. Opt. 26, 227 (1987).
11. Nonlinear Optical Properties of Organic and Polymeric Materials , D.J. Williams ed., ACS Symposium Series 233 (American Chemical Society), 1983.
12. K.D. Singer and A.F. Garito, "Measurements of Molecular Second-order Optical Susceptibilities Using DC Induced Second Harmonic Generation", J. Chem. Phys. 75, 3572 (1981).
13 S.J. Lalama and A.F. Garito, "Origin of the Nonlinear Second-order Optical Susceptibilities of Organic Systems", Phys. Rev. A 20, 1179 (1979)
14. C.C. Teng and A.F. Garito, "Dispersion of the Nonlinear Second-order Optical Susceptibility of an Organic System: p-nitroaniline", Phys. Rev. Lett. 50, 350 (1983).
15. K.D. Singer, M.G. Kuzyk, and J.E. Sohn, "Second-order Nonlienar Optical Processes in Orientationally Ordered Materials: Relationships Between Molecular and Macroscopic Properties", J. Opt. Soc. Am. B4, 968 (1987).
16. J. Thackara, M. Stiller, E. Okazaki, G.F. Lipscomb, and R. Lytel, "Optoelectronic Waveguide Devices in Thin-Film Organic Media", Conference on Lasers and Electro-optics, Baltimore, MD (1987), paper ThK29.
17. J. Thackara, M. Stiller, G.F. Lipscomb, A.J. Ticknor, and R. Lytel, "Poled Electro-optic Waveguide Formation in Thin-Film Organic Media", Appl. Phys. Lett. 52, 1031 (1988).
18. J. Thackara, M. Stiller, A.J. Ticknor, G.F. Lipscomb, and R. Lytel, "Poled Electro-optic Waveguide Devices in Thin-Film Organic Media", Conference on Lasers and Electro-optics, Anaheim, CA (1988), paper TuK4.
19. C.S. Willand, S.E. Feth, M. Scozzafava, D.J. Williams, G.D. Green, J.I. Weinshenk, H.K. Hall, and J.E. Mulvaney, "Electric-Field Poling of Nonlinear Optical Polymers", p. 107; and K.D. Singer, J.E. Sohn, and M.G. Kuzyk, "Orientationally Ordered Electro-optic Materials", in Nonlinear Optical and Electro-active Polymers, P.N. Prasad and D.R. Ulrich, ed., Plenum Press, New York (1988), p. 189.

ORIENTATIONALLY ORDERED NONLINEAR OPTICAL POLYMER FILMS

J.E. SOHN†, K.D. SINGER†, M.G. KUZYK†, W.R. HOLLAND†,
H.E. KATZ‡, C.W. DIRK‡, M.L. SCHILLING‡, and R.B. COMIZZOLI‡

† AT&T Engineering Research Center, P.O. Box 900, Princeton, NJ 08540 USA
‡ AT&T Bell Laboratories, Murray Hill, NJ 07974 USA

1. INTRODUCTION

Orientationally ordered polymer films continue to be an area of intense research as evidenced by contributions to recent symposia.[1] [2] Waveguide-quality materials with significant bulk optical nonlinearities can be assembled from organic molecules possessing large second-order nonlinear optical susceptibilities[3] by incorporating them into polymer glasses and poling with an electric field.[4] [5] [6] Orientational order is imparted by this poling, which consists of raising the temperature of the doped polymer above its glass-rubber transition, applying a strong electric field, and then cooling with the field still applied. The resulting polymers show sizable nonlinear optical properties as evidenced by electro-optic and second-harmonic generation measurements.[4][7]

The design of new materials possessing enhanced nonlinear optical susceptibilities can be guided by considering the model relating the molecular nonlinear optical susceptibilities of the molecular species to the bulk properties:[4][8]

$$\chi^{(2)} \sim N\beta\mu E_p. \tag{1}$$

This relationship shows that increasing the number density N, poling field strength E_p, molecular dipole moment μ, and molecular nonlinear susceptibility β, will increase the nonlinear optical susceptibility.

2. MOLECULAR SUSCEPTIBILITIES

The second-order nonlinear optical properties of donor- and acceptor-substituted aromatic compounds were determined using dc-induced second harmonic generation in the near-infrared.[9] [10] The study reveals the structure-property relationships that affect the molecular susceptibility β. Important parameters include the electronic properties of the donor and acceptor groups, the length and type of conjugated π-electron system bridging the donor and acceptor groups, and molecular configuration.

The experimentally determined quantity $\beta\mu$ are given in Table 1 for several disubstituted-benzene compounds and for extended aromatic systems in Table 2. A substantial increase in $\beta\mu$ is observed when the dimethylamino group (Compound A) is replaced by the dithiolylidenemethyl group (Compound B). The excited state properties of the dithiolylidenemethyl group are found to be important in this increase.[10] Substantial increases are also found with changes in the acceptor group. The increase in $\beta\mu$ on changing from nitro (Compound A) to dicyanovinyl (Compound C) to tricyanovinyl (Compound D) acceptor can be understood with the use of Hammett constants.[9] Combination of the better donor and acceptor groups (Compound E) yields a value of $\beta\mu$ nearly an order of magnitude greater than the prototypical nitroaniline A.

Similar trends in the effect of the electron-acceptor group as well as the effect of the length and type of conjugated π-electron system bridging the electron-donor and electron-acceptor groups is seen by comparing the susceptibilities of the compounds in Table 2. Increase in $\beta\mu$ is observed in these extended π-electron systems as the acceptor group is varied from nitro (Compound F) to dicyanovinyl (Compound G) to tricyanovinyl (Compound H). Comparing Compounds F and A (or G and C, or H and D), the expected increase in $\beta\mu$ with increasing

J. Messier et al. (eds.), Nonlinear Optical Effects in Organic Polymers, 291–297.

TABLE 1. $\beta\mu$ of disubstituted benzene compounds measured at $\lambda = 1.356\mu$m. Compound E measured at $\lambda = 1.58\mu$m

	Molecule	$\beta\mu$ $(10^{-30}\text{cm}^5\text{D}/\text{esu})$
A		138
B		358
C		271
D		846
E		1200

conjugation length is observed. It was found that azo and ethylene π-electron systems are nearly equivalent in determining the molecular susceptibility, when the effects of dispersion are taken into account.[10] Further increase in the length of the conjugated system was expected to increase $\beta\mu$ (Compound G → Compound K). However, a decrease was found, which may be attributed to a non-planar or cisoid π-electron system (confirmed by a higher energy of the first excited state and a lower ground state dipole moment for Compound K as compared to Compound G).[10] The importance of coplanarity is seen in comparing Compounds F and J, where a substantial decrease in $\beta\mu$ is found for the imine (Compound J), owing to the nonplanity of its extended π-electron system. A four-fold increase has been achieved (Compound H) as compared to the system used in demonstrating the potential of poled polymers to nonlinear optics.[7]

TABLE 2. $\beta\mu$ of extended aromatic compounds measured at $\lambda = 1.356\mu m$. Compound H measured at $\lambda = 1.58\mu m$

	Molecule	$\beta\mu$ $(10^{-30}cm^5D/esu)$
F		1090
G		2650
H		4110
J		500
K		1320

3. POLYMERIC AZO DYES

Copolymers of methyl methacrylate and azo dye-substituted methacrylate have been prepared.[11] These copolymers have a number density of optically nonlinear species three to four times greater than the guest-host systems previously demonstrated.[7] Preparation of these copolymers is shown schematically in Figure 1.

Methyl methacrylate and anilinoalkyl methacrylate are copolymerized to yield the random copolymer shown at the top in Figure 1. The azo coupling is performed using the appropriately substituted benzene diazonium salts. This procedure represents an improvement over polymerization of azo dye methacrylate monomers, since radical polymerization of methacrylates containing strong electron withdrawing groups (*e.g.* nitro, dicyanovinyl) yields low molecular weight polymers in low yields.

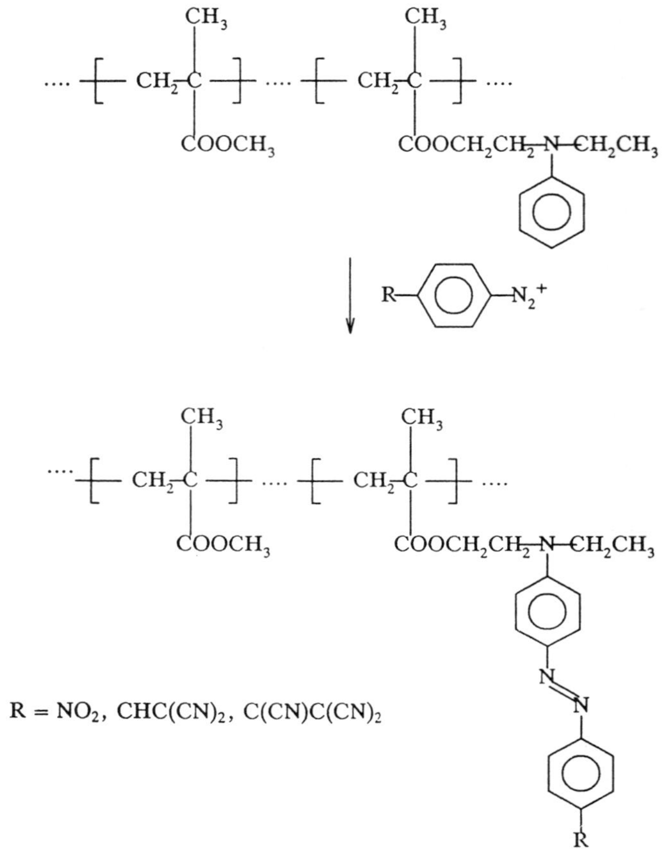

Figure 1. Preparation of copolymers of methyl methacrylate and azo dye-substituted methacrylate.

4. POLED POLYMER FILMS

The copolymers described above where R = CHC(CN)$_2$ or C(CN)C(CN)$_2$ represent increases in both the molecular quantity $\beta\mu$ and the number density N of the demonstration material of Disperse red 1 (Compound F) in PMMA. Larger poling fields E$_p$ are realized through the use of corona poling.[12] Thus the values of all parameters in Equation (1) have been increased.

Poling experiments were performed on films prepared by spin deposition onto indium-tin-oxide coated glass substrates. The nonlinear optical dyes used in these experiments are Disperse red 1 (Compound F) in PMMA,[7] DR1/PMMA, the dicyanovinyl azo dye Compound G in PMMA, DCV/PMMA, and a random copolymer with R = CHC(CN)$_2$ from Figure 1, DCV-MMA. The sample was then corona poled above its glass-rubber transition temperature (T$_g$ = 127°C for R = CHC(CN)$_2$ in Figure 1). Aluminum was deposited as the electrode material for the electro-optic measurements. The index of refraction was measured by determining the synchronous mode angles for prism-coupled waveguide excitation.[13]

The second-order nonlinear optical properties were determined by measuring the electro-optic and second-harmonic coefficients. The electro-optic measurements were performed in transmission in a Mach-Zehnder interferometer apparatus, using a variation of a previously described method.[14]

The second-harmonic coefficients were determined using a method previously described using both p- and s-polarized incident light (both with p-polarized harmonic light) at an incident wavelength of 1.58 μm.[7][15] This method allows for the determination of d_{33} and d_{31}.

The results of the electro-optic and second-harmonic measurements are given in Table 3 along with values of the refractive index for corona-poled films of the copolymer DCV-MMA, the guest-host systems DCV/PMMA and DR1/PMMA, and the electrode-poled film of DR1/PMMA previously reported.[7] The subscripts refer to measurements made immediately after poling (subscript 0) and those made after initial decay of the polarization (subscript f). The nonlinear optical coefficients of all corona-poled films are substantially larger than those previously reported in poled polymer systems[4] owing to the increased poling field E_p.

TABLE 3. Second-harmonic and electro-optic coefficients of poled films.

MATERIAL	N	n	d_0	d_f	r_f^{el}	r_f
(method)	$(10^{20}/cm^3)$	$(\lambda=0.8\mu m)$	$(10^{-9}esu)$	$(10^{-9}esu)$	$(10^{-12}m/V)$	$(10^{-12}m/V)$
DCV-MMA (corona)	~8	1.58	$d_{33}=51$ $d_{31}=17$	$d_{33}=46$ $d_{31}=15$	$r_{33}^{el}=15$ $r_{13}^{el}=5$	$r_{33}=18$ $r_{13}=6$
DCV/PMMA (corona)	2.3	1.53	$d_{33}=74$ $d_{31}=25$	$d_{33}=19$ $d_{31}=6$		
DR1/PMMA (corona)	2.3	1.52	$d_{33}=20$ $d_{31}=7$			
DR1/PMMA (electrode)	2.7	1.52	$d_{33}=6$ $d_{31}=2$			

0 subscript denotes coefficient prior to initial decay
f subscript denotes coefficient following initial decay

The increase in $\beta\mu$ observed for Compound G as compared to Compound F is translated into the bulk nonlinear optical properties of the corresponding films, roughly in agreement with the molecular measurements[9] and a linear model.[15] The increase in number density N of the copolymer (DCV-MMA) with respect to the guest-host system (DCV/PMMA) is not translated into the second-harmonic coefficient d_0. This may be attributed to restricted motion of the nonlinear optical moieties, which prevents maximal alignment with the current poling process. Another possibility is the formation of centrosymmetric aggregates by some of the dye residues. In this case, the residues in these aggregates do not contribute to the second-order nonlinear optical effects.

Hindered motion may be responsible for reducing the initial decay of the second-harmonic coefficient.[4][16] Corona-poled films of the copolymer show decay of the second-harmonic coefficient (d_f) to ~90% of the original value, while corona-poled films of the guest-host system decay to ~25%. Thus, use of these copolymers results in an increase in the stability of the nonlinear optical properties.

The ratio of the second-harmonic coefficients, as predicted by a simple thermodynamic model, [15] is $d_{33}/d_{31} = 3$, for all of the poled films. This strongly indicates that the poling potential is the only potential acting on the nonlinear optical dipoles during the poling process, and confirms the poling model for polymer glasses.[4]

The electro-optic effect can depend on contributions arising from electronic processes as well as acoustic and orientational processes. Only electronic processes are involved in second-harmonic generation. By comparing the electro-optic and second-harmonic coefficients, the processes contributing to the electro-optic effect can be analyzed. Using a two-level model, the bulk electronic electro-optic coefficient, r_{uv}^{el}, can be related to the second-harmonic coefficient, d_{vu} by accounting for dispersion and local field effects through[4]

$$r_{uv}^{el}(-\omega;\omega,0) = -\frac{4d_{vu}}{n_i^2(\omega)n_j^2(\omega)} \times \frac{f_{ii}^{\omega}f_{jj}^{\omega}f_{kk}^0}{f_{kk}^{2\omega}f_{ii}^{\omega'}f_{jj}^{\omega'}} \times \frac{(3\omega_0^2-\omega^2)(\omega_0^2-\omega'^2)(\omega_0^2-4\omega'^2)}{3\omega_0^2(\omega_0^2-\omega^2)^2}, \tag{2}$$

where ω' is the frequency of the fundamental used in measuring d_{kij}, and where the electro-optic coefficient is evaluated at frequency ω.

The electronic electro-optic coefficient after initial decay, r_f^{el} (reference wavelength 799nm, modulating frequency 5kHz), for the copolymer DCV-MMA is given in Table 3 along with the experimentally determined value of the electro-optic coefficient after initial decay r_f. These two coefficients are equal within experimental error, showing that even at this low modulating frequency, the electro-optic effect is probably dominated by the electronic contribution.

5. SUMMARY

The design of poled polymer systems for second-order nonlinear applications is guided by the expression relating the bulk second-order nonlinear optical susceptibility $\chi^{(2)}$ to the material parameters, N and $\beta\mu$, and the processing parameter E_p. Improvements in all of these parameters have been realized, as compared to those from the demonstration of optically nonlinear poled polymers. The combination of optimized electron-donor and electron-acceptor, covalent attachment of the dye to the polymer backbone, and processing with the more efficient corona poling technique, yields a polymeric material with substantially improved second-order nonlinear optical properties, as compared to the demonstration material. Further, the covalent attachment to the polymer backbone leads to improved stability of the nonlinear optical properties.

6. REFERENCES

1. *Nonlinear Optical Properties of Polymers*, A.J. Heeger, J. Orenstein, and D.R. Ulrich, eds., Materials Research Society Symposium Proceedings, **109** (Materials Research Society, Pittsburgh, 1988).

2. *Nonlinear Optical and Electroactive Polymers*, P.N. Prasad and D.R. Ulrich, eds. (Plenum, New York, 1988).

3. D. S. Chemla and J. Zyss, ed., *Nonlinear Optical Properties of Organic Molecules and Crystals* (Academic Press, New York, 1987).

4. K.D Singer, M.G. Kuzyk, and J.E. Sohn, *J. Opt. Soc. Am. B* **4**, 968 (1987).

5. J.I. Thackara, G.F. Lipscomb, M.A. Stiller, A.J. Ticknor, and R. Lytel, *Appl. Phys. Lett.* **52**, 1031 (1988).

6. C.S. Willand and D.J. Williams, *Ber. Bunsenges Phys. Chem.* **91**, 1304 (1987).

7. K.D. Singer, J.E. Sohn, and S.J. Lalama, *Appl. Phys. Lett.* **49**, 248 (1986).

8. S. Kielich, *IEEE J. Quant. Electron.* **QE-5**, 562 (1969).

9. H.E. Katz, K.D. Singer, J.E. Sohn, C.W. Dirk, L.A. King, and H.M. Gordon, *J. Am. Chem. Soc.* **109**, 6561 (1987).

10. K.D. Singer, H.E. Katz, J.E. Sohn, C.W. Dirk, L.A. King, and H.M. Gordon, *J. Chem. Phys.* (1988), submitted for publication.

11. H.E. Katz, D.I. Cox, M.L. Schilling, J.E. Sohn, and D.L. Fish, *Macromolecules* (submitted for publication).

12. R.B. Comizzoli, *J. Electrochem. Soc.* **134**, 424 (1987).

13. T. Tamir, ed., *Integrated Optics* (Springer-Verlag, New York, 1975).

14. M. Sigelle and R. Hierle, *J. Appl. Phys.* **52**, 4199 (1981).

15. K.D. Singer, M.G. Kuzyk, and J.E. Sohn, in *Nonlinear Optical and Electroactive Polymers*, P.N. Prasad and D.R. Ulrich, eds. (Plenum, New York, 1988).

16. H.L. Hampsch, J. Yang, G.K. Wong, and J.M. Torkelson, *Macromolecules* **21**, 526 (1988).

OVERVIEW-NONLINEAR OPTICAL ORGANICS AND DEVICES

DONALD R. ULRICH
Directorate of Chemical and Atmospheric Sciences
U.S. Air Force Office of Scientific Research
Bolling Air Force Base
Washington, DC 20332-6448 USA

ABSTRACT

Since 1982 the USAF has had a major basic research program in nonlinear optical organics with a major focus on polymers and some work on organometallics. Emphasis has been on both second and third order nonlinear effects. The Air Force laboratories as well as the Defense Advanced Research Projects Agency have been transitioning several of the basic research results to device application. Some of the current and projected directions for polymer development and device design are discussed.

INTRODUCTION

The use of organics for nonlinear optical processes has been gaining increased attention, particularly over the past two years. There has been a surge in the establishment of research programs in industry and academia in the U.S., Europe and Japan with the major focus on organic polymers. Some emphasis is also being placed on inorganic polymers and organometallics at a smaller level of effort.

Polymers are the subject of intense reserch because of the ability to tailor molecular structures which have inherently fast response times and large second and third order molecular susceptabilities. Polymers provide synthetic and processing options that are not available with the single crystal and multiple quantum well (MQW) classes of NLO materials, as well as excellent mechanical properties, environmental resistance and high laser damage thresholds.

In 1982 AFOSR initiated the first U.S. federally funded research program in nonlinear optical organics. In 1983 the Defense Science Office of the Defense Advanced Research Projects Agency and the US Army Night Vision Laboratory became coinvestors in the NLO organics research with AFOSR. Building on the work of AFOSR, in 1986 the Polymer Branch of the Air Force Wright Aeronautical Laboratory/Materials Laboratory, AFWAL/ML, in collaboration with the Frank J. Seiler Laboratory at the US Air Force Academy started an in-house and contract research program. By 1988 the Rome Air Development Center and the AFWAL/Avionics Laboratory were issuing requests-for-proposal (RFP). In April, 1988 the grantees, contractors and Air Force and other federal agency scientists and engineers met at the National Academy of Sciences to review the total program.

This paper reviews in part some of the results of the program and the current and projected directions for organic polymer and organometallic development and device design.

J. Messier et al. (eds.), Nonlinear Optical Effects in Organic Polymers, 299–325.
© 1989 by Kluwer Academic Publishers.

NONLINEAR OPTICAL APPLICATIONS FOR POLYMERS

The Air Force will have many uses for electrooptical and all-optical systems based on devices designed with materials having important optical properties. These include optical computing and optical storage, optical signal processing and optical sensor and vision protection against laser radiation.

Eighteen months ago there were four or five optical device groups in the U.S. which were either working on device architectures using NLO polymers or were doing some conceptualizing along these lines. They included the groups at Lockheed, University of Arizona, and the University of Southern California. These groups, the Air Force laboratories, DARPA and some industrial firms with an investment in telecommunications were surveyed in 1987 as to where NLO polymers would fit in.

The results are shown in Figure 1. Second order polymers were viewed as playing a major role in optical signal processing in such areas as spatial light modulators and neural nets. At the time optical communications were thought to be the domain of inorganic crystals, particularly lithium niobate or potassium dihydrogren phosphate. Now with the advances in polymers with properties commensurate with lithium niobate, this opinion is rapidly changing.

Third order polymers were going to play a major role in all-optical signal processing as well as tuneable filters, degenerate four wave mixing, phase conjugation and sensor protection. While it was accepted that polymers would find a role in all-optical signal processing, there was a majority position which predicted use in parallel processsing in polymer plane wave guides. A minority opinion stressed their role in serial processing in guided wave devices. All agreed that since the polymer area is embryonic considerable research has to be accomplished before the targets can be established.

At that time little role was seen in polymers in digital (optical) computing. While multiquantum well devices were thought to be the material of choice, it was stressed that this application is strongly dominated by electronics. However, with the promise of recent $\chi^{(3)}$ measurement of 10^{-8} to 10^{-9} esu in transparent polymers and proposed and theoretically calculated approaches to reach 10^{-7} or higher, this conclusion is being reconsidered.

ATTRIBUTES AND REQUIREMENTS OF POLYMERS FOR NONLINEAR OPTICS

Polymers are being investigated and developed as NLO materials for several reasons which are listed in Figure 2. The primary driver is the electronic origin of the nonlinear polarization. Organic polymers possess large, nonresonant optical susceptibilities whose origin lies in ultrafast lossless excitations of highly charged correlated pi-electron states. The optical behavior is nonresonant since the nonlinear optical polarization is electronic with little or no lattice phonon contribution.

Polymers offer time responses ranging over fifteen orders of magnitude, including the large nonresonant electronic nonlinearities (fsec-psec), thermal and motional nonlinearities (nsec-msec), configurational and orientational nonlinearities (μsec-sec), and photochemical nonlinearities. The contribution of each of these to nonlinear optical processes needs to be understood in order to design polymers with ultrafast response times. Measurement then becomes a critical issue.

Another major attribute of polymers is their low dc dielectric constants, being of the order of 3 as compared to 28 in the inorganic single crystal lithium niobate for example. In polymer waveguiding films the low dc dielectric constant means shorter time constants and small velocity mismatch. That is, the travelling wave device may be designed to achieve matching of optical and microwave velocities. The much longer dielectric constant in lithium niobate results in a loss of phase matching over shorter waveguide lengths. This results in higher drive voltages and power requirements, limiting frequencies accessible. There is some debate here with the inorganic crystal community since some advocates claim that comparison should be made with crystals such as $KTiOPO_4$, potassium titanium phosphate, rather than lithium niobate (KTP has higher SHG efficiency, but a lower laser damage threshold). However, the NLO polymer argument is supported by substantial device performance data whereas the inorganic crystal argument at this point rests more on opinion.

In second order polymers a large electrooptic effect r greater than 30 pm/V, the stability of electrically poled states, and low-optical loss of 0.1 dB/cm are required. For second harmonic generation a $\chi^{(2)}$ of 10^{-7} esu, a low birefringence of 0.1, no two-photon absorption and transparency for doubling especially at 0.85μm is required. These requirements were verified at this meeting (1).

While the switching time of conjugated polymers is several orders faster than hybrid lithium niobate and semiconductor devices, the power per bit is in the range of 1 watt, considerably higher than the other optical switching technologies. One focus of research is to reduce the switching element power requirement. According to the device design relationships, power requirements are inversely proportional to achieving a large intensity dependent index of refraction, n_2.

This requirement is reflected in the optical performance comparison of polymers to gallium arsenide/gallium aluminum arsenide multiple quantum wells (MQWs). The figure of merit for relative comparison is the ratio of the energy required to induce switching, given by n_2, to the product of switching time and the absorption coefficient associated with switching. The switch on/switch off time, or recovery period for polymers is femtoseconds compared to nanoseconds for the semiconductors; the adsorption coefficient of polymers is about 1/10,000 that of gallium arsenide. For these reasons the nonresonant polymer figure of merit is high compared to a moderate value for the resonance enhanced MQW case.

While the resonant enhanced n_2 of the MQW is high compared to the moderate nonresonant n_2 for the polymers, nonresonance means that the polymers show this over broadband transparency from the ultraviolet to the near infrared. The high semiconductor n_2 in contrast is wavelength specific, being limited to wavelengths close to the bandgap. Another important attribute of polymers is that they show all the NLO processes, while MQWs show NLO third order, but no optical amplification or third harmonic generation.

In devices based on third order response, nonlinearities of $n_2 >$ 10^{-16} m^2/w and an absorption coefficient of less 0.1 cm^{-1} are required over the next few years depending on application. In the long

302

term $n_2 \mathcal{V} 10^{-14}$ m^2/w at 1.3 and 1.55μm, no vibrational overtone absorption and improved thermal conductivity will be required. While off-resonance nonlinearities of 10^{-10} to 10^{-11} esu were reported for mainchain and side-chain polymers last year, they still imply devices with large operating intensities (2).

Ease of processing is another requirement, particularly for the deposition of optically clear films by nonvacuum techniques. Solubility is the key issue. The films must be able to transmit or guide light in addition to having large nonlinearities. Absorption and scattering losses need to be minimized, a loss of less than 0.1 dB/cm being required for wave guiding. Techniques for the fabrication of optically flat surfaces need to be refined.

NLO POLYMER CLASSES

The current classes of organic NLO polymers under investigation in the AFOSR program are listed in Figure 3. While inorganic polymers and organometallics are also under investigation, metal-containing delocalized electron polymers for third order will be discussed.

STATUS OF SECOND ORDER POLYMER AND DEVICE APPLICATION

Considerable advances have been made in the past year in isotropic polymers, which include glasses, alloys, and blends and composites for functions. Figure 4 shows the results of poled polymers where the active NLO unit is attached to the polymer backbone as a pendant side chain. Control of orientation and symmetry is achieved by poling in an external field at elevated temperatures resulting in second order susceptibilities larger than inorganic crystals (3).

The polymer had a second harmonic generation (SHG) after poling that is several times higher than lithium noibate. In addition, for first time in an NLO polymer the electrooptic coefficient is equal to that of an inorganic crystal. This is a major achievement since it was thought that an organic polymer with the same $\chi^{(2)}$ as an inorganic material would probably have a much smaller electrooptic coefficient r. The r in lithium niobate comes from lattice phonons and the contribution to r in organic polymers was essentially electronic. These polymers had a glass transiton temperature of 120°C. Accelerated life tests indicate that the second order activity should stay within 90% of the original value for five years. In April, 1988 it was reported that optically clear polymers with a loss less than 1dB/cm at 1.3μm and $\chi^{(2)}$ larger than 50 pm/V had been achieved (4). These are spin-coated from common solvents.

In another study on poled isotropic polymers by Marks, Carr and Wong, assemblies of appropriate nonlinear chromophores having noncentrosymmetry as well as high chemical stability and suitable processability are being constructed (5). A computationally efficient SCF-LCAO MECI π-electron theoretical approach has been developed to aid in chromophore design and to better understand molecular electronic/structure architectural features which give rise to high quadratic molecular optical nonlinearities (β). Selected high-β chromophores are then covalently linked via several synthetic procedures to robust, glassy film-forming chloromethylated or hydroxylated polystyrenes. By this procedure, it is possible to achieve very high chromophore densities in polymeric films with good optical transparency and stability characterisitics. Spin- coating of these polymers on ITO-coated conductive glass, followed by drying and poling above Tg yields robust films with high SHG efficiencies. As an example, in Figure 5 films of poly(p-hydroxystyrene) functionalized with N-(nitro-phenyl)-L-prolinol exhibit d_{33} = 18 x 10^{-9} esu at 1065 nm (ca. 16 times the corresponding value for KDP).

These factors which affect the temporal stability of the SHG capacity of the poled poymers are being investigated. For example, the hydrogen bonding network is a factor in extending lifetime. Figure 6 shows there is short decay which is not affected by annealing, the reduced amplitude arising from the unremoved polar solvent. After annealing tests reported after 325 days show that the long decay component is significantly improved.

Enhancement of the NLO effect is expected in pendant side chain structures by making the NLO polar group the side chain mesogen. Until two years ago there had been considerable interest in liquid crystalline polymers for third order NLO effects, particularly in rigid chain structures (6). It was postulated that the natural cooperative alignment of the liquid crystalline molecular structure in cooperation with the highly charge-correlated pi-electron state would lead to large nonresonant nonlinearities.

However, Garito and coworkers concluded in December, 1987 that χxxxx (-3w:w,w,w,) is much more sensitive to the length of the chain than the conformation. Using his recently developed many-electron theory of second and third order nonlinear optical susceptibilities, microscopic descriptions of χxxxx were compared for cis- and trans-polyenes (7).

Several groups in the U.S. and Western Europe have revived interest in liquid crystalline polymers for second order effects using NLO mesogens as backbone or pendant side chain groups, or in combination. At this meeting Garito extended his work to show that both βijk and γijkl of conjugated linear and cyclic chains orginate from electron correlation behavior during virtual pi-electron excitation processes in effectively reduced spatial dimensions (8). Since the nonlinear electronic excitations are naturally confined to quantum length scales, they are highly correlated and obey well-defined symmetry rules. Garito's conclusion that these excitation processes result in βijk values 100x times large than any previously reported values is expected to accelerate efforts dedicated to the synthesis of liquid crystalline polymers for second order effects.

Since 1985, work by Griffin has investigated the synthesis of liquid crystalline polymers where the pendant moiety in these side chain polymers is simultaneously the liquid crystalline (mesogenic) moiety and the nonlinear optical species having a pi-donor/pi-acceptor conjugated electronic structure. To this end polyester liquid crystalline side chain homo- and co-polymers have been synthesized. These materials can be made chiral by empolying a chiral diol in the polymerization step, avoiding the formation of a centrosymmetric system (9).

Griffin and Prasad have fabricated an electrically poled Langmuir-Blodgett film from one such copolymer, (NBSBV)n-(C*SBV)m, and have examined its SHG properties. Electrooptic modulation in a monolayer was observed in the surface plasmon geometry, the measured $\chi^{(2)}$ being approximately 10^{-7} (10). Griffin and Williams have made detailed dielectric measurements on one of these polymers and found it to be dual frequency addressable[11]. Polyamide side chain liquid crystalline polymers have been synthesized with improved retention of polar allignment.

The successful development of guest-host and side-chain polymer systems demonstrating large second order nonlinearities has made polymer thin films viable candiates for integrated optics. The state-of-the-art for second order devices is shown in Figure 7. The most fundamental electrooptic device is the modulator, a thin film waveguide electrooptic molulator employing one of three modulating systems. These are the Mach-Zehnder interferometer, directional coupler, and rotation of the optical polarization. The Mach-Zehnder interferometer is the most common design. This high-frequency traveling wave device has been designed to provide a match in the optical and electron phase velocities (12). Electrical modulating signals, operating at microwave frequencies, travel across electrodes at speeds which must be commensurate with the speed of light within the waveguide to achieve optimum performance.

Since the dielectric constant of the NLO polymers is on the order of 3, the traveling wave device may be designed to achieve close matching of optical and microwave velocities. The large dielectric constant of $LiNbO_3$ results in a rapid loss of phase matching over shorter waveguide lengths. This results in higher drive voltages and power requirements and limiting the frequencies accessible to 8 to 24 GHz. Because phase matching is possible for polymeric materials, the maximum frequency of single mode devices is limited only by electrode losses, being in the vicinity of 50 GHz using conventional electrodes.

Lytel and coworkers recently reported the first measurement of an electrooptic effect in a poled-polymer channel waveguide above a GHz using the pendant side chain polymer shown in Figure 4 (13). Little or no dispersion in the electrooptic coefficient was observed. The losses in these polymers are below a dB/cm, which allows path lengths of several cm, and consequently, half-wave voltages approaching TTL levels.

This remarkable demonstration of performance was achieved by the concurrent development by Lytel of the selective poling procedure (SPP), in which active, poled channel wavelengths are fabricated in the aforementioned polymers in a single step. Only those regions of the material defined by the electrode pattern on the substrate are poled. Three prototype devices have been fabricated: a travelling-wave phase modulator, demonstrating 1.3 GHz modulation; a directional coupler; and a Y-branch interferometer.

Results of the research are also being adopted to meet a US Air Force requirement for a compact laser diode source in the 450 to 500 nanometer range for improved tactical optical data storage (14). The high-efficiency frequency doubler is being developed also with one of the pendant side chain polymers of Figure 4. High speed radar signal processing with 20 GHz as a target is also being pursued with NLO polymeric materials.

In a later section the possibilities for second order polymers in implementing neural networks for optical computing are discussed.

STATUS OF THIRD ORDER POLYMER MATERIALS AND DEVICES

One of the goals of the current third order polymer research is to design multifunctional polymers having unique combinations of semiconductor, NLO and structural properties.

The objectives are to: (1) design polymers for maximum reliable nonlinear optical susceptibilities; (2) synthesize these electroactive polymers in a form exhibiting reasonable solubility in conventional solvents so that the polymers can be purified, characterized, and processed by convenient methods; and (3) explore the effects of charge transfer upon nonlinear optical activity.

Numerous theoretical calculations predict that optical nonlinearity in π electron systems increases with increasing π electron delocalization (or conjugation) (15). Electron delocalization depends upon orbital overlap between π-orbitals on adjacent atoms. Consideration of orbital overlap can be divided into two spatial factors. First, that overlap depends in an approximately exponential manner on the bond distance between atoms. Second, overlap depends upon the relative orientation of interacting orbitals with greater overlap observed for colinear π orbitals. From these considerations Dalton has pointed out that ladder polymers, with their planar configuration and uninterrupted π conjugation exhibit the molecular conformation necessary for optimum orbital overlap and hence maximum electron delocalization. These observations are supported by recent theoretical calculations. Also, ladder polymers exhibit symmetries appropriate for supporting polaron and bipolaron mid-gap state species. That is ladder polymers can be doped to charged lattices which will clearly exhibit different linear and nonlinear optical properties than the pristine polymers.

Because of the potential for optimum electron delocalization and because of the potential for supporting stable, charged lattices, ladder polymers were chosen as the most likely candidates to yield large nonlinear optical susceptibilities (15).

Because of strong interchain Van der Waals interactions deriving from extensive electron delocalization and because of the rigid (low entropy) nature of ladder polymers, these materials have traditionally exhibited poor solubility which in turn has resulted in low polymer molecular weights and difficulty in processing these polymers. It is important to destabilize polymer-polymer interactions relative to polymer-solvent interactions. A logical approach to this objective is to introduce steric interactions associated with substituent groups which prevent tight packing of the ladder polymers. Thus, as a means of achieving improved polymer solubility and processibility, ladder polymers are derivatized by synthesizing such polymers from appropriately derivatized monomers.

Ladder polymers have traditionally been prepared by polycondensation (15). The condensation reactions are required to complete each rung of the ladder and, in general, the two condensations occur at different rates (the condensations are thus referred to as being asymmetrical). To insure complete condensation and the fully-cyclized ladder structure yielding optimum electron delocalization, it is necessary to understand the polycondensation kinetics in detail. Moreover, the opportunity exists, by exploitation of the asymmetric polycondensation kinetics, to influence required polymer processing (e.g., fabrication of thin films) on the open-chain (or partially condensed) precursor polymer. Because of greater entropy and a conformation less favorable for polymer-polymer interaction, the precursor polymers are more soluble. Thus, an important route to processed ladder polymers, which have been purified and characterized by standard techniques, involves preparation and processing of precursor polymers and conversion to the final ladder polymer by thermal treatment.

Substituents can be used to fine tune the solubility of both precursor and ladder polymers in various solvents including both polar and nonpolar solvents. Solubility in water is influenced by terminating substituents with sulfonic acid or tetraalkylammonium groups; solubility in organic solvents can be influenced for alkyl, alkoxy, vinylamine, etc. groups.

Derivatization and precursor polymer synthetic methods represent the cornerstone of the approach (15). This is very likely the most successful and systematic approach to the preparation of electroactive polymers.

One of the most straightforward schemes involves the preparation of dichloroquinone manomers derivatized with vinylamine groups employing a Mannich reaction (15). Polycondensation of these monomers with aromatic amines can be carried out in simple solvents such as dimethylformamide (DMF) at ambient temperatures to yield highly processible open-chain precursor polymers. Optical quality thin films of the precursor polymer can be fabricated and converted to optical quality ladder polymer films by heating the films at approximately 250°C.

As-synthesized polymers yield third order susceptibilities as high as $3 \cdot 10^{-9}$ esu; these large values can be attributed to the combined effects of pi electron delocalization and charge transfer from the vinylamine substituent to the polymer backbone (Figure 8) (15). The experimentally observed dependence of nonlinear optical activity upon pi-orbital overlap is in agreement with recent calculations of Medrano and Goldfarb of Air Force Wright Aeronautical Laboratory/Materials Laboratory for various lengths and conformations of rigid-rod macromolecules (16). Consideration of trends observed to date suggest that nonlinear susceptibilities may be increased an additional one to two orders of magnitude by careful attention to optimizing electron delocalization. Third order susceptibilities in the range 10^{-8} to 10^{-11} esu were observed for pristine polymers. At the time of observation, these were the largest values of third order susceptibility observed for any organic material. Recently, comparable values have been observed for related systems by a number of groups including Prasad (17) and Garito (18).

The greatest variability in measured third order susceptibilities is observed for vinylamine-derivatized ladder polymers and can be attributed to interruption of pi electron conjugation associated with defects in the ladder structure. Defects have been identified as arising from air oxidation of imine nitrogens or incomplete condensation (Figure 8); air oxidation disrupts pi delocalization by introducing sp^3 centers. This work is unique in representing the first quantitative characterization of the perfection of the lattice for an electroactive polymer. The results suggest approaches for realizing the optimum optical nonlinearity predicted for ladder polymers.

Preliminary studies suggest that it is possible to enhance nonlinear optical susceptibility by more than an order of mangitude by chemical or electrochemical doping (15). Moreover, in some cases, doping leads to improved transparency at visible light frequencies.

Dynamics of the dialkyl portions of the vinylanmine substituents appear to modulate coupling of the vinylamine pi-orbitals with the pi-system of the polymer backbone influencing the magnitudes of both the electronic and thermal components of the degenerate four wave mixing (DFWM) signal. Alicyclic substituents appear to be more sterically hindered leading to large electronic components of the third order susceptibility and to reduced thermal tails in the temporal response.

Very preliminary studies on oxidized and reduced (charged) lattices suggest that third order susceptibilities approaching 10^{-7} esu may be realized for doped organic materials (15). Some support for this result is provided by preliminary studies of stacked organic metals (19). However, the frequency dependence of the optical nonlinearity needs to be carefully defined and the nature of the modified polymer lattice characterized before doped materials can be considered for device application.

When ladder polymers are systematically cycled through their voltammetry, dramatic color changes are observed. Magnetic and optical measurements suggest that these changes can be associated with the generation of polaron and bipolaron states. For a number of systems, the optical phenomenon can be characterized as bleaching of the (visible)π-π* interband transition together with the generation of (infrared) intraband transitions (Figure 9). Not only does this process produce a new optical lattice (which may or may not have greater optical nonlinearity but will most certainly exhibit different optical nonlinearity) but the phenonemon can immediately be applied to develop an optical switch (although the speed will be limited by the electrochemical response time).

Ladder polymers studied in this work exhibit unusually high optical damage thresholds (approximately 1 GW/cm^2) and show no saturation of the nonlinear response before damage (15). Absence of saturability suggests nonresonant optical nonlinearity while linear absorption suggest some resonance absorption at measurement frequencies (e.g., 532 nm); the existence of both localized and delocalized electronic state and the existence of a variety of delocalized states (e.g., exciton, polaron, bipolaron, conduction band) may result in a decorrelation of linear and nonlinear effects.

The demonstrated concept of Dalton and Hellwarth for designing laser resistant polymers with χ^3 of 10^{-8} to 10^{-9} esu and transparency in the visible has set the direction for a new US Army - US Air Force - Defense Advanced Research Projects Agency (DARPA) program for sensor and vision protection with importance to the NATO alliance. The envisioned way of protecting high gain optical military systems such as rangefinders, visual sensors, cameras, and the human eye against a rapidly tunable, pulsed, visible and near IR laser threat is predicated on demonstration of hybrid device concepts, potentially employing multiple NLO polymers. The device requirements dictate that the polymers should possess subnanosecond response times and millisecond recovery speeds, and should have a broadband wavelength response (over visible and near IR). The most vigorous performance parameters are for eye protection, demanding responses in the picosecond/picowatt domain or smaller, 100% transparency in the unradiated state, and should be able to dissipate 100% of the over-threshold energy incident upon the divice. An example is an optical limiter.

In another advance, organometallic polymer research by Garito, Dalton, Buckley and Prasad as well as Wegner has shown that very significant third order effects are obtained using pthalocyanine metal organic derivatives.[15,20-22] Garito has shown that planar structures with extended pi-orbitals will have high activity and speeds through saturable absorption (Figure 10) (23). Effective n_2 equal to AlGaAs mulitquantum structures are potentially achievable. Dalton has shown that lattice charge in addition to electron delocalization (interorbital spacing optimization) is another important factor to enhancing χ^3 in metal containing delocalized electron polymer systems.[15]

A variety of organometallic macrocycle materials have been prepared and preliminary nonlinear optical measurements carried out. The observed behavior for pthalocyanines shows considerable variablitlity in optical nonlinearity with minor changes in chemical structure of the metallomacrocycle. These phenomena my be related to interstack dipole coupling.

Recent values of third order susceptibility to 10^{-7} esu have been realized in isolated cases. Dalton has shown that ultimately it will be important to extend consideration of oxidation/reduction (lattice charging effect) to the consideration of metal containing delocalized electron polymer systems as the variable redox states of metal should permit greater lattice change to be introduced and should permit realization of mixed valence centers.

Recent results by Singh, Prasad and Karasz, and Drury and Lusignea have developed high optical quality, highly oriented uniaxial poly(p-phenylene vinylene) (PPV) (24,25). Since the third order nonlinear optical susceptibility is a fourth rank tensor, a large orientational anisotropy has been observed. Device structures based on this anisotropy have been proposed using the concept of orientational bistability (26). Femtosecond degenerate four wave mixing at 602 and 580 nm on a 10:1 stretch-oriented film shows a χ^3 value along the draw direction of 4×10^{-10} esu with a subpicosecond response.

The change in χ^3 value (actually the square root of the DFWM signal) as a function of film rotation with respect to the incident electric field vector yields the polar plot shown in Figure 11. The highest value of χ^3 is obtained when the electric vectors of all the four waves are parallel to the draw direction. The minimum value for χ^3 is for the orientation when all the electric vectors are perpendicular to the draw direction. The $\chi^{(3)}_{\parallel}/\chi^{(3)}_{\perp}$ ratio is 37 which is a very high degree of orientational anisotropy. As shown by Xray, there is a high degree of polymer chain alignment along the draw direction (27). This research then confirms that the largest component of χ^3, and therefore the microscopic nonlinearity tensor γ, is along the chain direction as is predicted from theoretical calculations of microscopic nonlinearity in pi-conjugated polymeric and oligomeric structures.

The PPV can be conveniently processed into various shapes using a water soluble precursor route. The oriented fibers have good mechnical strength such that even a two micron thick free-standing film can be prepared. Doping can produce high electrical conductivity along the draw direction, indicating a high effective pi-conjugation in this polymer--a criterion need for large χ^3. The film exhibits high optical damage threshold under picosecond and femisecond pulse illumination.

The state of the art for $\chi^{(3)}$ devices is shown in Figure 12. Internationally there has been far less progress in $\chi^{(3)}$ device development than in $\chi^{(2)}$ devices. The reason for this is that most of the research efforts in the United States, Europe and Japan have been in second order polymers.

While nonresonant third order nonlinearities of 10^{-10} to 10^{-11} esu had been reported one year ago, two orders of magnitude higher had to be achieved for practical device application. As reported in this section, polymers have now been achieved which demonstrate $\chi^{(3)}$ of 10^{-8} to 10^{-9} esu. While 10^{-10} esu was on the periphery for all-optical waveguide devices, these polymers could still find use in bistable optical switches and optically-controlled modulators and switches.

The new third order polymers will now make all-optical waveguide devices a realistic possibility. Third-order integrated optical devices will include distributed couplers, Mach-Zehnder interferometers and Bragg reflectors. The optical Mach-Zehnder modulator will utilize $\chi^{(3)}$ polymers which exhibit intensity dependent refractive indices (29). In a typical Mach-Zehnder switch, the input beam is divided into two channels. When a high intensity modulating beam is introduced into a channel, the refractive index of the channel changes and it creates a phase difference of the input beam in the channel relative to the beam in the other channel. If the phase difference is equal to pi, the two cancel each when they are recombined. A fast optical Mach-Zehnder switch may be used for digital signal processing.

Optical bistability has been observed in a poly-4-BCMU polydiacetylene polymer quasi-wavelength interferometer by Singh and Prasad (30). Biaxial NLO polymers as demonstrated with the rigid rod aromatic heterocyclic polymer, poly(p-phenylene-2, 6-benzobisthiazole), or PBT, and PPV offer additional device design options based on polarization bistability (31).

Thin film devices such as etalons are now possible with the improved $\chi^{(3)}$ polymer (32). A requirement defined at this workshop for etalons over the next 3-5 years were polymers with $\chi^{(3)} \chi 10^{-7}$ esu and $\alpha \stackrel{<}{=} 0.1$ cm.

There will be large payoff in fast optical multiplex switching, based on high n_2 polymers, which combines optical signals from many channels converted to a single temporal signal for easy transmission (33). As a high intensity pulse from a picosecond mode-locked laser transverses through a high n_2 medium, it sequentially opens the channels and thus creates a temporal signal. A similar design is used to demultiplex the transmitted signal to retrieve the individual signals. This can lead to digital optical information processing and other applications cited in Figure 12.

Device design with $\chi^{(3)}$ materials is often guided by the architecture or electrical parameters. Thus careful investigation of polymer properties concurrent with their operational performance in device prototypes should lead to parallel analog processing of 2-D images and other applications cited in Figure 12. For example the optical Kerr cell design can be used also for analog optical signal processing such as optical scanning.

NLO POLYMERS IN OPTICAL NEURAL NETWORKS

At AFOSR work is starting which will investigate the role of second order polymers in nonlinear optical neural nets and architectures. Neural networks are models of computation that are based on the way brains perform their computation. The human brain has somewhere on the order of 10^{11} to 10^{12} neurons, massive numbers of computational units. Each neuron is locally connected to 10^3 to 10^4 other neurons. They differ from serial networks, illustrated by digital computers, in interconnectivity because they are able to perform certain tasks very efficiently by parallel signal processing.

Optical implementation is particularly attractive (34). In comparison to electronic devices, optical devices are inherently parallel and thus not limited by wire interconnections and cross-talk. Figure 13 compares the Figure of Merit for an optical neurocomputer with that of the brain and Very Large Scale Integration. As Malloy and Giles at AFOSR point out, 10^6 neurons is reasonable to achieve for an optical array, 10^{12} as in the human brain being very difficult.

There are several ways to implement optical interconnections, but most are limited by the dimensionality of interconnection medium. For the basic structure of an optical neural network (Figure 14), the interconnection cloud could be a two-dimentsional spatial light modulator or a three-dimensional volume holographic element. The restrictions for the optical implications of neural nets have yet to be defined. Architectures, dynamics and representation of information have to be expressed optically; models are starting to emerge. The architecture-imposed requirements on device design is an unknown which impacts the material properties requirement.

For the basic structure of an optical neural network (Figure 14), the interconnection cloud could be a two-dimensional volume holographic element (34). In the near-term there is a need one- and two-dimensional arrays with 10^3 and 10^6 elements respectively with response times in the microsecond to nanosecond range. Multiple quantum well self-electrooptic effect devices (SEED) have a variety of desirable functions for optoelectronic neural net device arrays including wavelength compatibility with semiconductor sources and detectors, voltage levels compatible with electronic circuitry, transparent substrates, and the ratio of fast index change for switching to resultant thermal index change. (Malloy and Giles point out that in its simplest form the SEED switches off when a threshold intensity of light is reached).

Malloy and Giles see the ability to understand, control and tailor the nonlinear excitonic mechanisms in the III-V semiconductor multiple quantum wells as compared with the current state of understanding in polymers as the primary point of this section. Material requirements are still difficult to delineate because of the vague and equivocal device requirements. In the long term tailored polymeric materials are seen as viable for designing arrays; however, it remains to be determined how $\chi^{(2)}$ and $\chi^{(3)}$ polymers can mimic neuron slabs and arrays and how they can optically implement interconnections in optical, optoelectronic and hybrid optical/electronic architectures.

Malloy and Giles point out that if the optical device works on resonant transitions sensitive to temperature, many applications are ruled out (34). The most recent approaches for $\chi^{(2)}$ and $\chi^{(3)}$ polymers in a nonresonant or near off-resonance mode offer some promise here provided that the calculated large low n_2 and absorptions can be achieved. In addition to being competitive with the aforementioned attributes for SEED optics, the electronic subpicosecond response times and other temporal responses of polymers offer considerable latitude for designing large element arrays working in the 10 nanosecond to 100 picosecond individual element range. Second order polymers in photo-addressable spatial light modulators (SLMs) operating at room temperature or third order polymers in optically bistable polymer etalons offer appraoches for implementation of optical neural net designs.

FUTURE DIRECTIONS FOR RESEARCH

The first generation of NLO polymers, developed over the period from 1983 to 1989, focused on homopolymers (figure 15). Large second order effects have been developed in poled isotropic and oriented polymers. Third order effects have been developed in bond alternation polymers, including ladder, ordered, conducting and semiconducting polymers.

The next generation of second order polymers (figure 15) are expected to be based on self-induced poling and polarization and self-organization. Third order effects will be developed in polymers based on near-off resonance phenomena and planar structures. Second and third order polymers are expected to take the form of molecular blends and alloys and molecular composites. The AFOSR and Air Force Wright Aeronautical Laboratory/Materials Laboratory have planned a major initiative, which starts on 1 October 1989, to address polymer-polymer interactions for NLO and polydispersity. Emphasis will be on organometallics and inorganic polymers and theoretical calculations and modeling.

Work in our program is also proceeding for the development of polymer quantum well structures as shown in Figure 16. The tailoring of macromolecular architecture emphasizes in-situ polymerization and ordering, liquid crystallinity, guest-host blends and rigid/flexible chain conformations.

SUMMARY

Some of the recent advances made in the synthesis and characterization of NLO polymers have been discussed. Poled isotropic polymers now perform as well as inorganic crystals. The chemistry which contributes to long term poling stability is starting to be understood. Considerable advances have been made in third order polymer design, synthesis and purity. Optical transparent ladder polymers with $\chi^{(3)}$ commonly measured at 10^{-9} esu and some samples with 10^{-8} esu are being reported. The mechanisms for improving visible transparency through doping have been delineated in these polymers. They are now the subject of serious investigation for sensor and vision protection against pulsed laser threats. Mechanisms for synthesizing metal containing delocalized electron polymer systems with n_2 that can theoretically approach MQWs have been demonstrated through saturable absorption and lattice charging. A few isolated measurements of $\chi^{(3)} \sim 10^{-7}$ esu have recently been reported. High optical quality, stretch oriented PPV with a large anisotropy value in the $\chi^{(3)}$ now makes orientational bistability devices a reality.

Results of the $\chi^{(2)}$ polymer research are already being adopted to meet Air Force and DOD requirements for high-efficiency frequency doublers, travelling-wave phase modulators, optical shutters, waveguide multiplexers, and electro-optic Bragg cells. While $\chi^{(3)}$ polymer research is moving rapidly into sensor protection devices, $\chi^{(3)}$ polymer development is just now starting to approach device quality in terms of $\chi^{(3)}$ of 10^{-7} to 10^{-8} esu. Significant work still needs to be done to achieve optically transparent polymers with ≥ 0.1 cm^{-1}.

Both the second and third order polymer systems are long term candidates for optical implementation of neural net architectures. Material requirements are still difficult to delineate because of the vague and equivocal device requirements. Polymers are seen as viable for designing arrays because of their nonresonant mode promise.

ACKNOWLEDGMENTS

Special appreciation is expressed to Ms. Donna Proctor for her help in the preparation of this paper as well as to Dr. Donald L. Ball for his encouragement and support. We are indebted to the dedication, achievement and enthusiastic inspiration of all the investigators sponsored by AFOSR in this area, many of whom could not be listed here.

REFERENCES

1. NATO Workshop, Nonlinear Optics in Polymers, NICE-Sophia Antipolis, June, 1988.

2. D. R. Ulrich, "Nonlinear Optical Polymer Systems and Devices," MOL. CRYST. LIQ. CRYST., 1988 (in press); p. 8

3. J. Riggs and J. Stamatoff, private communications; R. DeMaritino, et. al., NONLINEAR OPTICAL PROPERTIES OF POLYMERS, Materials Research Society Proceedings, Volum 109, A. J. Heeger, J. Orenstein, and D. R. Ulrich, Eds. Materials Research Society, Pittsburgh, Pennsylvania, 1988: pp. 65-77, J. Stamatoff, AFOSR Contract F49620-87-C-0115

4. J. Stamatoff, AFOSR Program Review, National Academy of Sciences, April 20-21, 1988; A. Buckley and J. Stamatoff, NATO Workshop, Nonlinear Opticas in Polymers, NICE-Sophia Antipolis, June 1988; J. Stamatoff, AFOSR Contract F49620-87-C-0115.

5. T. Marks, S. Carr and G. Wong, AFOSR Program Review, National Academy of Sciences, April 20-21, 1988; NATO Workshop, Nonlinear Optics in Polymers, NICE-Sophia Antipolis, June, 1988; T. Marks, AFOSR Grant AFOSR-86-0105.

6. D. R. Ulrich, SPIE OPTICAL ENGINEERING REPORTS, pp. 5A-7A, No. 43, July, 1987; D. R. Ulrich, POLYMER, Vol. 28, pp. 533-542, 1987, D. N. Rao, et. al., APPL. PHYS. LETT. Vol. 48, pg. 1187 (1986); A. F. Garito, et. al., MOLECULAR AND POLYMERIC OPTOELECTRONIC MATERIALS: FUNDAMENTALS AND APPLICATIONS, Proceedings SPIE, 482, G. Khanarian, 1986; pp. 2-11.

7. A. F. Garito, NONLINEAR OPTICAL PROPERTIES OF POLYMERS, Materials Research Society Proceedings, Volume 109, A. J. Heeger, J. Orenstein, and D. R. Ulrich, Eds., Materials Research Society, Pittsburgh, Pennsylvania, 1988; pp. 91-101; A. F. Garito, AFOSR Contract F49620-85-C-0105.

8. A. F. Garito, "Recent Developments in Nonlinear Optical Properties of Polymers," Abstract 2.2, International Conference on Organic Materials for Non-Linear Optics, Oxford, England, June 29-30, 1988; A. F. Garito, AFOSR Program Review, National Academy of Sciences, April 19-20, 1988, A. F. Garito, AFOSR Contract F49620-85-C-0105.

9. A. C. Griffin, AFSOR Program Review, National Academy of Sciences, April 20-21, 1988; A. C. Griffin, AFOSR Grant AFOSR-84-0249.

10. M. M. Carpenter, P. N. Prasad, and A. C. Griffin, "The Characterization of Langmuir-Blodgett Films of a Nonlinear Optical, Side Chain Liquid Crystalline Polymer," Thin Solid Films (in Press); P. N. Prasad, AFOSR Contract F49620-87-C-0042?; A. C. Griffin, AFOSR Grant AFOSR-84-0249.

314

11. A. C. Griffin and C. Williams, private communication; A. C. Griffin, AFOSR Grant AFOSR-84-0249; C. Williams, AFOSR Contract F49620-87-C-0111.

12. G. I. Stegeman, C. Seaton, and R. Zononi, NONLINEAR OPTICAL PROPERTIES OF POLYMERS, Materials Research Society Proceedings, Volume 109, A. J. Heeger, J. Orenstein, and D. R. Ulrich, Eds., Materials Research Society, Pittsburgh, Pennsylvania; pp. 53-64; R. Lytel, G. F. Lipscomb, and J. I. Thackara, ibid.: pp. 19-28; J. Stamatoff and J. Riggs, private communication; O. K. Kwon, F. R. W. Pease and M. R. Beasley, IEEE ELECTRON DEVICE LETTERS, USA, Vol EDL-8, No. 13, pp 582-585, December 1987

13. R. Lytel, NATO Workshop, Nonlinear Optics in Polymers, NICE-Sophia Antipolis, June, 1988; R. Lytel, "Advances in Organic Integrated Optic Devices," Abstract 3.5, International Conference on Organic Materials for Nonlinear Optics,: June 29-30, 1988 Oxford, England; R. Lytel, AFOSR Program Review, Natonal Academy of Sciences, April 20-21, 1988.

14. Recent Research Accomplishments of the Air force Office of Scientific Research, 1988, Bolling Air Force Base, Washington, D.C. 20332-6448, U.S.A. (in press).

15. L. R. Dalton and R. Hellwarth, AFOSR Contract F49620-87-C-0010; L. Dalton, AFOSR Contract F49620-85-C-0096; L. Dalton, NATO Workshop, Nonlinear Optics in Polymers, NICE-Sophia Antipolis, June, 1988; L. Dalton, AFOOSR Program Review, National Academy of Sciences, April 20-21, 1988; L. Dalton, NONLINEAR OPTICAL PROPERTIES OF POLYMERS, Materials Research Society Proceedings, Volume 109. A. J. Heeger, J. Orenstein, and D. R. Ulrich, Eds., Materials Research Society, Pittsburgh, Pennsylvania, 1988: pp. 301-312; L. R. Dalton, NONLINEAR OPTICAL AND ELECTROACTIVE POLYMERS, P. N. Prasad and D. R. Ulrich, Eds., Planum Press, New York, 1988; pp. 243-272.

16. J. Medrano and I. Goldfarb, NATO Workshop, Nonlinear Optics in Polymers, NICE-Sophia Antipolis, June, 1988.

17. P. N. Prasad, private communication.

18. A. F. Garito, private communication.

19. P. G. Huggard, W. Blau, and D. Schweitzer, APPL. PHYS. LETT., Vol. 51, p. 2183 (1987).

20. A. F. Garito, AFOSR Contract F49620-85-C-0105.

21. P. N. Prasad, AFOSR Contract F49620-87-C-0097.

22. A. Buckley, AFOSR Contract F49620-86-C-0129.

23. A. F. Garito, AFOSR Contract F49620-85-C-0105.

24. B. P. Singh, P. N. Prasad, and F. E. Karasz, POLYMER (in press); F. E. Karasz, AFOSR Contract F49620-87-C-0027; P. N. Prosad, AFOSR Contract F49620-87-C-0097.

25. M. Druy and R. Lusignea, AFOSR Contract F49620-88-C-0065.

26. K. Otsuka, J. Yumoto, and J. J. Song, OPT, LETT., Volume 10, p. 508, 1985.

27. D. R. Gagnon, F. E. Karasz, E. L. Thomas, and R. W. Lenz, SYNTH. METAL., Volume 20, p. 85, 1987.

28. D. R. Ulrich "Nonlinear Optical Polymer Systems and Devices," MOL. CRYST. LIG. CRYST., Volume 160, pp 1-32, (1988).

29. J. Riggs, private communicaton; G. Stegeman, private communication.

30. b. P. Singh and P. N. Prasad, J. OPT. SOC. AM. B5, 453 (1988).

31. R. Lytel, private communication; R. Lusignea and L. Domash, private communication; P. N. Prasad private communication; D. N. Rao, et. al., APPL. PHYS. LETT., Vol. 48, p. 1187 (1986); A. F. Garito, et. al., MOLECULAR AND POLYMERIC OPTOELECTRONIC MATERIALS: FUNDAMENTALS AND APPLICATIONS, Proceedings SPIE, 682, 1986: pp. 2-11.

32. R. Lytel, et. al. NONLINEAR OPTICAL AND ELECTROACTIVE POLYMERS, P. N. Prasad and D. R. Ulrich, Eds., Planum Press, New York (1988) pp. 415-426.

33. J. Riggs, private communications G. B. Kushner and J. A. Neff, NONLINEAR OPTICAL PROPERTIES OF POLYMERS, Materials Research Society Proceedings, Volume 109, A. J. Heeger, J. Orenstein, and D. R. Ulrich, Eds., Materials Research Society, Pittsburgh, Pennsylvania, 1988; pp. 3-17.

34. K. J. Malloy and C. L. Giles, NONLINEAR OPTICAL PROPERTIES OF POLYMERS, Materials Research Society Proceedings, Volume 109, A. J. Heeger, J. Orenstein, and D. R. Ulrich, Eds., Materials Research Society, Pittsburgh Pennsylvania, 1988; pp. 77-87.

35. K. J. Malloy and C. L. Giles, private communication.

36. D. Psaltis, special issue of APPLIED OPTICS on "Neural Networks," Vol. 26, 1987; N J. Malloy and C. L. Giles, private communication.

316

Why NLO Polymers?

- Subpicosecond Response Times
- Large, Nonresonant Nonlinearities
- Low DC Dielectric Constants
- Low Switching Energy
- Broadband
- Low Absorption
- Absence of Diffusion Problems
- Potential for Resonant Enhancement
- Ease of Processing and Synthesis Modification
- Room Temperature Operation
- Environmental Stability
- Mechanical and Structural Integrity

Fig. 2   The advantages of NLO polymers.

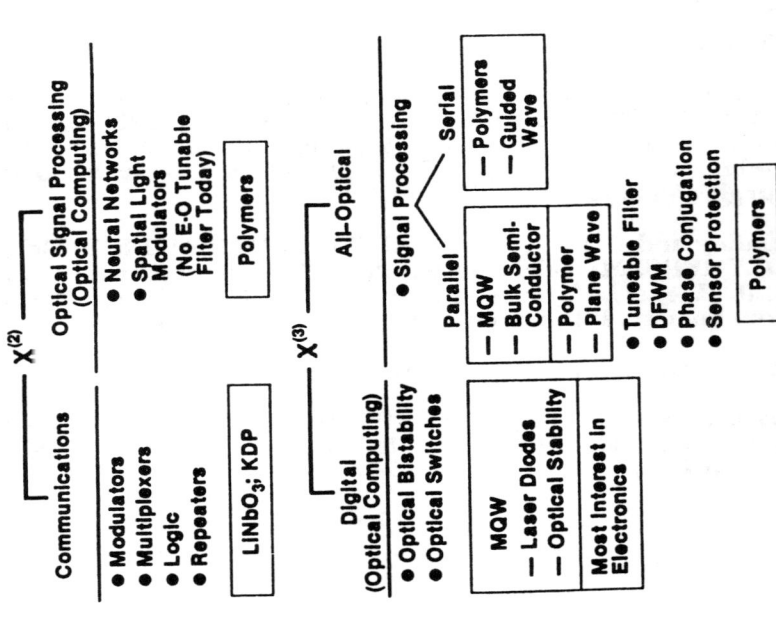

Where Will NLO Polymers Fit In?

Fig. 1   The role of polymers in nonlinear and electrooptical applications.

Status NLO Polymer Classes

Class	Examples	NLO Function
ISO Tropic	Glasses Alloys Composites	$\chi^{(2)}$, $\chi^{(3)}$
Bond-Alternation	Ladder Polymers PTL, PQL Polyacetylene Polythiophene	$\chi^{(3)}$
Liquid Crystalline Polymers (LCP)	Side Chain LCPs	$\chi^{(2)}$
Rigid Rod Aromatic Heterocyclics	PBT LCPs PBO BBL	$\chi^{(3)}$
Polydiacetylenes		Mostly $\chi^{(3)}$ Some $\chi^{(2)}$

Fig. 3   Status of nonlinear optical and electrooptical polymers.

Second Order Polymers for Electrooptical Devices

● **Pendant Side Chain Structure**

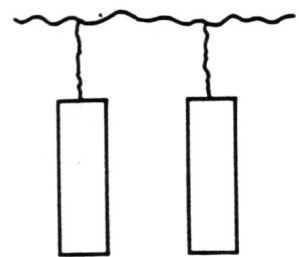

● **High Activity**	Polymer	LINbO$_3$
For SHG	$\chi^{(2)}$ = 120 pm/V	10 pm/V
For Electrooptics	r = 35 pm/V	30 pm/V
FOM = $\frac{r}{\varepsilon}$	10	1

● **Excellent Secondary Properties**

   Spin Coatable for Thin Film Waveguides,
   2-4 Micron

   Low Dielectric Constant ($\varepsilon_{Polymer}$ = 3;
   ε_{LINbO_3} = 30)

   Low Loss (<1 db/cm at 830 nm)

   Melt Processable for Optics

   Tg ~ 120°C

Fig. 4   Comparison of poled second order polymers with the inorganic crystal lithium niobate.

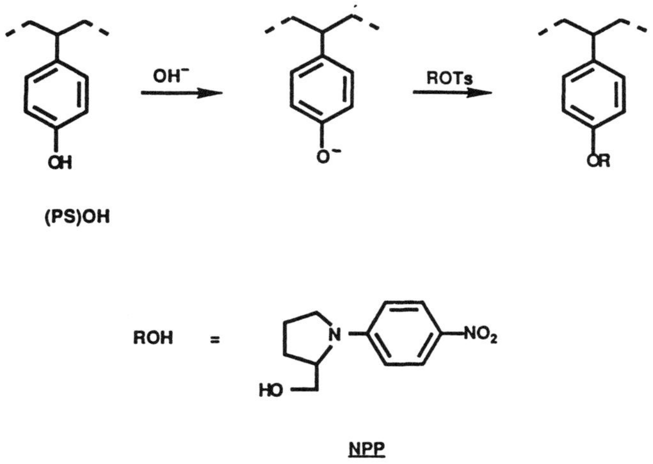

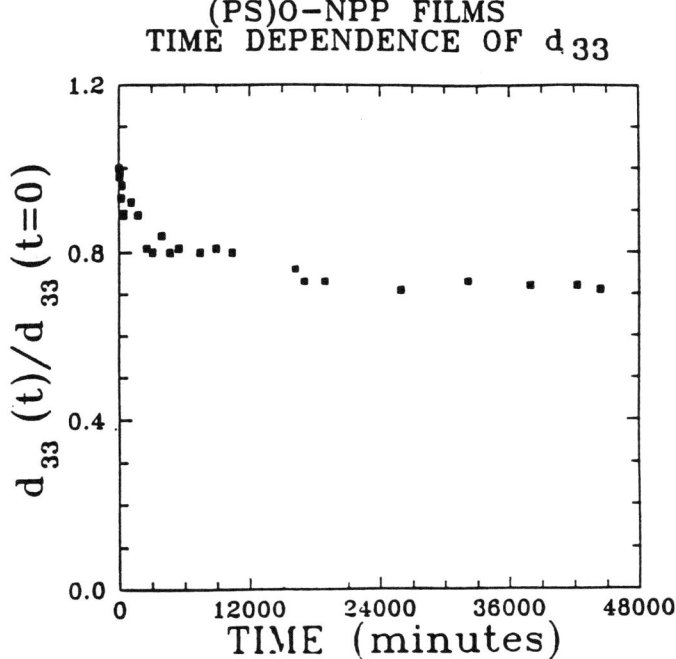

Fig. 5 **Functionalization** of **poly(p-hydroxystyrene)** with **NLO** chromophores.

Fig. 6 **Temporal stability of poled poly(p-hydroxystyrene) with NLO chromophores.**

● Traveling-Wave Electrode Mach-Zehnder Electrooptic Modulator

	LiNbO₃ Modulator	Polymer Modulator
Switching Voltage (V)	3 1/2–10 1/2	1.3 (0.7 With Higher $X^{(2)}$)
Power Requirement (W)	0.6–5	0.03
Maximum Frequency (GHz)	8–24	>50

— No Expected Velocity Mismatch Implies Higher Frequency
  Devices are Possible With Polymers

● Electrooptic Bragg Cell
  — High Speed Radar Signal Processing
  — 20 GHz as a Target

● Second Harmonic Generation
  — High Efficiency Doubling of a Diode Laser

Fig. 7   Device state-of-the-art with second order polymers.

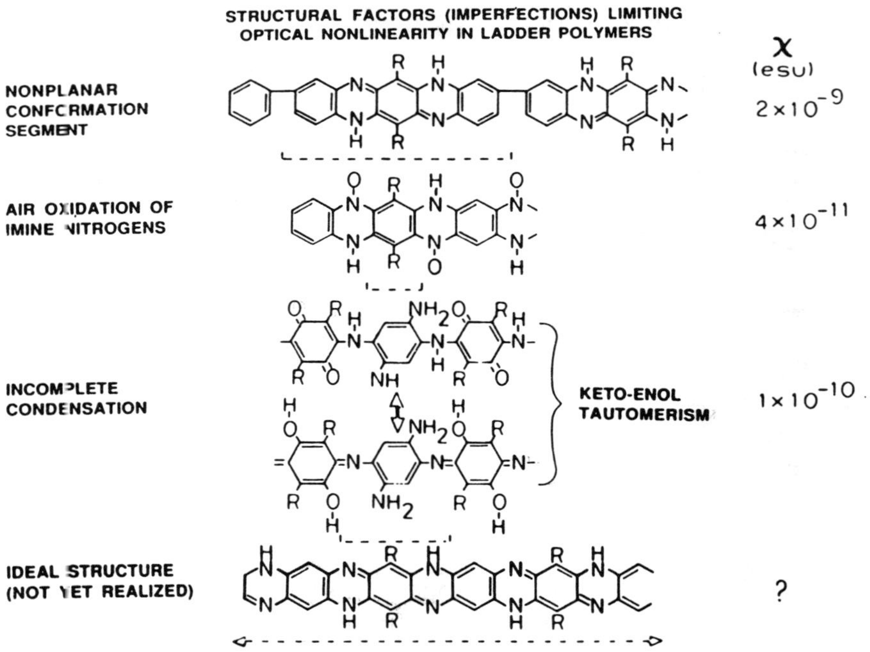

Fig. 8   Structural design factors of NLO ladder polymers.

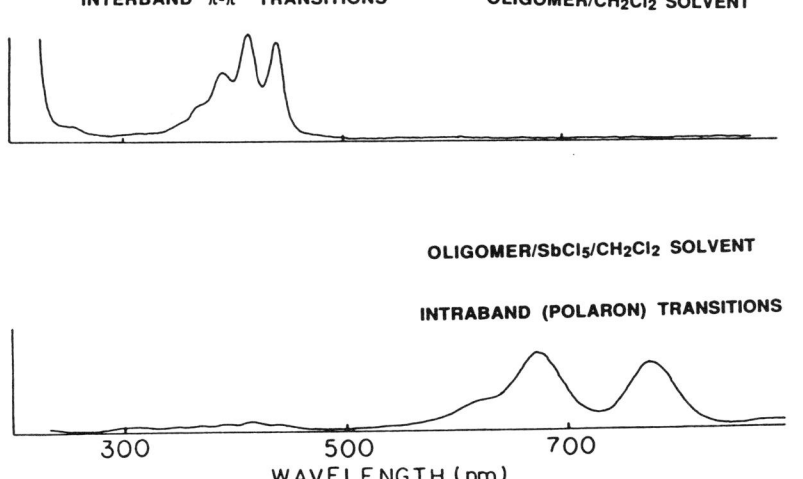

EFFECT OF CHEMICAL DOPING ON THE LINEAR OPTICAL
SPECTRUM OF A DELOCALIZED ELECTRON OLIGOMER:
INTRODUCTION OF VISUAL TRANSPARENCY AND
GENERATION OF A CHARGED (POLARON) LATTICE

Fig. 9 Introduction of visible spectrum transparency in NLO doped ladder polymers.

● Planar Structure With Extended π Orbitals

M = Metal Atom

● Shift in Focus From Conjugated Off-Resonance to Systems on Resonance With Narrow Molecular Extinction Coefficients

● High Activity Through Saturable Absorption
— Effective N_2 Equals AlGaAs MQW Structures

● Excellent Secondary Properties
— Spin Coatable For Thin Film Structure Applications

Optical Bistability
Parallel Processing

Fig. 10 Third order saturable absorption polymers.

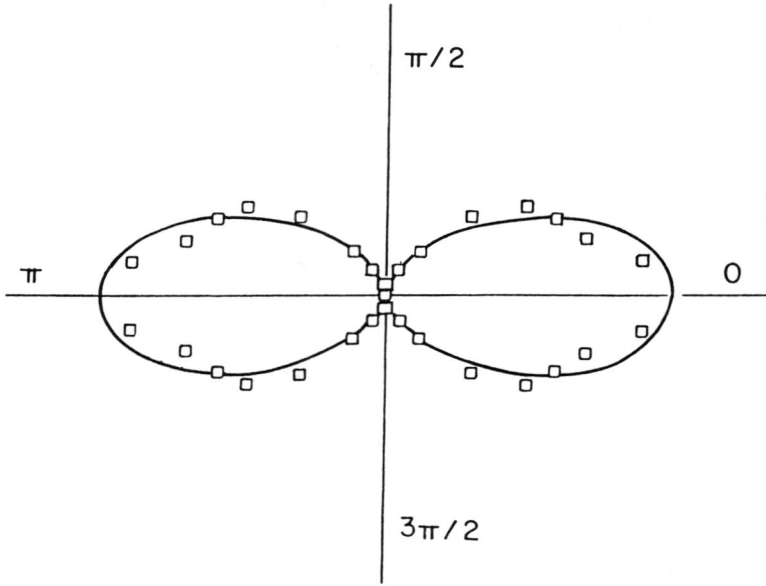

Fig. 11 The polar plot (orientational anisotropy) of the square root of the degenerate four wave mixing signal intensity (proportional to $\chi^{(3)}$) for the 10:1 stretch-oriented PPV film. The squares are the observed data points; the solid curve is the theoretical fit.

- Far Less Progress Than $X^{(2)}$

- Large Payoff in Multiplexing and Demultiplexing
  - High Speed
  - Handle Many Inputs
  - Very High $X^{(3)}$

- Can Lead to:
  - All Optical Interferometer and Optical Switch
  - Optical Bistability and Digital Optical Information Processing
  - Optically Induced Dynamic Grating and Real-Time Holography

- Device Performance Depends on Material Properties in a Complicated Way
  - Determined by Device Architecture
  - Parallel/Analog Processing of 2D Image
  - Phase Conjugate Optics and Image Processing
  - Self Focusing/Defocusing Applications

Fig. 12 Device state-of-the-art with third order polymers.

SIZE OF NEUROCOMPUTERS

(OR WHY OPTICS)

Figure of Merit for Neurocomputer Performance

$$F = F\ (\ N,I,S\ )$$

N = # of neurons
I = # of interconnections/neuron
S = speed of neuron update

$$F = (\ N\ I\ )\ x\ (\ N\ I\ S\ )\ =\ N^2\ I^2\ S$$

System	N	S (sec^{-1})	I	NxI	F
Brain	10^{12}	10^3	10^3	10^{15}	10^{33}
Optical Neurocomputer	10^6	10^9	10^3	10^9	10^{27}
VLSI (fully connected)	10^3	10^9	10^3	10^6	10^{21}
VLSI (nearest neighbor)	10^6	10^9	$O(1)$	10^6	10^{21}

OPTICS OFFERS A LARGE # OF REAL (NOT SIMULATED) INTERCONNECTIONS (JUST LIKE BIOLOGY)

Fig. 13   The role of optics in neurocomputers.[35]

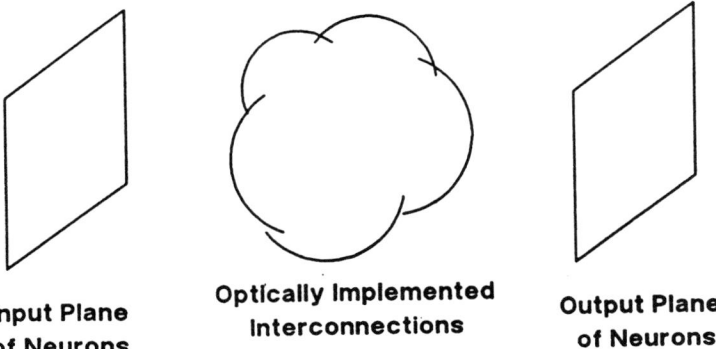

Input Plane
of Neurons

Optically Implemented
Interconnections

Output Plane
of Neurons

Fig. 14   Basic structure of an optical neural network. The planes of neuron could be connected to other planes not depicted.[36]

ADVANCED NLO POLYMERS

GENERATION	SECOND ORDER NLO $\chi^{(2)}$	THIRD ORDER NLO $\chi^{(3)}$
	HOMOPOLYMERS	
I (1983 - 1989)	o ISOTROPIC o POLED o ORIENTED	o BOND ALTERNATION o LADDER o ORDERED
	POLYMER - POLYMER INTERACTIONS BLENDS - ALLOYS MOLECULAR COMPOSITES	
II (1990 - ?)	o SELF-INDUCED POLARIZATION o SELF-ORGANIZING	o NEAR-OFF RESONANCE o PLANAR

Fig. 15   Advanced NLO polymers

POLYMER MULTIPLE QUANTUM WELLS

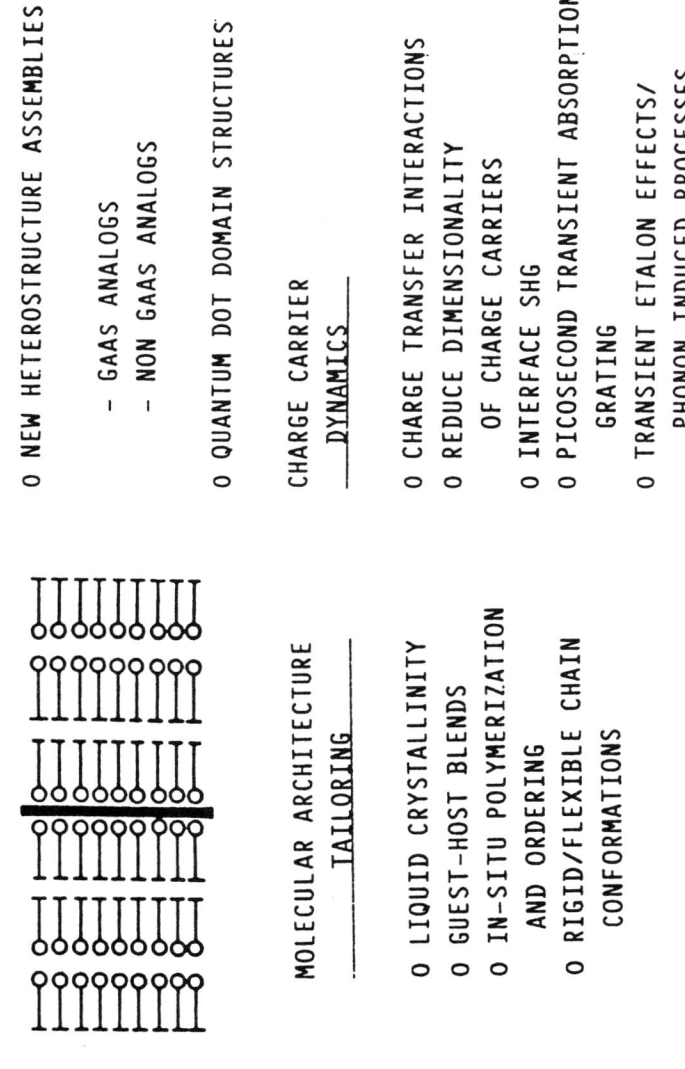

O NEW HETEROSTRUCTURE ASSEMBLIES

- GAAS ANALOGS
- NON GAAS ANALOGS

O QUANTUM DOT DOMAIN STRUCTURES

CHARGE CARRIER
DYNAMICS

O CHARGE TRANSFER INTERACTIONS
O REDUCE DIMENSIONALITY
  OF CHARGE CARRIERS
O INTERFACE SHG
O PICOSECOND TRANSIENT ABSORPTION/
  GRATING
O TRANSIENT ETALON EFFECTS/
  PHONON INDUCED PROCESSES

MOLECULAR ARCHITECTURE
TAILORING

O LIQUID CRYSTALLINITY
O GUEST-HOST BLENDS
O IN-SITU POLYMERIZATION
  AND ORDERING
O RIGID/FLEXIBLE CHAIN
  CONFORMATIONS

Fig. 16 Polymer multiple quantum wells

NON LINEAR OPTICAL POLYMERS
FOR
ACTIVE OPTICAL DEVICES

A. Buckley and J. B. Stamatoff

Hoechst Celanese Research Division
Robert L. Mitchell Technical Center
86 Morris Avenue, Summit, New Jersey 07901, USA

ABSTRACT
    Progress in the field of NLO polymeric materials has been
exceptionally rapid. The promise of these materials based upon
measurements from organic single crystals has been translated
into real properties of significant commercial value. Device
prototyping activities are well underway for a host of unique
designs. Device results suggest rapid transitioning of this
technology to advanced systems for communications, signal
processing, data storage, and optical computing.

THE PROMISE OF ORGANICS
    Basic research on the second order nonlinear optical
properties of organics has been intensely pursued (1,2,3). The
results were remarkable and indicated a class of materials with
properties of high value. Organics were shown to be extremely
fast due to nonlinear polarization of loosely bound electron
clouds. The crystals showed very high activity (e.g. $\chi^{(2)}$ =
500 pm/V for 2,4-methylnitro-aniline). Optical damage
thresholds exceeded 1 gigawatt/cm^2 for the nonresonant process.
Finally the polarization optic coefficient and the Miller's
delta coefficient indicated that, as a class, organics
possessed higher figures of merit for both electrooptical and
all optical processes than inorganics.
    Rapid application and improvement of organics was barred due
to serious materials limitations. Organic crystals have high
vapor pressures, are very difficult to grow with the required
optical perfection, and cannot be readily fabricated into forms
required for device applications (e.g. thin films or fibers).
Of equal but less obvious importance is the barrier which
crystallization creates between molecular tailoring and
material properties. Uncontrolled crystallization habits
essentially eliminate the possibility of controlling the
symmetry of the resulting materials. Thus one of the most
promising aspects of organics, the ability to design and
synthesize highly active molecules is very restricted for
crystalline materials.
    Polymeric NLO materials offered the true promise of
organics. Polymers could maintain the NLO advantages of
organics and avoid the disadvantages of organic crystals.
Polymers are readily synthesized in large volume. They may be
designed to be stable and have low vapor pressures. Polymers
are readily formed into thin films and fibers as required for
device applications. Importantly, symmetry of the NLO active
parts of the polymer can be imposed by externally applied

327

J. Messier et al. (eds.), Nonlinear Optical Effects in Organic Polymers, 327–336.
© *1989 by Kluwer Academic Publishers.*

328

fields (i.e. by electrical poling). Thus, highly active
molecules which are designed at the molecular level may be
processed to achieve the required symmetry and, thus, result in
highly active materials.

REALIZATION OF POLYMERIC NLO MATERIALS
Through a series of molecular design, synthesis, and
fabrication steps, highly active NLO polymers have been
developed at Hoechst Celanese. The properties of these
polymers are listed in figure 1.

NLO Polymer Properties

$\chi^{(2)}$ at 1.3 μm	> 50 pm/V
r (calculated)	> 14 pm/V
n (1.3 μm)	1.57
ε (DC)	3.5
ε (n^2)	2.6
waveguide loss at 1.3 μm	0.9 dB/cm

FIGURE 1. Properties of NLO polymers

The properties are optically roughly equivalent to LiNbO$_3$
with some important differences. First, the dielectric
constant is approximately an order of magnitude smaller than
that of LiNbO$_3$. Thus, these materials achieve a high
polarization optical coefficient and permit designing high
frequency electrooptical devices. Second, the polymer may be
formed into high quality thin films by spin coating. This
makes the fabrication of waveguided devices very attractive.
Third, the electrooptical activity and index of refraction may
be controlled by electrical poling.
Measurement of $\chi^{(2)}$ was accomplished using the apparatus
shown in figure 2. Tunable nanosecond lasers pulses are split
into two beams for the reference and sample cells. Thin poled
NLO polymer films are rotated in the laser beam under computer
control and the generation of light at 2ω is monitored. Figure
3 shows a typical Maker fringe pattern from a thick polymer
poled film.

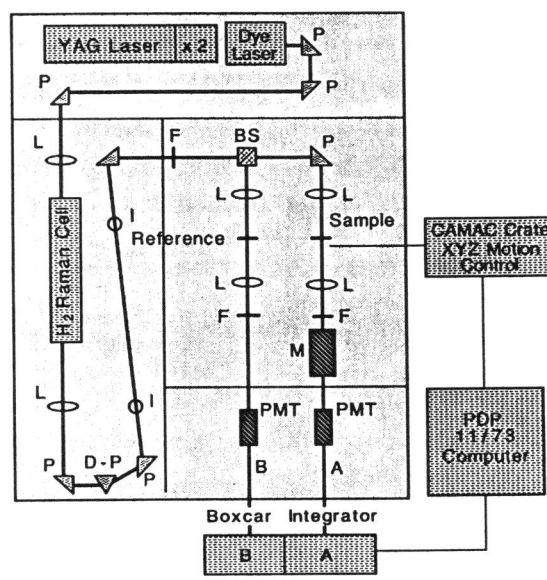

Schematic of 2nd and 3rd Harmonic Generation Experiment

FIGURE 2. Laser Apparatus

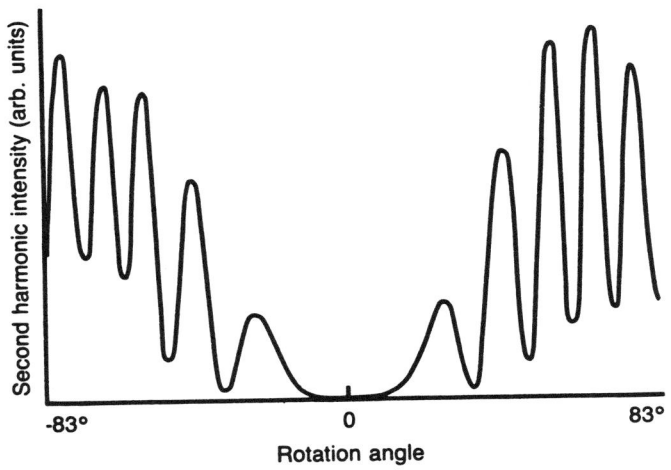

FIGURE 3. Maker Fringe Pattern

Electrooptical activity is also directly measured. Figure 4 shows an optical arrangement which we use to measure electrically induced birefringence as detected by polarization rotation. The device uses a lock-in amplifier to reduce noise and increase sensitivity. Poled films are measured using 45° linear polarization as indicated in figure 5 for various incident angles.

330

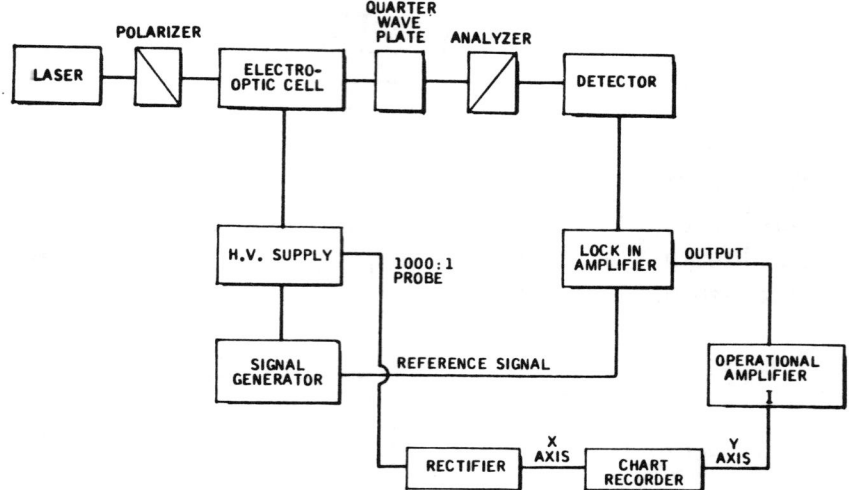

FIGURE 4.   Electrooptical Measurement Apparatus

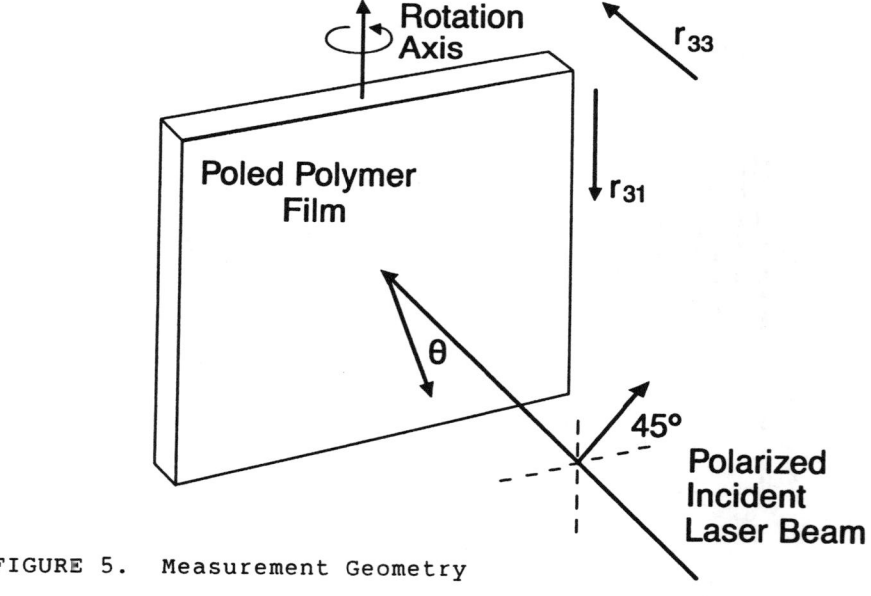

FIGURE 5.   Measurement Geometry

The linear electrooptical response of these $\chi^{(2)}$ polymer films is displayed in figure 6. The linear E/O effect is a second order NLO effect. We have correlated the magnitude of the effect with the strength of the poling field used to establish a noncentric structure. Further, motional effects would give rise to quadratic electrooptical activity which is easily distinguished from this data.

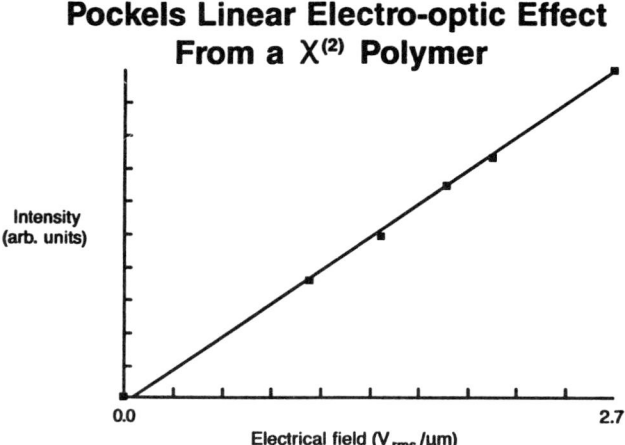

FIGURE 6. Linear Electrooptical Effect

High frequency electrooptical measurements are made in waveguided formats with thin low capacitance electrodes using interferometry. Figure 7 is the schematic of a table top or external Mach Zehnder interferometer which is used to measure the Pockels constant in slab or 2d waveguide format.

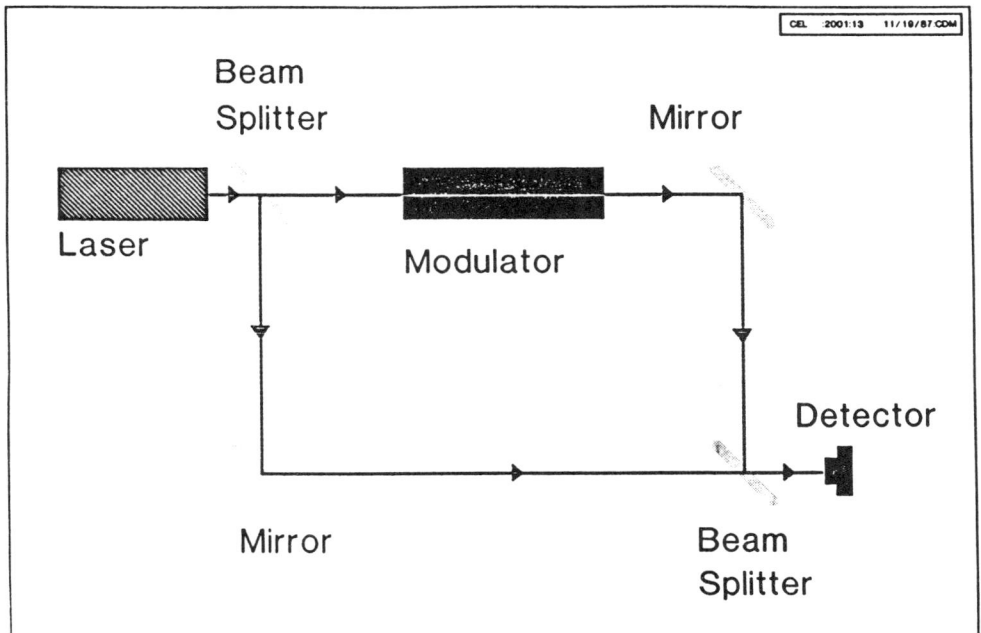

FIGURE 7. Table Top Mach Zehnder Apparatus

Thus, the activity of films are measured by both optical and electrooptical techniques. Assuming that,

$$\chi^{(2)}(-2\omega;\omega,\omega)=\chi^{(2)}(-\omega;\omega,0)$$

and using the relationship,

$$r=-2\chi^{(2)}/n^2\varepsilon$$

results from both techniques may be compared. This comparison is given in Figure 8. Note that all r values have been corrected for dispersion using a simple two state model. Generally results are quite comparable. Differences may be due to the finite width of the absorption band which is not considered in the simple two state dispersion correction.

FIGURE 8.

Electronic Contributions to r

Material	pm/V		
	$\chi^{(2)}$	$r^{2\omega}_{exp}$	$r^{e\text{-}o}_{exp}$
PMMA:pna	0.9 (1.06)	0.45 (.633)	0.5
PMMA:MNA	0.67 (1.06)	0.33 (.633)	.32
HCC #1622	8.4 (1.06)	1.75 (.830)	3.75
HCC #1238	36.5 (1.34)	10.0 (.830)	20.0
HCC #1232	65.0 (1.34)	16.5 (.83)	?

APPLICATION TO ACTIVE OPTICAL DEVICES

Demonstration of many device prototypes is possible using NLO polymers. A comparatively simple device is the slab modulator which was designed and constructed at Hoechst Celanese (Figure 9). For this device, light is coupled into one mode of a slab waveguide using prisms. If the incident light is polarized at 45° to the plane of the slab waveguide, then light is coupled into both the TE and TM modes. Using E/O active waveguiding films, the the phase of the TE mode may be selective retarded relative to the TM mode. At sufficiently high modulation voltage the emerging light is plane polarized at 90° to the original incident light. Through the use of a polarizer the device becomes a modulator. Modulators of this type were constructed at Hoechst Celanese and represent the first organic based active optical devices.

333

At Hoechst Celanese modulators requiring only four volts to achieve V_π (i.e to achieve 100% modulation) have been constructed. Recently an analog modulator based upon the same design was demonstrated at two trade shows. In this demonstration, an amplitude modulated audio signal was impressed upon an optical beam. Similar analog modulation can be used to convert high frequency microwave signals to optical signals for transmission along optical fiber.

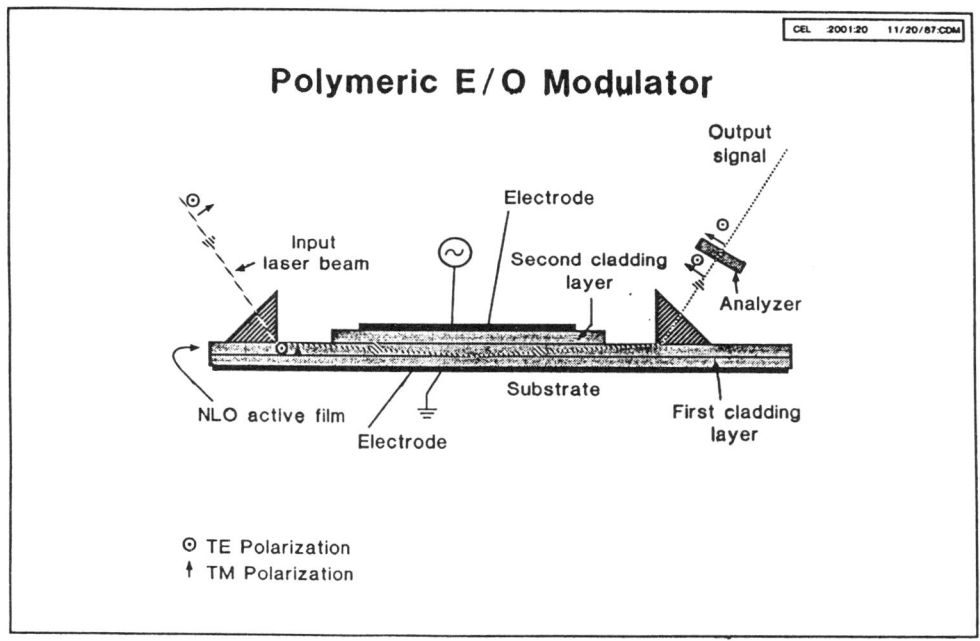

FIGURE 9: Slab Waveguide Device

These devices are far simpler to construct than waveguides requiring 2d confinement of light (i.e. linear waveguides). However linear guides permit using very small electrodes and ultimately designing very high bandwith devices. One of the most fundamental and useful is the traveling wave Mach Zehnder modulator. A central feature of this class of devices is the ability to match the phase velocity of the modulating microwave signal with the guided optical mode. As a result very high bandwith devices may be achieved with modulation frequencies limited only by electrode losses. Hoechst Celanese is teaming with GEC Marconi Laboratories to develop this device. Figure 10 details the designed performance and shows the key steps including design by finite element modeling, pattern formation in the NLO active polymer, and end face preparation of the NLO polymer.

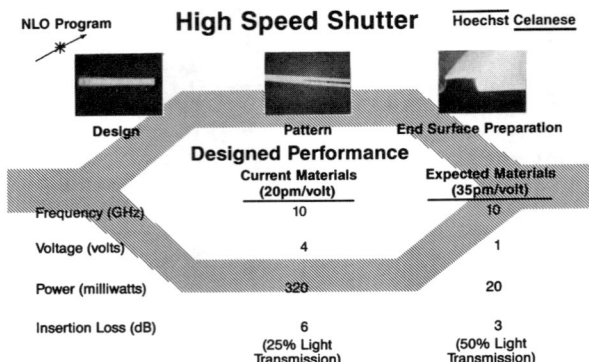

High Speed Shutter

NLO Program   Hoechst Celanese

Design   Pattern   End Surface Preparation

Designed Performance

	Current Materials (20pm/volt)	Expected Materials (35pm/volt)
Frequency (GHz)	10	10
Voltage (volts)	4	1
Power (milliwatts)	320	20
Insertion Loss (dB)	6 (25% Light Transmission)	3 (50% Light Transmission)

FIGURE 10.   Designed Performance of a Waveguide Interferometer

Design studies have shown that organic devices have very high bandwith exceeding 10 GHz. This attribute is in addition to other more obvious advantages such as low switching voltages, low insertion loss, and high expected device yield.

Figure 11 shows an enlarged view of the end surface of a linear waveguide. The rib guide is 2μm thick and 5μm wide. It was formed from a spin coated film of the NLO polymer by reactive ion etching. Smooth end surfaces were obtained by cleaving and are quite suitable for end fire coupling.

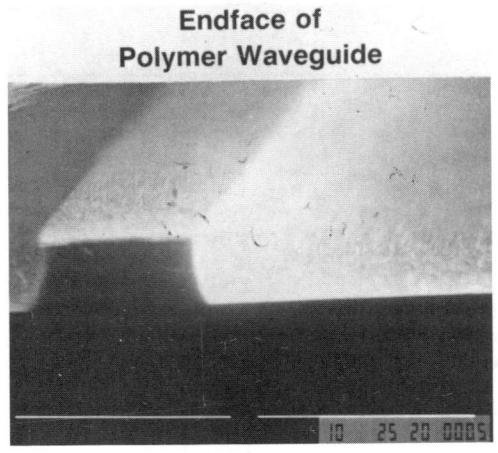

Endface of Polymer Waveguide

FIGURE 11.   End Surface of an NLO Waveguide

A second class of waveguided devices permit efficient frequency conversion. Figure 12 shows a schematic of this type of device. Alternating regions on the polymers are poled to form periodic strips across the waveguide. Light of frequency w is guided as a single mode and progressively is converted into light at frequency 2ω. Alternating regions of $\chi^{(2)}$ active and inactive polymer produce effective phase matching despite the different velocities of the primary and harmonic beams. (This is accomplished by allowing wavefronts which become out of phase to catch up in the unpoled regions). This design has been tested for inorganic materials but fabrication is very tedious. NLO polymers films may be selectively poled and are thus ideally suited to this application.

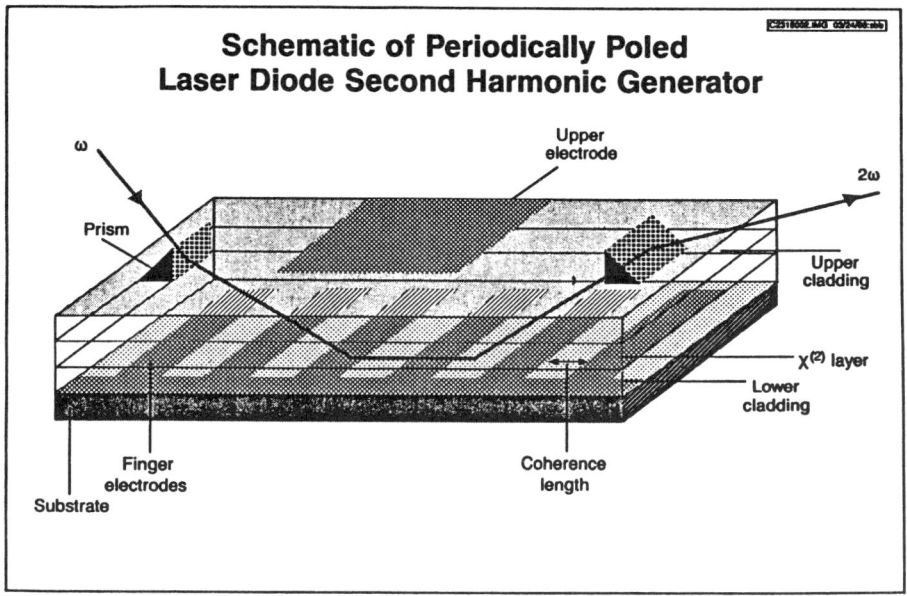

Schematic of Periodically Poled Laser Diode Second Harmonic Generator

FIGURE 12. Waveguide Harmonic Generation Device

FUTURE PROSPECTS
We believe that these materials are currently of high commercial value. Device applications are being extended to include very high frequency modulators, analog modulators for microwave signal processing, directional couplers for single mode local area networks, and spatial light modulators for optical computing. Improvements of an order of magnitude over the current NLO activities (both r and $\chi^{(2)}$) is entirely possible. With such improvement, multifunctional integrated electooptical waveguided designs can be achieved.

Similar development of $\chi^{(3)}$ materials is also underway at Hoechst Celanese. One example, shown in Figure 13, is a yarn of NLO fibers $6\mu m$ in diameter. Reasonably high $\chi^{(3)}$ activites (on the order of 10^{-11} esu) have been achieved for a polymer of high molecular weight which may be readily formed into fibers. Using fiber waveguide techniques, picosecond optical devices have been constructed in optical glasses of far lower activity (4).

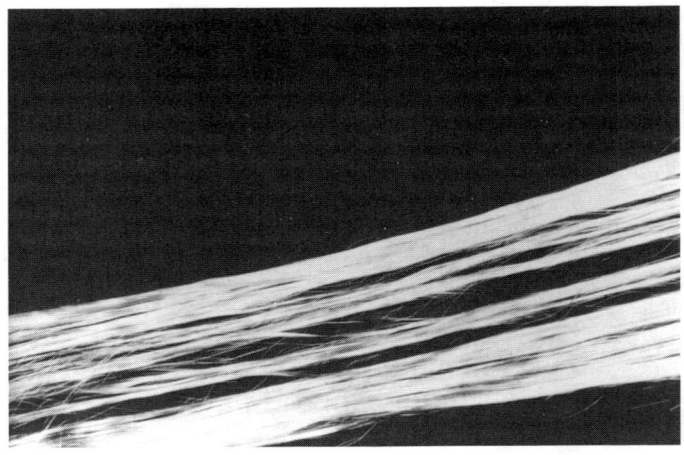

FIGURE 13. NLO Polymer Yarn

   With such robust processing options, NLO polymers will
likely play a central role in developing all optical and
ultrafast switches, amplifiers, and multiplexers.

ACKNOWLEDGEMENTS

   The authors are deeply indebted to Prof. A.F. Garito of the
University of Pennsylvania. We thank Drs. G.W. Calundann, G.E.
Gillberg and G.V. Nelson for stimulating conversations. We
gratefully acknowledge the timely support and direction given
to our program by Drs. D. Ulrich and J. Neff.
   This work was partially funded by DARPA/DSO through AFOSR
under Contract No. F49620-86-C-0129 and AFOSR under Contract
No. F49620-86-C-0115.

REFERENCES

1. Garito A.F., Teng C.C., Wong K.Y., and Zammani'Khamiri O.,
   Mol. Cryst. Liq. Cryst. 106, 219-258, 1984.

2. Zyss J., J. Mol. Electronics 1, 25-45, 1985.

3. Sauteret C., Hermann J.P., Frey R., Pradere F., Ducuing J.,
   Baughman R., and Chance R., Phys. Rev. Lett. 36, 956-976,
   1976.

4. Frieberg S.R., Smith P.W., J. Quantum Electronics QE23,
   2089-2094, 1987.

LINEAR ELECTROOPTIC COEFFICIENT OF A FERROELECTRIC POLYMER

Y. LEVY*, V. DENTAN*, M. DUMONT*, P. ROBIN**, E. CHASTAING**

* INSTITUT D'OPTIQUE, Université Paris-Sud, Bât. 503, 91406 ORSAY
** THOMSON-LCR, Domaine de Corbeville, 91406 ORSAY

Ferroelectric copolymers of vinylidene fluoride and trifluorethyle-ne, P(VDF-TrFE), commercialised by ATOCHEM are well known for their piezo-electric and pyroelectric applications [1]. Nonlinear optical properties have been studied by several authors and it has been shown that second harmonic generation could be achieved in those polymers [2]. It is also possible to enhance the second harmonic generation signal by doping the ferroelectric polymer films with non-linear optical active molecules. To evaluate the applications of ferroelectric polymer in the nonlinear optical field, it is very important to measure the electrooptic coeffi-cient. The Attenuated Total Reflection configuration (ATR) is a suitable experiment for this measurement [3].

As it will be shown later on, the ATR measurements associated with a simplified model of piezoelectricity provide the two contribution parts in the electrooptic coefficient. The first is directly produced by the piezoelectric effect for which a modification of the refractive index is related to a change of the thickness. The second is due to a pure molecu-lar electrooptic effect. We show in this paper that the piezoelectric constant of the film can also be deduced from a precise analysis of data.

Attenuated Total Reflection method (ATR)

The Kretschmann ATR prism configuration is sketched in Figure 1.A. A glass prism is coated with a multilayered structure constituted by a sil-ver thin layer, a Ferroelectric Polymer Film (FEPF) and an aluminium layer. Through the prism, a p polarized Transverse Magnetic (TM) He-Ne laser beam is reflected on the multilayered structure. By rotating the prism about an axis perpendicular to the plane of incidence, we varied the incidence angle (labeled ϕ for the external angle and Ψ for the inter-nal angle in figure 1) which changes the longitudinal component of the wave vector (parallel to the base). The laser beam excites the surfaces plasma wave (SPW) when this longitudinal component matches the SPW wave vector. In the same way, the TE and TM polarized guided modes can be excited but naturally at different angles of incidence. These modes are seen as sharp minima in the reflectivity versus angle curves [4]. The presence of both TE and TM modes is a consequence of the thickness of the dielectric film which is much larger than the penetration depth of the SPW mode at the silver/FEPF interface. By applying a low voltage between the two electrodes (Ag and Al layers), a modification of the optical thickness is induced which produces an angular shift of the reflectivity minima [5]. This angular variation of the guided modes can be related to the electrooptic coefficient of FEPF. With 4V applied across few microns thick film, the reflectivity change is close to 10^{-3}. Therefore, we used

337

J. Messier et al. (eds.), Nonlinear Optical Effects in Organic Polymers, 337–342.
© *1989 by Kluwer Academic Publishers.*

338

a modulated voltage ($\simeq$ 2 kHz) and a lock in amplifier for the detection.

Figure 1-B shows the complete arrangement used for recording simultaneously the reflectivity $R(\phi)$ and differential reflectivity $\Delta R(\phi)$. If the modification is due to a refractive index change, then $\Delta R(\phi)$ can be expressed as :

$$\Delta R(\phi - \phi_o) = \frac{\partial R}{\partial n} \frac{\partial n}{\partial E} \Delta E = - \left[\frac{\partial R}{\partial \phi_o} \frac{\partial \phi_o}{\partial n}\right] \frac{\partial R}{\partial E} \Delta E \qquad (1)$$

where ΔE is the peak to peak electric field amplitude and $\partial n/\partial E$ is proportional to the electrooptic coefficient of the material. An approximate formulation of $\Delta R(\phi)$ may be expressed by remarking that, for a non absorbing layer, a small change of the refractive index n produces only a shift of the angular position ϕ_o of the minimum. With this approximation, the differential reflectivity is proportionnal to the derivative of R versus ϕ. With the ATR configuration, $\partial \phi_o/\partial n$ is always negative. Therefore $\partial R/\partial n$ is negative for $\phi < \phi_o$ and positive for $\phi > \phi_o$. In regard of eq. 1, the shape of ΔR on Figure 1.B corresponds to a positive electrooptic coefficient. In fact this choice is arbitrary because the sign depends on the orientation of electrooptic axes with respect to the substrate.

Experimental results on a ferroelectric polymer

Samples were prepared on thin glass slides which were silver coated (420 Å) by thermal evaporation in vacuum. Copolymer films were then deposited by spin coating using a solution of 120 g/l in methylethyl cetone. By using a rotation speed comprised between 2000 r.p.m. and 4000 r.p.m., film thicknesses of 1,8 μm to 2.6 μm were obtained. The samples were dried for 1 hour at 100°C, then 1/2 hour at 130°C to achieve a complete crystallization. An upper aluminium electrode was then thermally evaporated. The samples were poled at room temperature using a field of 80 MV/m. The back face of the substrate was then put in optical contact with a low refractive index prism (n_p = 1.5163 at λ_o = 633 nm ; angle A_p = 60°) through an index matching oil in order to minimize the interface losses.

After the deposition of the FEPF and of the aluminium electrode on the silver layer, we verified the ferroelectric properties of the material, for a TE or TM resonance mode. The shape of the differential reflectivity $\Delta R(\phi)$ and hence of the electrooptic coefficient can be changed at will by applying a positive or negative polarization voltage ($|V| > 180$ V)during a short time. In an other experiment, we recorded the TM, $R(\phi)$ and $\Delta R(\phi)$ curve (figure 2) on the largest possible range (16° $< \phi <$ 60°). Each minimum correspond to the excitation of a guided mode since the SPW mode cannot be recorded because of the low refractive index of the prism. The TE recorded reflectivity, not represented in figure 2, exhibits a similar set of minima shifted with respect to the TM peaks. We calculated the refractive index n and the thickness e of the FEPF, by using a least squares fitting procedure with the Fresnel formulae. We found n = 1.416 and e = 26375 A. The differential reflectivity $\Delta R(\phi)$ shown in figure 2, exhibits a surprising feature : the shape of ΔR resonance curves is inverted for modes located both apart of an incidence angle close to 44°. This shape inversion cannot be explained by considering only the refraction index change, as discussed above with equation 1. We have to take in account the ferroelectric character of the FEPF which involves a modification of the thickness (piezoelectric effect). Therefore, when a modulated electric field is applied, both n and e modify the reflectivity. With this assumption, $\Delta R(\phi)$ can be expressed as :

$$\Delta R(\phi) = \left[\frac{\partial R}{\partial n} \frac{\partial n}{\partial E} + \frac{\partial R}{\partial e} \frac{\partial e}{\partial E} \right] \Delta E \qquad (2)$$

With the least squares procedure, we obtained a good fit between experimental and theoretical curves. The corresponding values of $\Delta n = 1.75 \ 10^{-5}$ and $\Delta e = -1.07$ A are so determined and provide the piezoelectric coefficient $(1/e)\Delta e/\Delta E = -13.4$ pm/V and the electrooptic coefficient $\chi^{(2)} = 2n(\Delta n/\Delta E) = 16.3$ pm/V. The signs are determined for an electric field in the same direction as the initial polarization field.

For explaining the results, we propose a simplified model which provides an insight into the physical interpretation on the shape inversion of the curve $\Delta R(\phi)$. The TE and TM modes for which we observe the minima of the reflectivity curves are governed by the transverse resonance condition.

$$(2\pi/\lambda_0) \ ne \cos \theta_m - \psi_1 - \psi_2 = 2 \ m\pi \qquad (3)$$

where λ_0 is the wavelength in vacuum, θ_m is the propagation angle of the waves in the film, ψ_1 is the phase shift of the reflected wave at the FEPF/Aℓ interface, and ψ_2 is the phase shift at the film/Ag interface including the small perturbation due to the prism. The integer m characterizes the order of the mode and the resonance condition depends on the state of the polarization through the phase shifts ψ_1 and ψ_2.

Using the previous approximation for non absorbing layers (cf. equation (1)), equation 2 for the mth mode may be written as :

$$\Delta R(\phi)_m = - \frac{\partial R}{\partial \phi} \left[\frac{\partial \phi_m}{\partial n} \frac{\partial n}{\partial E} + \frac{\partial \phi_m}{\partial e} \frac{\partial e}{\partial E} \right] \Delta E \qquad (4)$$

The external incidence angle ϕ_m of the mth mode is related to the propagation angle θ_m and to the intermediate angles φ'_m and φ_m defined in figure 1. Through the Snell-Descarte formulae one can write :

$$d \phi_m = \frac{\partial \phi_m}{\partial e} de + \frac{\partial \phi_m}{\partial n} dn = \kappa_e \frac{de}{e} + \kappa_n \frac{dn}{n} \tag{5}$$

where κ_e and κ_n can be expressed from equation 3.

The index of refraction of a dense isotropic material is given by the Lorentz-Lorenz formula :

$$n^2 - 1 = \chi = N\alpha / (1 - N\alpha/3) \tag{6}$$

where α is the polarizability of molecules and N their density. From this formula it is evident that the index variation due to an applied electric field must be written as :

$$dn = dn_1 + dn_2 \tag{7}$$

where dn_2 is the molecular electrooptic effect due to the variation of α and dn_1 is related to the density variation induced by the piezoelectric deformation of the material. As the FEPF is clamped to the substrate, the volume per unit area is proportional to the thickness e then N is inversely proportional to e. Using this property in equation (6) one obtains :

$$de/e = b \, dn_1/n \quad \text{with} \quad b = - 6 \, n^2 / (n^2 + 2) (n^2 - 1) \tag{8}$$

The equation (5) becomes :

$$n \, d\phi_m = (\kappa_n + b \, \kappa_e) \, dn_1 + \kappa_n \, dn_2 = \kappa_1 \, dn_1 + \kappa_n \, dn_2 \tag{9}$$

The figure 3 shows the variation of κ_e, κ_n and κ_1 with the mode position ϕ_m, for values of n and e corresponding to our sample. As b is negative, dn_1 and de are of opposite sign and κ_1 changes of sign for ϕ_m close to 47.5°.

From experimental records shown in figure 2, the inversion point, ϕ_o, can be estimated and, by equating eq. 9 to zero, one obtains :

$$dn_2/dn_1 = - \kappa_1(\phi_o)/\kappa_n(\phi_o)$$

This discussion explains very simply that the sign inversion of $\Delta R(\phi)$ is due to the piezoelectric properties of the FEPF. It allows a rapid estimation of the ratio dn_2/dn_1 . The function $\kappa = [\kappa_1(\phi_m) \kappa_2(\phi_o) - \kappa_2(\phi_m) \kappa_1(\phi_o)] [\kappa_2(\phi_o) - \kappa_1(\phi_o)]^{-1}$, defined by : $n \, d\phi_m = \kappa(\phi_m) \, dn$, is drawn on figure 3. This function is proportional to the maxima amplitude of the $\Delta R(\phi)$ resonances in fig. (2). Let us notice that the κ functions are very sensitive to the value of n but much less to e : for a given material the behaviour is similar with samples of different thicknesses.

Of course the previous fitting procedure is necessary for the determination of the piezo electric and of the overall electrooptic $\chi^{(2)}$ coefficients. Our simplified model allows also to calculate Δn, the piezoelectric contribution Δn_1 from Δe (with eq. (8) and to extract the molecular electrooptic contribution Δn_2 from eq. (7). On finds the electrooptic coefficients :

$$\chi_1^{(2)} = 2 \, n \, \Delta n_1 /E = 17.9 \text{ pm }/V \quad ; \quad \chi_2^{(2)} = 2 \, n \, \Delta n_2 /E = -1.6 \text{ pm }/V$$

Nevertheless, this decomposition of $\chi^{(2)}$ into $\chi_1^{(2)}$ and $\chi_2^{(2)}$ cannot be considered as totally accurate since it is based on a simplified theory. The first objection comes from the anisotropy of the material. Similar calculations made on TE reflectivity measurements lead to an anisotropy close to 2.10^{-2}. Moreover we must emphasize the tensorial nature of $\chi_2^{(2)}$ which has been ignored in the above discussion. Wich TM modes, we measure a combination of $\chi_{zzz}^{(2)}$ and $\chi_{xxz}^{(2)}$ which varies from one resonance to another (the decomposition of the optical electric field on x and z directions is a function of θ). On the other hand with TE modes we measure $\chi_{yyz}^{(2)}$.

The second objection is the use of the Lorentz-Lorenz formula to take into account the index variation induced by mechanical deformations. The local field effects are probably more complicated in a ferroelectric polymer material than in the homogeneous medium assumed in the Lorentz-Lorenz theory : a correction to the b factor in (8) could be necessary. In addition, mechanical stress in the clamped film may induce a further anisotropy.

In conclusion, we have shown that the piezoelectric properties of our material explain the surprising inversion of sign of the apparent electrooptic effect observed by ATR method. In the P(VDF-TrFE) copolymer the piezoelectric contribution to Δn is dominant and the method for extracting the purely molecular electrooptic effect is not accurate due to the oversimplified theory. We will soon be able to determine the anisotropy of the material and the tensorial components of $\chi^{(2)}$ by fitting simultaneously TE and TM modes, but the most difficult problem is the improvement of the Lorentz Lorenz expression within the special case symmetries of our material (determination of the b factor).

BIBLIOGRAPHIE

[1] The application of ferroelectric polymers - Edited by T.T. WANG, J.M. HERBERT and A.M. GLASS BLACKIE.
[2] H.SATO and H.GAMO - Japanese Journal of Applied Physics 25, (1986).
[3] Electromagnetic Surfaces Modes, edited by A.D. BOARDMAN, John WILEY and sons LDT (1982).
[4] Y. LEVY, M. JURICH and J.D. SWALEN, J. Appl. Phys., 57, (1985).
[5] G.H. CROSS, I.R. GIRLING, I.R. PETERSON, N.A. CADE, Elect. Lett., 21, (1986).

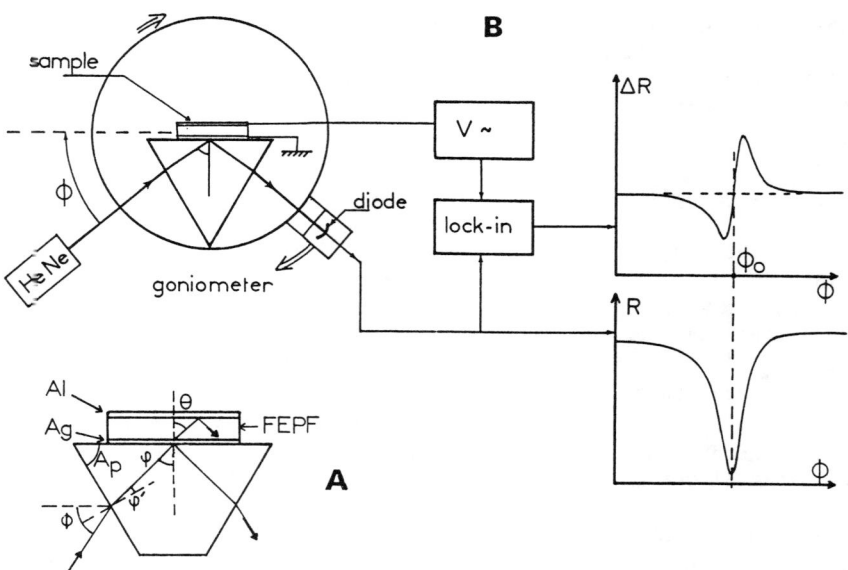

Figure 1 : A> The Kretschmann ATR configuration with the definition of angles. The prism is covered with the multilayered Ag/dielectric film/Aℓ structure. B> Experimental arrangement.

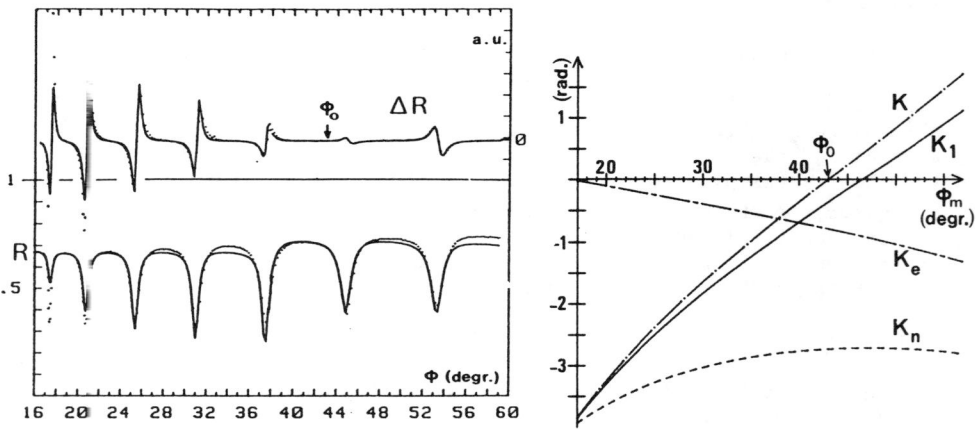

Figure 2 :
Experimental and theoritical
curves R(φ) and ΔR(φ).
..... recorded data.
_____ theoritical curves.

Figure 3 : Plots of the functions
calculated with the optical constants of our sample.
. $K_n(\phi_m) = n\ \partial\phi_m/\partial n$
. $K_e(\phi_m) = e\ \partial\phi_m/\partial e$
. $K_1(\phi_m) = K_n + b\ K_e$
. $K(\phi_m)$ as defined in the texte.

APPLICATION OF THIRD-ORDER NON-LINEARITIES OF DYED PVA TO REAL-TIME HOLOGRAPHY

R.A. LESSARD, J.J.A. COUTURE, P. GALARNEAU

LROL-Département de Physique, Université Laval, Sainte-Foy, Québec, G1K 7P4, Canada

1. INTRODUCTION

Actually, many experimental investigations for new optical recording materials seem to be very important to permit applications of holography and four-wave mixing techniques. Construction of holographic optical elements (HOE) and filters, aberration corrections and non-destructive testing as well as optical data storage require recording materials of high optical performance.

Although many publications[1-11] reported studies on organic compounds, KAKICHASHVILI[1] was the first to introduce "Polarization Holography Applications" obtained with photochromic trimethylspirane-benzopyran. Five years laser, TODOROV, NIKOLOVA and TOMOVA[2-8] pointed out that methyl-orange (M.O.) and methyl-red (M.R.) respectively introduced in polyvinyl alcohol (PVA) and poly-(methylmethacrylate) matrix may record polarization holograms by TRANS-CIS photoisomerization. Macroscopically, the photo-induced anisotropy is achieved by photodichroism of dyes molecules and birefringence of PVA under action of linear polarized light. Todorov's films of M.O./PVA gave 35% diffraction efficiency for holograms recorded with two orthogonal circular polarized light beams onto samples that were heated during 30 minutes at 80°C beforehand. These samples showed an uniform response in the 280-2800 cycles/mm spatial frequency domain. We have studied the spatial frequency response[10] of azo-dyes in PVA; those azo/PVA samples showed a uniform response in the range of 500 to 3600 cycles/mm and a spatial bandwidth of about 4000 cycles/mm.

Furthermore, a study[11] of DFWM of two azo-dyes in solution, chrysoidin and benzopurpurin 4B, points out that reflectivities up to 80% may be achieved with pump intensity of the order of 6 MW/cm². The influence of solvent, dye concentration and pump intensity on the conjugate beam have also been studied.

In the study of photodichroism properties of AZO-dyes, the most probable mechanism involved in real-time recording and erasing processes is the TRANS↔CIS photoisomerisation[12] as illustrated in figure 1.

J. Messier et al. (eds.), Nonlinear Optical Effects in Organic Polymers, 343–349.
© 1989 by Kluwer Academic Publishers.

The recording process can be seen as induced by a n→π[*]
transition occurring under light illumination when azo molecules absorb polarized light. After that, a rotation over the σ
bond is achieved because the π bond is strongly disturbed by
one electron of the π[*] state changing a TRANS isomer into CIS
isomer[12,13,14]. Moreover, an erasure process is achieved when
azo samples are heated or placed in darkness. The thermal
CIS-TRANS isomerization[12] takes place and the most stable
TRANS form appears. Consequently, for a given temperature and
under specific polarized light action, two isomers species
exist in equilibrium. However, steric arrangement, local viscosity and good choice of solvent promote this TRANS↔CIS photoisomerization. Hence, the studied azo samples must be able
to photoisomerize but, at the same time, they are placed in a
solid matrix in order to avoid molecular motion.

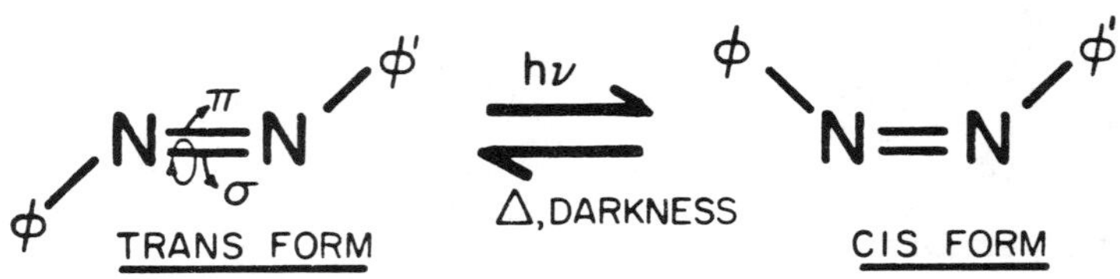

FIGURE 1. Basic photoisomerization mechanism.

2. PREPARATION OF STUDIED PHOTOSENSITIVE FILMS

Azo-dyes are attractive because of their low cost and many
of them are soluble in water. Moreover, as water is a good
solvent for PVA, preparation of samples can be greatly simplified. For the preparation of photosensitive films we chose
PVA as host material because we can easily change its viscosity; also its solubility in water and acetic acid is very good.
Moreover, adding acetic acid in aqueous azo dye/PVA solution
gives a bathochromic shift resulting in strong absorption in
the 450 to 580 nm range and very good transmittance at 632,8
nm.

The studied AZO DYES (Mordant Yellow 3R, Methyl-Orange and
Chrysoidin) were dissolved in aqueous solution (6 wt %) in
order to obtain maximum dye concentration. We prepared films
of 15 μm and 30 μm of thickness by pooring 1,1 ml or 2,2 ml of
AZO-DYE/PVA solution on clean microscope slides placed in horizontal position. After twenty-four hours, those samples
lost a large part of their solvents and they can be used as
solid photosensitive materials. Thickness measurements (15
and 30 μm) were made with a SLOAN DEKTAK IIA apparatus. The
effective thickness of samples were evaluated by angular se-
lectivity bandwidth technique[10,15] and they were respectively
12 ± 1 μm and 24 ± 1 μm.

3. EXPERIMENTAL

The holographic characterization was performed with a two
beams arrangement which permits the study of real-time holo-

graphic responses of our samples for one of the three polarization states: linear parallel (↑↑), orthogonal linear (↑→)
and orthogonal circular (↻ ↺). Our experimental set-up[16],
illustrated by the figure 2, was specifically designed to ensure rapid variations of the recording angle (2θ) without modification of the recording conditions (pathlength difference,
superposition of the recording beams and fringes stabilization
conditions).   The interbeam angle domain was 18°<2θ<160°.
Also four-wave mixing studies can be easily conducted with
this special apparatus.   The reading helium-neon laser beam
and the fringes stabilization system are not represented in
figure 2. All kinetic measurements of diffraction efficiency
during the grating growth were performed with fringes stabilization.

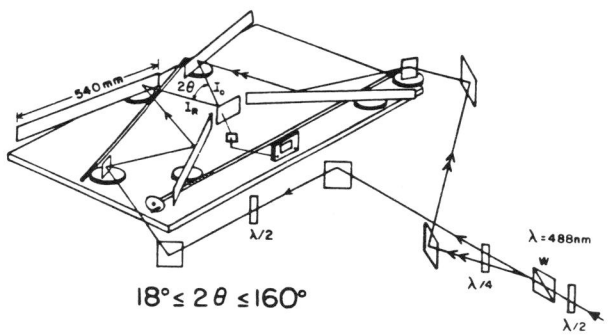

FIGURE 2. Stabilizated holographic set-up.

4. RESULTS
     The kinetic recording and erasing processes for the three
AZO DYE/PVA systems under orthogonal circular polarizations
illumination is shown in figures 3, 4 and 5. In those experiments the interbeam angle (2θ) was kept at 18°, the writing
beams came from an Ar[+] at 488 nm and the reading was performed with a 10 mW He-Ne laser (632,8 nm) under circular polarization illumination.

     For each AZO-DYE/PVA system, the growth rate of the grating
(dη/dt) increases with the power of the two beams used. The
best diffraction efficiencies are obtained with a power level
of the writing beams greater than 5 mW (18 mW/cm^2) in the case
of Mordant-Yellow 3R and Methyl-Orange.   For Chrysoidin/PVA
layers, the minimum power required to achieve best efficiencies is 20 mW (72 mW/cm^2). The most effective AZO-DYE/PVA
system is Mordant-Yellow 3R. Aside from the dynamic grating,
a more persistent grating is formed when high intensities are
used as observed in the previous figures 3, 4, 5.

     The holographic response of Mordant Yellow 3R/PVA layers
under orthogonal linear polarizations of the two writing beams
is illustrated in figure 6. Those results are quite similar
to the previous obtained with orthogonal circular polarizations states.   The kinetics under orthogonal circular polarizations with respect of the beam ratio has been investigated.

In this case the total power was kept at 15 mW and the beam ratio varied from 1 to 100. As expected the maximum response was achieved for K ratio near 1 as shown in figure 7.

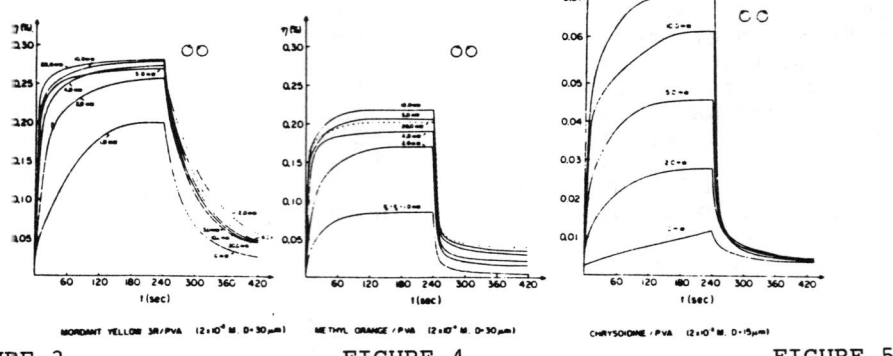

FIGURE 3   FIGURE 4   FIGURE 5

Kinetic recording and erasing processes under orthogonal circular polarizations illumination.

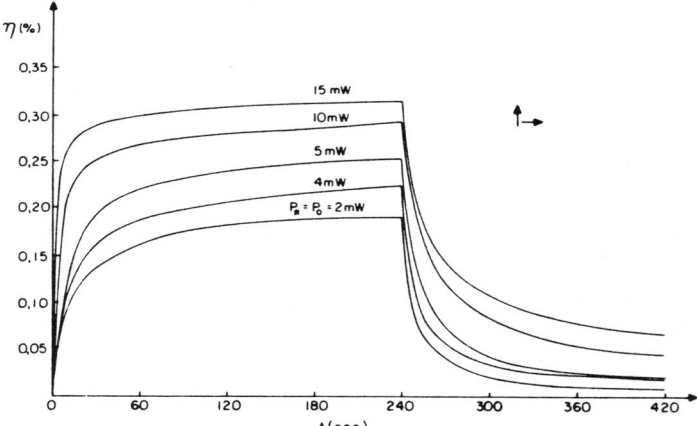

FIGURE 6. Kinetics under linear orthogonal polarizations illumination.

To complete the present study, the spatial frequency response of AZO-DYE/PVA photosensitive films have been considered. Figures 8, 9 and 10 present the M.T.F. of those films for the three polarization states.

One may notice uniform spatial frequency response up to 3300 cycles/mm for each system. Similar holographic responses is observed in the case of orthogonal linear and circular polarizations states. The spatial frequency bandwidth is about 4000 cycles/mm as required in many applications.

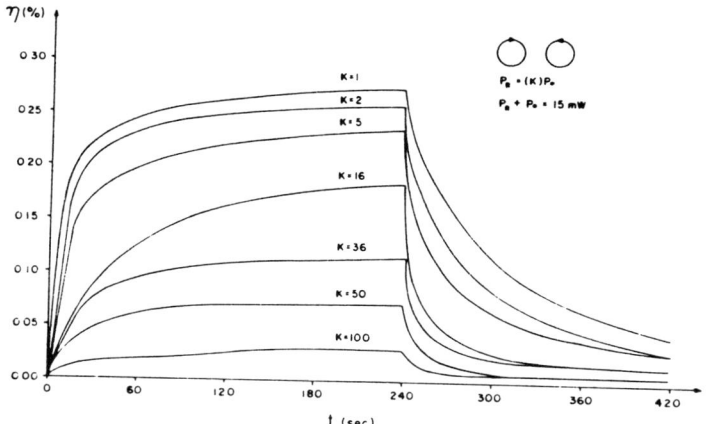

FIGURE 7. Kinetics under orthogonal circular polarizations
with respect of the beam ratio (K = P_R/P_O).

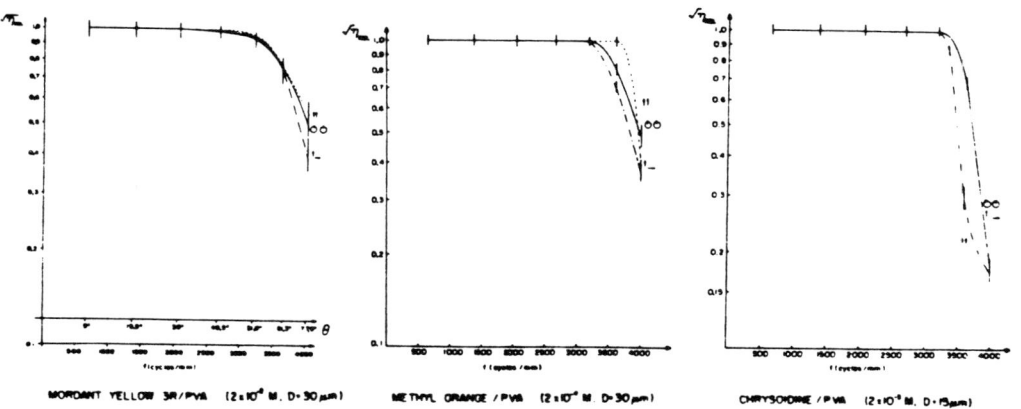

FIGURE 8   FIGURE 9   FIGURE 10

Spatial frequency response (M.T.F.) of the three studied AZO-
DYES/PVA systems with respect of the three polarizations recording states.

5. EVALUATION OF THE THIRD ORDER NONLINEARITY

Considering a pure absorption grating with an absorption
described by

$$\alpha = \alpha_o + (\alpha_1) \cos \vec{K} . \vec{r} \quad , \tag{1}$$

and following Kogelnik's theory[17] for volume holograms, one
may obtain

$$\eta \simeq \left[\frac{\alpha_1 D}{2 \cos \Theta} \right]^2 . \qquad (2)$$

Using the relation[18]

$$\chi^{(3)} = \frac{2 n c \alpha_1}{\omega E^2}$$

which relates the $\chi^{(3)}$ and the modulation depth, the $\chi^{(3)}$ values were evaluated at a reference incident power of 2 mW for orthogonal circular polarizations, they are given in Table 1.

DYE	CHRYSOIDINE	METHYL ORANGE	MORDANT YELLOW 3R
Concentration	2×10^{-3} M	2×10^{-4} M	2×10^{-2} M
Thickness of the film	15 μm	30 μm	30 μm
η at 2 mW	0.028%	0.171%	0.310%
$\chi^{(3)}$	1.9×10^{-8} $\frac{m^2}{V^2}$	2.4×10^{-8} $\frac{m^2}{V^2}$	3.2×10^{-8} $\frac{m^2}{V^2}$

TABLE 1. Evaluation of $\chi^{(3)}$ for orthogonal circular polarization.

6. CONCLUSION

Three erasable and reusable AZO-DYES/PVA systems were studied for their holographic recording properties. Mordant Yellow 3R/PVA films were found to be more interesting when good diffraction efficiency and write/read cycle are involved. When high repetition rate is required, Methyl-Orange/PVA system seems to be more appropriate since its rising time is shorter than the two other studied systems.

REFERENCES

1. Kakichashvili, Sh.D.: Polarization Recording of Holograms. Opt. Specktrosk, 33, 324 (1972).
2. Nikolova, L. and Todorov, T.: Volume amplitude holograms in photodichronic materials, Optica Acta, 24, 1179 (1977).
3. Todorov, T., Nikolova, L., Tomova, N., Dragostinova, V.: Photochromism and dynamism holographic recording in a rigid solution of fluorescein, Optical and Quantum Electronics, 13, 203 (1981).
4. Todorov, T., Tomova, N., Nikolova, L.: High sensitivity material with reversible photo-induced anisotropy, Opt. Comm., 47, 123 (1983).
5. Todorov, T., Nikolova, L., Tomova, N.: Polarization holography. 1.: A new high-efficiency organic material with reversible photo-induced birefringence, Appl. Opt., 23, 4309 (1984).

6. Todorov, T., Nikolova, L., Tomova, N.: Polarization holography. 2.: Polarization holographic gratings in photoanisotropic materials with and without intrinsec birefringence, Appl. Opt., 23, 4588 (1984).
7. Todorov, T., Nikolova, L., Stoyanova, K., Tomova, N.: Polarization holography. 3. Some applications of polarization holographic recording, Appl. Opt., 23, 785 (1985).
8. Todorov, T., Nikolova, L., Tomova, N., Dragostinova, V.: Photoinduced Anisotropy in Rigid Dye Solutions for Trensient Polarization Holography, IEEE J. Quant. Elec. QE-22, 1262 (1986).
9. Gehrtz, M., Pinsl, J., Brächle, C.: Sensitive Detection of Phase and Absorption Gratings: Phase-Modulated, Homodyne Detected Holography, Appl. Phys., B 43, 61 (1987).
10. Couture, J.J.A. and Lessard, R.A.: Modulation transfer function measurements for thin layers of Azo dyes in PVA matrix used as optical recording material, Appl. Opt., 27, August 15, (1988).
11. Mailhot, S., Galarneau, P., Lessard, R.A., Denariez-Roberge, M.M.: Degenerate four-wave mixing in organic AZO-DYES, CHRYSOIDIN and BENZOPURPURIN 4B, Appl. Opt., 27, August 15, (1988).
12. Zollinger, M.: Color Chemistry, VCH Verlagsgesellschat mbH, D-6940 Weinheim (Federal Republic of Germany), pp. 103-114 (1987).
13. Noller, C.R.: Chemistry of organic compounds, Second edition, (W.B. Saunders 1957), p. 663.
14. March, J.: Advanced organic chemistry, Third Edition (Wiley-Interscience (1985)), pp. 210, 215.
15. Couture, J.J.A., Lessard, R.A.: Effective thickness determination for volume transmission multiplex holograms, Can. J. Phys., 64, 553 (1986).
16. Maksymyk, I.: Automatisation de la mesure de la F.T.M. de milieux enregistreurs, Thèse M.Sc., Université Laval, Ste-Foy, Québec, 1987.
17. Kogelnik, H.: Coupled Wave Theory for Thick Hologram Gratings, Bell Syst. Teck. J., 48, 2909 (1969).
18. Mailhot, S.: Conjugaison de phase par mélange à quatre ondes dans les colorants azoïques: la benzopurpurine 4B et la chrysoïdine, Thèse M.Sc., Université Laval, Ste-Foy, Québec, 1988.

ULTRAFAST THIRD-ORDER NON-LINEAR OPTICAL PROCESSES IN POLYMERIC FILMS

PARAS N. PRASAD

Department of Chemistry, State University of New York at Buffalo,
Buffalo, New York 14214, USA

1. INTRODUCTION

Nonlinear optical processes have received international attention
because of their importance in optical signal processing and computing
(1). For all optical processing, one needs to utilize third-order
nonlinear optical interaction which makes the refractive index intensity
dependent. Compared to second order nonlinear optical interactions, our
understanding of structure-property relation for third order effect is
very limited. Recognizing the importance of third-order nonlinear optical
effects in all optical signal processing and that organics have the
largest non-resonant $\chi^{(3)}$, the interest recently has shifted heavily
towards the study of third-order nonlinear optical interaction in organics
(1,2).

This paper includes selective results from our comprehensive program in
nonlinear optical effects in organic molecules and polymers. The focus of
our study has been on the third order effect. We have calculated
microscopic nonlinearities of organic molecules in several series of
conjugated structures using ab-initio SCF approach coupled with the finite
field method. The effects of increase in the Π-electron conjugation
length and molecular conformation, as well as heavy atom effect and the
role of substituents have been investigated in order to derive an
understanding of molecular structure-property relation so that structural
parameters associated with enhanced optical nonlinearities can be
identified. This theoretical study has been complemented with the
measurements of optical nonlinearities in sequentially built and
systematically derivatized structures.

Femtosecond degenerate four wave mixing has been used to measure
resonant and nonresonant $\chi^{(3)}$ in several polymeric systems and Langmuir-
Blodgett films. In the non-resonant case the response is limited by the
laser-pulse width. In the resonant cases, the response time has been
found to vary over many orders of magnitude depending on the dynamics of
the resonant process. In the case of conducting polymers, the observed
optical nonlinearities are found to be consistent with the mechanism
involving solitonic and polaronic contributions. In the same group of
polymers, when the neutral form of the polymer is doped and consequently
oxidized (positively charged), the conductivity increases by many orders
of magnitude but $\chi^{(3)}$ is found to dramatically reduce.

Nonlinear optical processes in polymer waveguides have been
investigated. Intensity-dependent phase-shifts leading to intensity-
dependent coupling angles as well as intensity dependent coupling
efficiencies and consequently, limiter action have been observed. In a
Fabry-Perot etalon geometry as well as in a quasi-waveguide interferometer
containing a nonlinear optical polymer, optical switching and bistable
behavior have been observed. These studies also provided an opportunity
to determine the sign of $\chi^{(3)}$.

J. Messier et al. (eds.), Nonlinear Optical Effects in Organic Polymers, 351–363.
© 1989 by Kluwer Academic Publishers.

2. MICROSCOPIC THEORY OF OPTICAL NONLINEARITY

Organic systems consist of molecular units (molecules or polymeric chains), which in the absence of any charge or intermolecular charge transfer interact only weakly. In such cases one often uses an oriented gas model to relate the molecular properties to the corresponding bulk properties. This oriented gas model has also been used to relate the molecular optical nonlinearity to the bulk nonlinearity. For the third order effect, the microscopic (molecular) nonlinearity is defined by the second hyperpolarizability γ and the corresponding bulk nonlinearity by the third order susceptibility $\chi^{(3)}$. They can be related under oriented gas approximation by the following equation (3)

$$\chi^{(3)}(\omega_4; \omega_1, \omega_2, \omega_3) = F(\omega_1)F(\omega_2)F(\omega_3)F(\omega_4) \sum_n \langle \gamma^n \rangle \qquad (1)$$

In the above equation $F(\omega_i)$ is the local field at optical frequency ω_i and has been assumed to be independent of the site index n. One often uses Lorentz approximation for the local field (3). The term $\sum_n \langle \gamma^n \rangle$ represents a site average sum. Both $\chi^{(3)}$ and γ are fourth rank tensors. Clearly under this model, the optical nonlinearity of an organic system is primarily determined by the microscopic nonlinear optical properties of the constituent molecular units. It is, therefore, of fundamental importance to creat an understanding of the microscopic nonlinearity and its dependence on molecular structure. Only through such understanding can one use molecular engineering to optimize the nonlinear optical behavior of organic materials.

For the calculation of microscopic nonlinearities, the two different approaches are: (i) The derivative method and (ii) the sum-over-states method. The derivative method is based on the Taylor expansion of the energy or the dipole moment as a function of the applied electric field as follows:

$$\epsilon(E) = \epsilon^o - \sum_i \mu^o E_i - 1/2 \sum_{i,j} \alpha_{ij} E_i E_j - 1/3 \sum_{i,j,k} \beta_{ijk} E_i E_j E_k$$

$$- 1/4 \sum_{i,j,k,l} \gamma_{ijkl} E_i E_j E_k E_l \qquad (2)$$

$$\mu_i(E) = \mu^o_i + \sum_j \alpha_{ij} E_j + \sum_{jk} \beta_{ij} E_i E_j + \sum_{jkl} \gamma_{ijkl} E_j E_k E_l \qquad (3)$$

In the electric dipole approximation the radiation field is assumed as an electric field providing Stark dipolar interaction. The third-order nonlinear optical coefficient γ_{ijkl}, therefore, is simply given by the fourth derivative of the energy or third derivative of the induced dipole moment (4). With a proper choice of the wavefunction, both the energy derivative or the dipole moment derivative methods should yield the same value. The derivatives can be evaluated either by a finite field (numerical) method or analytically. Again, the two methods of obtaining derivatives are equivalent, except that in the finite field method, to avoid numerical instability, one has to be careful in choosing the range of field strengths (4).

The sum-over-states method is based on the perturbation expansion of the Stark energy term in which nonlinearities are introduced as a result of mixing with excited states (5). For example the expression for $\gamma(-3\omega; \omega,\omega,\omega)$ which will be responsible for third harmonic generation is given as

$$\gamma_{ijkl}(-3\omega;\omega,\omega,\omega) = \frac{e^4}{4h^3} \sum_{m_1 m_2 m_3} \frac{\langle g|r_i|m_3\rangle\langle m_3|r_j|m_2\rangle\langle m_2 r_k|m_1\rangle\langle m_1|r_l|g\rangle}{(\omega_{m_1 g}-\omega)(\omega_{m_2 g}-2\omega)(\omega_{m_3 g}-3\omega)}$$

$$+ \ldots$$

In this calculation one computes the energies and various expectation values of the dipole operator for various excited states. These terms are then summed to compute γ. If one does an exact calculation, in principle both the derivative and the sum-over-states methods should yield the same result. However, such exact calculations are not possible. Important choices, one has to make, are the form of hamiltonian and the basis functions. One can use an ab-initio method or take a semi-empirical approach. At this point, it is our feeling that each method has its own merit and limitation since it involves a different set of approximations, choice of hamiltonians and basis functions.

The derivative method is conceptually sound in the sense that one can easily relate the different derivatives to the different anharmonic terms of the electronic potential energy well. Furthermore, in the simplest (time-independent) form, the derivative method demands that only the ground state properties be computed. However, this type of calculation only yields static hyperpolarizabilities. Nonetheless, the static hyperpolarizabilities are good estimates of the nonlinearities away from any electronic resonances and can provide a better handle on understanding the role of structural variations. Therefore, these types of calculations can be very useful in understanding the structure-property relationship.

The sum-over-states method requires that not only the ground states but all excited state properties be computed as well. One often truncates the sum-over-states to include only a few excited states. Furthermore, electron correlation effects become considerably important to describe the excited state properties. One advantage of this method is that it can be conveniently used to calculate dynamic (frequency dependent) hyperpolarizabilities, and, therefore to see the effect of electronic resonances.

We have adopted the derivative method to calculate the second hyperpolarizability γ and create an understanding of the structure-property relationship. We have used an ab-initio approach at the SCF level. The details of this procedure are described elsewhere (4). Here only the important steps are qualitatively described. We define the molecular hamiltonian including field dependent Stark term. A single LCAO-MO Slater determinant is constructed with a choice of AO basis set assuming a single configuration closed shell. In other words, configuration interactions are not included but, as was discussed above, electron correlation may not be as important in describing the ground state property as it is for the excited states. The next step is optimization of molecular geometry. Then the dipole moment is calculated as a function of field. The various derivatives of the dipole moment with respect to the applied field are calculated. We have used both the finite field and analytical method (the latter using coupled perturbed Hartree-Fock approach (4)). To get different tensor components, the field directions are changed. To get insight into various orbital contributions, we have used an orbital transformation method which transforms an occupied orbital into a corresponding virtual orbital in the presence of field. This transformation also allows us to conveniently separate the σ- and π-contributions to optical nonlinearity. In addition, to generate some understanding of the regional contribution (eg a particular functional group) to nonlinearity, we investigate the contour plots of various derivatives of electron density. For example, the third derivative of the electron density with respect to field provides information on regional contribution to γ.

We find that the choice of basis functions is very crucial in the calculation of molecular nonlinearities. The split valence basis sets describe the geometry of the molecules adequately and should be used for geometry optimization. However, it is the tail portion of the wavefunction which describes the anharmonic behavior of the electron and to describe this portion of the wavefunction one must include diffuse polarization functions. For small molecules, the exclusion of diffuse polarization function leads to even a wrong sign for γ. We find that a compromise between computational complexity and quality of calculation can be reached by utilizing a 321-G basis set to optimize the geometry and include diffuse polarization functions to calculate α, β and γ.

We have computed α and γ for various finite chain oligomers in the polyene (alternate single and double bonds), polyyne (alternate triple and single) and cumulene (all double) series. The details of our results are discussed elsewhere (4). Here we summarize only some important conclusions of our computation. In comparing our ab-initio SCF calculation with those from the sum-over-states calculations using the same orbital basis, we find that the two methods yield agreement on polarizability but not on second hyperpolarizability, γ. As expected on the basis of a simple free electron model, the second hyperpolarizability γ is anisotropic, with the longitudinal component γ_{zzzz} (along the chain direction) growing very rapidly for all three series, polyenes, polyynes and cumulenes, as the chain length grows. The γ values are positive in sign. The corresponding orbital analysis shows that the sigma orbital contribution to the longitudinal hyperpolarizability tends to partially cancel the large π-contribution. Contour maps of the third derivative indicates that the second hyperpolarizability is a cooperative effect with substantial charge transfer along the entire length of the chain.

Often a power law is used to describe the dependence of α and γ on the chain length (effective conjugation length). We have taken the computed values of α and γ for oligomers of trans-polyene and polyyne series and fitted the values with the equation

$$F = A + B(N-\delta)^C$$

where F = α or γ; A is included to take into consideration the end effect and δ is incorporated to adjust for an effective conjugation length. N is the number of repeat units and C describes the power law. Our ab-initio calculation yields C in the range 1.3 to 1.5 for α and 3.2 to 3.4 for γ. In contrast, a free electron model predicts a N^3 power dependence for α and N^- power dependence for γ (6). Earlier semiempirical calculations using Hückel or PPP hamiltonian predict C = 2.3 for α and C = 5.3 for γ (7). A more recent calculation by de Melo and Silbey predicts C = 4.5 for γ (8). The sum-over-states approach have been used by Garito and co-workers. They have calculated using a semiempirical approach, γ for both cis- and trans-polyenes. Their calculation yields C = 5 for γ for trans-polyenes (9).

To create insight into conformational changes and the effect of increased π-electron density, we have computed α and γ for trans-butadiene, cis-butadiene and thiophene. In this group, cis and trans-butadiene differ in the conformation. Thiophene can be envisioned as derived from a cis-butadiene structure by replacing two hydrogens by a sulfur atom which increases the π-electron density in a five-membered ring structure. We find the values of γ follow the order: trans-butadiene > thiophene > cis-butadiene.

With an objective to understand the role of substituents, we have computed α, β and γ for various acetylene derivatives R_1-C≡C-R_2. At this time, the calculation has been done using only the split valence set 321-G. Without the inclusion of diffuse polarization functions, these results may not provide any reliable measure of the actual trend. However, the calculation shows some interesting trend. With the substitution of electron withdrawing groups (eg R_1, R_2 = NO_2), the γ value increases.

We have begun to correlate the theoretical predictions by measurement of α and γ on sequentially synthesized and systematically derivatized structures. One series we have investigated is that of thiophenes (10). We have measured orientationally averaged α by measuring the refractive indices of the thiophene monomer and oligomers (N = 2-6) in THF solutions of various compositions. By using the Lorentz-Lorenz approximation of additive molecular polarizations, $\langle \alpha \rangle$ was calculated using a least square fit. Figure 1 shows the dependence of $\langle \alpha \rangle$ on the number of repeat unit of thiophene. The orientationally averaged γ $[\langle \gamma \rangle = \frac{1}{5}(\gamma_{zzzz} + \gamma_{xxxx} + \gamma_{yyyy} + 2\gamma_{yyzz} + 2\gamma_{xxzz} + 2\gamma_{xxyy})]$ was calculated from $\chi^{(3)}$ measurements in THF solutions by using degenerate four wave mixing as described below in the measurement section. Again, the assumption of additivity of the molecular hyperpolarizabilities of the solute and solvent along with the Lorentz approximation for the local field is used. From this measurement we have computed both the magnitude and the sign of γ. The sign of γ is found to be positive in each case. The $\langle \gamma \rangle$ values for the monomer and various oligomers are also plotted as a function of the number repeat unit N in Figure 1. Clearly, the $\langle \gamma \rangle$ value increases much more rapidly with the increase of N. Now we compare the results of our experimental studies with the predictions of our ab-initio calculations. We have calculated both α and γ tensor components of thiophene with a 321-G basis set augmented with diffuse polarization functions of S and P type on carbon and P and D type on sulfur. The calculated values of $\langle \alpha \rangle$ and $\langle \gamma \rangle$ are

8.21×10^{-24} cm^3, 1×10^{-36} esu; the corresponding experimental values are

9.8×10^{-24} cm^3 and 3.5×10^{-36} esu. The agreement for $\langle \alpha \rangle$ is quite good but for $\langle \gamma \rangle$, the values do not agree as well. For bithiophene, due to the computational complexity, we have at this time only the α value which is

1.45×10^{-23} cm^3; the experimental value is 2.5×10^{-23} cm^3. As our calculation for various π-electron structures predicts, the sign of γ is positive. A fit of the power law to our experimental values yields C = 1.69 for $\langle \alpha \rangle$ and C = 4.05 for $\langle \gamma \rangle$. The $\langle \alpha \rangle$ power law is in better agreement with that predicted by our ab-initio calculation, but again the observed power law for $\langle \gamma \rangle$ does not agree as well with our ab-initio calculation.

We have also determined $\langle \alpha \rangle$ and $\langle \gamma \rangle$ for 2-nitro substituted terthiophene and found them to be larger than that of terthiophene. In case of $\langle \gamma \rangle$, the values for 2-nitroterthiophene is an order of magnitude larger than that for terthiophene. Therefore, this result is in agreement

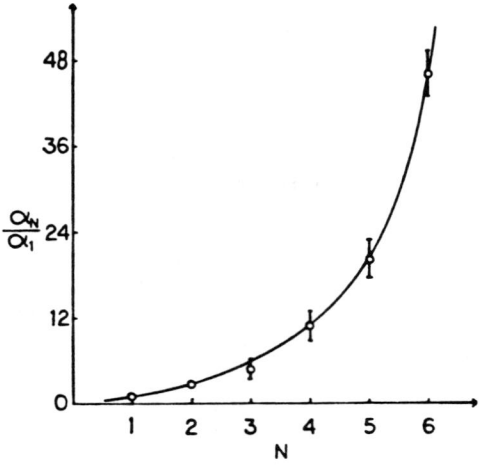

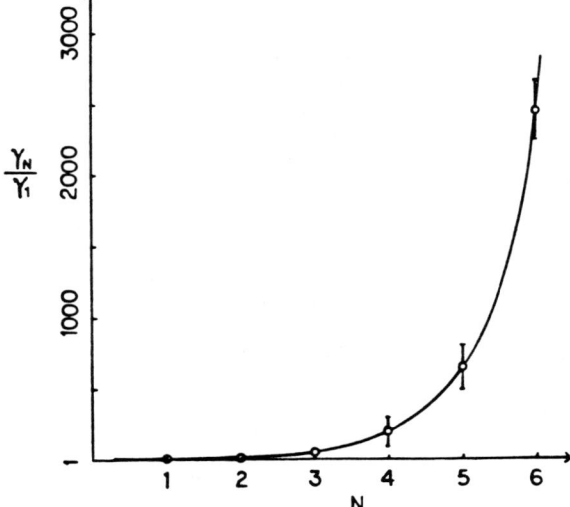

FIGURE 1. top) Dependence of the ratio $\dfrac{\langle\alpha\rangle_N}{\langle\alpha\rangle_1}$ on the number, N, of the thiophene repeat unit. The terms $\langle\alpha\rangle_N$ and $\langle\alpha\rangle_1$ represent the orientationally averaged polarizabilities for the N-mer and monomer respectively. Bottom) Dependence of the ratio $\dfrac{\langle\gamma\rangle_N}{\langle\gamma\rangle_1}$ on the number, N, of the thiophene repeat unit. The terms $\langle\gamma\rangle_N$ and $\langle\gamma\rangle_1$ represent the orientationally averaged second hyperpolarizabilities for the N-mer and the monomer respectively.

with the trend predicted by our ab-initio calculation on substituted
acetylenes.

3. MEASUREMENTS OF THIRD-ORDER OPTICAL SUSCEPTIBILITIES

In the measurement of $\chi^{(3)}$, our interest is not just the magnitude, but
also the sign of $\chi^{(3)}$ and its response time. Furthermore near resonance
$\chi^{(3)}$ can be complex and show a strong frequency dispersion. In an
anisotropic medium $\chi^{(3)}$ is expected to behave like a fourth rank tensor.
If there are important phonon contributions, $\chi^{(3)}$ can also be expected to
show a pronounced temperature dependence. We have used a variety of
techniques for measuring these relevant parameters characterizing $\chi^{(3)}$.
These methods include degenerate four wave mixing, nonlinear optical
waveguide, nonlinear Fabry-Perot, surface plasmon nonlinear optics and
third harmonic generation. The method we use most extensively is
degenerate four wave mixing (DFWM).

DFWM can be envisioned as Bragg-diffraction from a transient grating.
This method of measurement has been discussed in detail elsewhere (1).
Two beams $I_1(\omega)$ and $I_3(\omega)$ cross in the sample at an angle θ and set up an
intensity grating. In the case of a non-resonant third order electronic
process, this intensity grating results in a refractive index grating due
to the intensity dependence of the refractive index derived from $\chi^{(3)}$.
When a third beam $I_2(\omega)$ is incident on the sample counterpropagating to
beam $I_1(\omega)$ it is Bragg-diffracted from this grating to produce a signal
$I_4(\omega)$ which is the phase-conjugate to beam $I_3(\omega)$ and, therefore, is
counterpropagating. It is split from $I_3(\omega)$ by a beam splitter. When all
input beams $I_1(\omega)$, $I_2(\omega)$ and $I_3(\omega)$ are time coincident in the sample the
magnitude of the DFWM signal $I_4(\omega)$ is related to $[\chi^{(3)}]^2$. Generally, one
uses a comparison with CS_2 as the reference sample. This method has been
described in detail elsewhere. However, it is worth remarking that
various values of $\chi^{(3)}$ have been reported for CS_2. We have taken a value
of 6.8×10^{-13} esu for $\chi_{1111}(-\omega;\omega,-\omega,\omega)$ for CS_2 as reported by picosecond
measurements (11). The change in the DFWM signal $I_4(\omega)$ as a function of
time-delay of the backward beam $I_2(\omega)$ gives the decay of the refractive
index grating and hence the time-response of the optical nonlinearity.
Obtaining the time-response readily is an important advantage of DFWM in
measurement of $\chi^{(3)}$. Furthermore, in the counterpropagating geometry as
discussed above, the phase matching condition is automatically satisfied.
Since all the input and output optical frequencies are of the same value
(hence degenerate), one measures $\chi^{(3)}(-\omega;\omega,-\omega,\omega)$. This $\chi^{(3)}$ value is an
important parameter for the design of devices utilizing optical switching
and bistability. We have also demonstrated that DFWM can conveniently be
used to measure the anisotropy of $\chi^{(3)}$ in an anisotropic medium (12).

There are many other processes such as thermal effects (thermal and
ultrasonic grating) and excited state gratings which also contribute to
degenerate four wave mixing. These processes have different time-response
which has been discussed elsewhere (13). To separate the purely
electronic $\chi^{(3)}$ from other types of gratings, femtosecond time-resolution
is of significant value.

For degenerate four wave mixing studies in our laboratory, the IR
output of a mode-locked Nd:YAG laser (Spectra-Physics, Model 3800) is
first compressed in a grating-fiber compressor (Spectra-Physics, Model
3690) and is frequency doubled to synch-pump a CW dye laser (Spectra-
Physics, Model 375). The pulses are amplified in a 3 stage amplifier
(Spectra-Physics, Model PDA) pumped by a frequency doubled 30Hz Q-switched
Nd:YAG laser (Spectra-Physics, Model DCR-2A) to generate ~ 400 fs nearly
transform-limited amplified pulses. The output of the amplifier is split

into three beams using beam splitters such that the intensities of beams 1
and 2 are ~ 0.5 times that of the parent beam, while the intensity of the
probe beam (beam 3) is only 0.1 times that of the parent beam. Each of
these three beams goes through a delay line so that their arrival times at
the sample can be synchronized. The beams are focused into the sample
cell to a spot size of ~ 400 μm in such a way that beams 1 and 2
counterpropagate and beam 3 is incident at a small angle of ~4° with beam
2. The conjugate signal is picked through a beam splitter arranged in the
path of beam 3. The intensity of I_4 was measured by a photodiode and
boxcar combination.

We have investigated the $\chi^{(3)}$ behavior for a number of organic
polymeric structures. The results will be discussed here in two separate
categories: (i) non-resonant case where the optical frequency used for
the $\chi^{(3)}$ measurement is not within the electronic resonance of the
materials and (ii) resonant-cases where electronic resonance plays an
important role. Since conjugated polymeric structures in general have a
distribution of conjugation length, even the non-resonant cases actually
do show residual weak absorption. The non-resonant $\chi^{(3)}$ values have been
measured for many conjugated polymeric structures which include aromatic
heterocyclic polymers such as poly-p-phenylene-bisbenzathiazole (PBT),
poly-p-phenylene-bisbenzaoxazole (PBO); soluble polydiacetylenes such as
poly-4-BCMU, polyimides such as LARC-TPI, polyphenylacetylene, and poly-p-
phenylene vinylene (PPV). Another system is oligomers of thiophene as
discussed above. Important conclusions of our investigation of nonresonant $\chi^{(3)}$ are as follows:

(i)   The response time of the non-resonant $\chi^{(3)}$ in organics are the
        fastest nonlinear response which in our case has been limited by the
        laser pulse-width (350 fs). Therefore, femtosecond response is
        clearly demonstrated.
(ii)   The maximum non-resonant value of $\chi^{(3)}$ in organic conjugated
        polymeric structures investigated in our laboratory has still been
        less than 10^{-7} esu.
(iii) $\chi^{(3)}$ shows a strong dependence on the π-electron conjugation. Two
        types of results demonstrate this behavior. In the first example,
        poly-4-BCMU in the film undergoes a conformational transition from a
        more conjugated red form to the yellow form in which π-electron
        delocalization is reduced. We find a corresponding drastic
        reduction in the $\chi^{(3)}$ value. The second example is provided by a
        systematic study of the thiophene oligomers (Figure 1) as discussed
        above. In going to higher oligomers, the effective π-electron
        delocalization length increases leading to a rapid increase of Y
        (and hence $\chi^{(3)}$).
(iv)   We have investigated oriented polymer films such as biaxial PBT and
        uniaxial PBO films. The measured $\chi^{(3)}$ shows considerable anisotropy
        which can be theoretically fitted by using the fourth rank tensor
        properties of $\chi^{(3)}$.
(v)   The temperature dependence of $\chi^{(3)}$ investigated so far only for a
        limited number of cases, is found to be insignificant between 80K
        and 373K, provided there is no structural change. This implies that
        optical phonon and orientational contributions are not significant.
(vi)   For the cases where we have been able to determine the sign, $\chi^{(3)}$
        under non-resonant condition has been found to be positive.

Organic systems exhibit a rich variety of excitations which can be
photogenerated. To examine the effect of electronic resonance on $\chi^{(3)}$ we
have studied a variety of systems in which various types of excitations
can be photoinduced. Table I lists these systems with the type of

TABLE I

Resonant $\chi^{(3)}$ by Four Wave Mixing

Systems	Mechanism	Response Time	$\chi^{(3)}$
PVK:TNF	correlated electron-hole pairs	>100 ps	10^{-11} esu
polyacetylene-PMMA	soliton/polaron dynamics	fs	10^{-9} esu
phthalocyanines (LB films)	exciton dynamics	ps	10^{-8} esu
polythiophene (electrochemically deposited)	polaron dynamics	fs	10^{-9} esu
polyalkylthiophene (solution cast films, L-B films)	polaron dynamics	fs	10^{-9} esu

photoinduced excitations. The observed $\chi^{(3)}$ along with its response time is also listed. Both the magnitude of resonant $\chi_{eff}^{(3)}$ and its response time vary over many orders of magnitude depending on the dynamics of the excitation. Even under resonance conditions, the response time of the nonlinearity can be in femtoseconds. Polythiophenes are the examples where the response time is again found to be limited by the laser pulse width (~350 fs). The resonant nonlinearity is sufficiently large that we have been able to see DFWM signal even from a bilayer (~80A thick) Langmuir-Blodgett film of phthalocyanines. However, the resonance enhancement is not as large as is observed for inorganic quantum well semiconductors. We have also found that depending on the dynamics of excited states, the resonant $\chi^{(3)}$ value varies with the pulse width of the laser used. For example, $\chi^{(3)}$ in phthalocyanine increases by a factor of 3 when the laser pulse width is increased from 400 fs to 2 ps.

An interesting group of conjugated polymers is provided by those which show bond alternation and can, therefore, support conformational defects such as solitons and polarons (14). These excitations have been suggested as charge carriers to give rise to large conductivity in polymers (14). In the neutral state of the polymer it is a semiconductor but when electrochemically or chemically doped by an electron acceptor the polymer backbone oxidizes and carries a positive charge along with a conformational defect such as soliton, polaron, or bipolaron. Therefore, it provides an opportunity to examine the effect of these conformational defects on the $\chi^{(3)}$ behavior. We have investigated the $\chi^{(3)}$ behavior in polythiophene where at a very low doping level, polaronic defects exist to give rise to spin as detected by EPR. As the doping level is increased, the spin goes through a maximum and then drops when polarons start condensing (pairing) to form bipolaronic defects. The electrical conductivity increases by many orders of magnitude. We have conducted an in-situ electrochemical doping study of $\chi^{(3)}$ in a polythiophene film deposited electrochemically using bithiophene. We find that even a small amount of doping (~3%) which barely produces any change in the linear absorption leads to a drastic reduction of $\chi^{(3)}$. This behavior is reversible and highly reproducible through the redox cycle.

We have also conducted an in-situ chemical doping experiment in which a multilayer (40 layer, approximately 1500Å) Langmuir-Blodgett film of 3-dodecyl substituted polythiophene was doped with iodine and its effect on $\chi^{(3)}$ was investigated. Again we find a similar behavior that when the polymer is oxidized, $\chi^{(3)}$ drops to less than 20% of its original value even though the electrical conductivity increases by over 8 orders of magnitude. Since in the chemical doping by iodine, very high doping levels can be reached, even at heavily doped levels the $\chi^{(3)}$ magnitude stays low. This behavior again is reproducible and reversible through many cycles. At this point, our feeling is that the doping affects the π-electron correlation to influence the $\chi^{(3)}$ behavior.

4. DEVICE PROCESSES

The optical signal processing will utilize the phenomenon of optical switching and optical bistability. We have investigated these processes in three different geometries: (i) surface plasmon, (ii) Fabry-Perot etalon and (iii) planar waveguide. In the surface plasmon geometry we have observed electro-optic modulation in an electrically poled Langmuir-Blodgett film of a side-chain liquid crystalline polymer. Using the $\chi^{(3)}$ process we have observed intensity dependent phase shift (change in coupling angle) in poly-4-BCMU polydiacetylene Langmuir-Blodgett films in a surface plasmon geometry. However, our view is that the surface plasmon geometry is not appropriate for investigating optical switching derived from electronic $\chi^{(3)}$. The reason is that the metal film used in the surface plasmon geometry has a complex dielectric constant for most part of the spectrum. Hence damping of the electromagnetic wave occurs which leads to thermal effects that often dominate intensity-dependent shift of the coupling angles due to electronic $\chi^{(3)}$.

In the Fabry-Perot geometry we have investigated both absorptive and dispersive bistability. A bistable behavior implies that for certain input range, the system exhibits two output states (15). In increasing the input intensity, the switching from a low output state to a high output state occurs at one intensity level. However, in the reverse cycle when the input intensity is decreased, the switching from the high to low state occurs at a different input level. Therefore, by a proper choice of the input hold level, one can use the optical bistability behavior for latching or memory operation. In absorptive bistability, one uses the saturable absorption behavior of the nonlinear medium. The transmission of the Fabry-Perot cavity is initially low due to absorption. As the input intensity increases, at certain level, the medium bleaches due to saturation of absorption. The cavity then switches from the low to high input state. In the reverse cycle, as the input is lowered, the local field within the cavity remains high to keep the cavity in the high output level well below the initial input switching level thus generating a hysteresis.

In dispersive bistability in the Fabry-Perot geometry, one takes advantage of the intensity dependent refractive index. The Fabry-Perot is detuned so that the transmission is low. Now by increasing the intensity of the input signal, the refractive index of the nonlinear medium is changed to bring the cavity in resonance condition at which point the device switches to the high output level. Again in the reverse cycle due to the high field inside the cavity, a hysteresis results. We have observed dispersive bistability in a film of a polyimide.

We have also conducted studies of optical switching in a polymer waveguide. We have earlier reported on optical switching in a poly-4-BCMU quasi-waveguide (16). A quasi-waveguide is formed when the wave is guided at only one interface but the other interface is leaky because the refractive index of the film (the guiding medium) is less than that of the cladding medium (in our case the prism substrate). The physics of the quasi-waveguide process is similar to an interferometer as described above in case of a Fabry-Perot device. Recently we have also successfully observed optical nonlinear processes in an optical waveguide of polyamic acid in which propagation distances over 5 cm have been easily achieved (17). The process of intensity-dependent phase shift (coupling efficiency changes due to change of refractive index as the input intensity changes) has been observed using nanosecond, 80-picosecond and 400 femtosecond pulses. The observed corresponding limiter action behavior in the input-output characteristics is shown in Figure 2 using 400 fs pulses. Our analysis shows that the observed switching with the femtosecond and picosecond pulses is due to electronic nonlinearity but that observed with

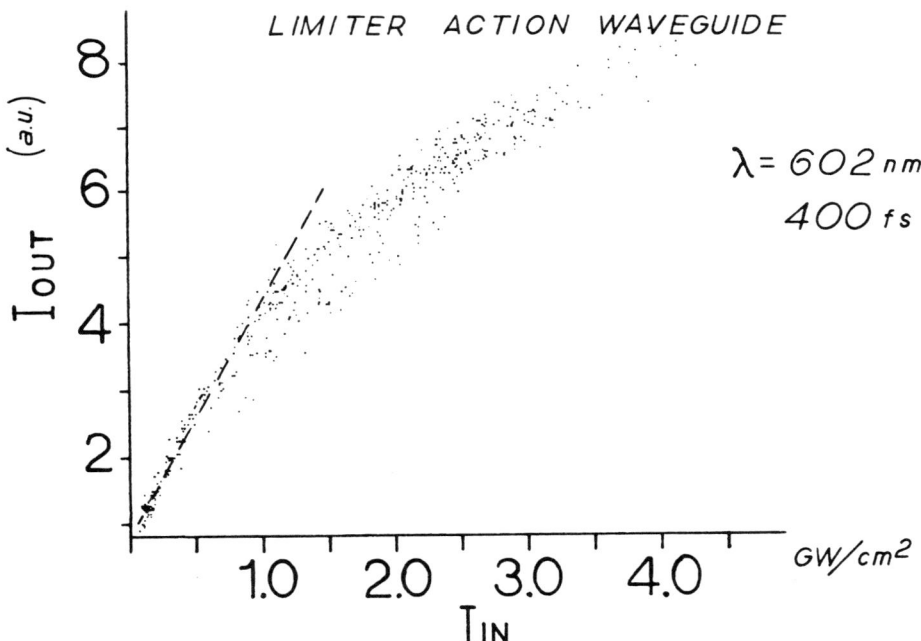

FIGURE 2. The output intensity (I_{OUT}) plotted vs. the input intensity (I_{IN}) for the polyamic acid nonlinear waveguide. The data points shown by dots reveal a limit action behavior. The dashed line represents the linear relation.

nanosecond is mostly dominated by thermal nonlinearity. Our studies of optical switching also yields a valuable fundamental parameter, that is, the sign of $\chi^{(3)}$. From the direction of shift of the coupling angle as a function of the intensity, one can obtain the sign of $\chi^{(3)}$. For the cases investigated we have found a positive sign for $\chi^{(3)}$. Our theoretical study of microscopic optical nonlinearity as discussed above also yields a positive sign for the third order nonlinear coefficient γ (and hence $\chi^{(3)}$).

5. FUTURE DIRECTIONS

Although we have begun to build some microscopic understanding of third-order optical nonlinearity, it is still in its infancy. We need to considerably improve upon our understanding of structure-property relationship in order to project structural requirements for enhanced third-order optical nonlinearity. The currently achieved $\chi^{(3)}$ in organic systems is the highest for non-resonant cases but still need to be increased to be compatible with the diode laser technology. Electronic resonance enhancement does not seem to increase the $\chi^{(3)}$ value as dramatically as is observed in inorganic multiple quantum well semiconductors.

In order to improve understanding of the structure-property relation we must study the role of substituents in addition to that of effective conjugation. For this objective, further studies of theoretical calculation and experimental measurements of γ on sequentially built and systematically derivatized structures are necessary. We are taking this approach. In addition, we must increase our data base of $\chi^{(3)}$ on novel polymeric structures and macrocycles.

The role of bulk effects has not been investigated. As discussed above the Lorentz-approximation along with the oriented gas model has been used. The contribution due to intermolecular charge-transfer states, exciton bands, polariton coupling etc. are topics which we also plan to investigate.

A problem in the past has been the optical quality of the $\chi^{(3)}$ polymeric materials. Most likely device structures will utilize the organic $\chi^{(3)}$ material in the guided wave forms. Organic polymeric films with large $\chi^{(3)}$ often tend to be unprocessable and extremely lossy due to both residual absorption and refractive inhomogeneities. Our research program also plans to work on structural control, characterization and material processing to address this issue.

6. ACKNOWLEDGEMENTS

The author wishes to thank his group members Drs. J. Swiatkiewicz, B. P. Singh, R. Burzynski, J. Pfelger, M. Samoc, A. Samoc; Miss P. Chopra, Mr. L. Carlacci, Mr. M. T. Zhao, Mr. Y. Pang, Mr. X. Huang, Mr. P. Logsdon, Mr. M. Casstevens, Mr. M. Carpenter and Mr. He whose works have contributed to this review. The author also thanks Dr. D. Ulrich of AFOSR for his encouragement. This work has been supported by the Air Force Office of Scientific Research through contract number F4962087C002. Partial supports from NSF-Solid State Chemistry Program Grant number DMR8715688 and Eastman Kodak are also acknowledged.

7. REFERENCES

1.   "Nonlinear Optical and Electroactive Polymers" Eds. P. N. Prasad and
     D. R. Ulrich, Plenum Press (New York, 1988).

2.   "Nonlinear Optical Properties of Polymers" Symposium Proceedings,
     Vol. 109, Materials Research Society, Fall 1987 meeting, Boston,
     Mass.

3.   "The Principles of Nonlinear Optics" Y. R. Shen, Wiley & Sons (New
     York, 1984).

4.   P. Chopra, L. Carlacci, H. F. King and P. N. Prasad, J. Phys. Chem.
     (in press).

5.   J. F. Ward, Rev. Mod. Phys. $\underline{37}$, 1 (1965).

6.   M. P. Boggaard and B. J. Orr, Int. Rev. Sci. Phys. Chem. Ser. $\underline{2}$, 149.

7.   E. F. McIntyre and H. F. Hameka, J. Chem. Phys. $\underline{68}$, 3481 (1978); O.
     Zamani-Khamiri and H. F. Hameka, J. Chem. Phys. $\underline{72}$, 5903 (1980).

8.   C. P. de Melo and R. Silbey, Chem. Phys. Lett. $\underline{140}$, 537 (1987).

9.   A. F. Garito, A Topical Workshop on Organic and Polymeric Nonlinear
     Optical Materials, ACS Polymer Chemistry Division, May 16-19, 1988,
     Virginia Beach.

10. M. T. Zhao, B. P. Singh and P. N. Prasad, submitted to J. Chem. Phys.

11. N. P. Xuan, J-L Ferrier, J. Gazengel, and G. Rivoire, Opt. Commun.
     $\underline{51}$, 433 (1984).

12. D. N. Rao, P. Chopra, S. K. Ghoshal, J. Swiatkiewicz and P. N.
     Prasad, Appl. Phys. Lett. $\underline{48}$, 1187 (1986).

13. P. N. Prasad in Reference 1, p. 41.

14. "Handbook of Conducting Polymers" Vols. 1 and 2, Ed. T. A. Skotheim,
     Marcel Dekker (New York, 1986).

15. "Optical Bistability: Controlling Light with Light" H. M. Gibbs,
     Academic Press (New York, 1985).

16. B. P. Singh and P. N. Prasad, J. Opt. Soc. Am. $\underline{B5}$, 453 (1988).

17. R. Burzynski, B. P. Singh, P. N. Prasad, R. Zanoni and G. I.
     Stegeman, to be published.

PICOSECOND PHASE CONJUGATION IN YELLOW POLYDIACETYLENE SOLUTIONS

J.M. NUNZI, Labo. de Physique Electronique des Matériaux.
CEA - Saclay IRDI/D.LETI/DEIN   91191 GIF SUR YVETTE FRANCE
J.L. FERRIER and R. CHEVALIER, Labo. d'Optiques des Fluides, ERA CNRS 780
Université d'Angers, 49045 ANGERS,   FRANCE

INTRODUCTION
    Polydiacetylene solutions have been reported to exhibit a very strong
degenerate non linearity near the excitonic absorption band. This led to
phase conjugation efficiencies greater than unity at 1 GW.cm^{-2} pump
powers[1]. In addition, using a crossed polarizations configuration,
intensity gratings could be avoided, and fast, picosecond response, could
be measured[2]. Intensity gratings gave a slowly decaying signal due to
thermal effects which hid fast medium response[1]. However, those experiments left some facts unexplained:
    All the explored polarization configurations gave a signal of the same
magnitude. At first sight, this could indicate that the signal had the
same origin in each configuration because we are concerned with polymers
which should behave one dimensionally[3].
    As concerns the slow thermal response, it can have two different ori-
gins[3]. One which is due to thermal expansion of the solution at constant
temperature and which can also give rise to acoustic oscillations when the
pulse duration is lower than the acoustic period. The other which is due to
polymer temperature induced structural changes at constant density. The
former provides hydrodynamical informations on the solution, while the
latter provides informations characteristic of the polymer nature (coupling
between electronic states and internal vibrations, effect of excited
species, etc.).

    Since previous experiments where performed using 180 ps pulses, and
since calculated and observed slow responses had the same magnitude as fast
response, a proportionality rule states that with 30 ps pulses, "thermal"
effects should fall at least one order of magnitude below "fast response",
allowing us to answer the above problems. In that study, we show that this
is indeed the case. Additionally, we prove that efficient picosecond phase
conjugation can be performed in polydiacetylene solutions, with reflectivities close to unity, without worrying about polarizations condition, and
without damage to the material.

EXPERIMENTS
    Sample is a 4-BCMU yellow solution in DMF, prepared according to ref. 3.
Concentration is close to 2 g/l (0,2 % solution) and is adjusted to obtain
a 50 % transmission of the 1 mm thick sample cell, as required for optimum
coupling efficiency[4].

    The experimental setup for phase conjugation has been described
elsewere[5]. The passively mode locked Quantel Nd-YAG laser provides 5 mJ,

J. Messier et al. (eds.), Nonlinear Optical Effects in Organic Polymers, 365–368.
© *1989 by Kluwer Academic Publishers.*

30 ps, 532 mm single pulses at 1 Hz repetition rate. The beam is divided into 3 parts:

Two strong pumps with intensities $I_1 \approx 140$ MW.cm^{-2} and $I_2 \approx 100$ MW.cm^{-2}, and a weaker probe, with intensity $I_3 \approx 13$ MW.cm^{-2}. Probe and pump 1 make a $10,75°$ angle.

Polarizations of the 4 beams are selected with glan polarizers. Acquisition, averaging and analysis of the signal is under microcomputer control.

RESULTS

The first studied configuration (I) is the one with all polarizations parallel (fig. 1), which is relevant of $\chi^{(3)}_{xxxx}$. An exchange of sample cells reveals that conjugate reflectivity is 30 times larger than in CS$_2$. This allows calibration of nonlinear susceptibilities [5]. We also observe a 25 % conjugate reflectivity when all beams coincide. Analysis of signal dependence with laser intensity around mean values confirms the regular third power proportionality of the effect[1,2]. Thus, we can deduce the nonlinear degenerate susceptibility $\chi^{(3)}_{xxxx} = 10^{-11}$esu, with 20 % accuracy, for this yellow solution, at 532 mm.

Fit of the temporal response with a sech function gives a full width at half maximum (FWHM) of 50 ps, which is an intensity autocorrelation function of the laser pulses. At 100 ps delay, there is a slow – thermal contribution which is 6 % of the initial response.

In the configuration corresponding to the large spaced intensity grating contribution (II), which is relevant of $\chi^{(3)}_{yxyx}$ in this case, the response is the same as for configuration (I). In the configuration corresponding to the small spaced grating contribution (III), relevant of $\chi^{(3)}_{yyxx}$ in this case, the magnitude of the signal is the same as in both configurations (I) and (II). However, we do not resolve any slow contribution to the signal in our 100 ps probing region. Erasure of this small grating contribution also gives a response with 50 ps FWHM (fig. 2).

In the polarisation grating configuration (IV) which is relevant of $\chi^{(3)}_{yxxy}$, the signal is 4 times weaker than in the other configurations. The temporal response also fits a sech function with 50 ps FWHM, and without visible slow effect.

DISCUSSION

As concerns the 50 % absorbed energy, it has to be converted to heat, and in the narrow intensity grating configuration (III), the first maximum of acoustic oscillations of the signal should appear at 68 ps "read" pulse delay[3]. But, as this time is comparable to the pulse duration, such oscillations are averaged to a continuous background signal, as in ref. 1, but which is weak (< 6 %) in our experiments. In the large intensity grating configuration (II), the first acoustic oscillation should appear at 1 ns read pulse delay. But it should also remain weak and be at most 4 times larger than the continuous background that we could not resolve in configuration (III), one part of which being the average of those oscillations. Hence, the continuous 6 % slow effet of figure 1, which cannot be due to thermal expansion, may be due to polymer characteristic changes of the index of refraction (n) expressed by a term $(\partial n / \partial T)_\rho$ [3].

However, erasure of the small grating contribution gives a 50 ps FWHM (fig. 2), as for grating reading (fig. 1), and this is characteristic of an initial response faster than laser coherence time[7]. Hence, in our

experiments, both components of the signal (slow and fast) have different origins, but are characteristic of the polymer nature.

Additionally, an interesting result comes from the relative $\chi^{(3)}$ moduli
measured in the 4 studied polarization conditions. We found:

$$\chi^{(3)}_{xxxx} \approx \chi^{(3)}_{xyxy} \simeq \chi^{(3)}_{xxyy} \simeq 2\chi^{(3)}_{xyyx} \ .$$

For a one dimensional isotropic mixture of polymers[3], all perpendicular
polarizations $\chi^{(3)}$ components should equate $\chi^{(3)}_{xxxx}/3$. Therefore, near
the excitonic absoption, yellow polydiacetylene solutions does not behave
as a one dimensional material; perhaps because the polymer chains do not
extend linearily[8], as this has been seen from phase conjugation in the
4-BCMU red gel form[9].

CONCLUSION

In conclusion, we have observed a strong picosecond response in a
material which can withstand gigawatt laser powers at 1 Hz repetion rate
during days without observable damage. Since those experiments have been
performed in a 0,2 % solution, and since concentrations up to 15 % are
easily obtained[3], larger and still fast responses are now within hand
reach.

Extrapolation of the measured susceptibility to a 100 % densely
packed yellow 4-BCMU polymer leads to $\chi^{(3)}_{xxxx}$ 10^{-8} esu.

As concerns the relative magnitude of $\chi^{(3)}$s, since the possibility of
bad addition of fast and slow responses due to laser coherence is not ruled
out, experiments with a fully transform limited, picosecond laser are now
in progress.

ACKNOWLEDGEMENTS
We thank J. Messier and F. Charra for helpfull discussions and comments.

REFERENCES
1. W.M. DENNIS, and W. BLAU,
   Opt. Com. 57, 371 (1986).
2. W.M. DENNIS, W. BLAU and D.J. BRADLEY,
   Appl. Phys. Lett. 47, 200 (1985).
3. J.M. NUNZI and D. GREC,
   J. Appl. Phys. 62, 2198 (1987).
4. R.G. CARO and M.C. GOWER,
   IEEE J. Quant. Electr. QE18, 1376 (1982).
5. N. PHU XUAN, J.L. FERRIER, J. GAZENGEL, G. RIVOIRE,
   Opt. Com. 51, 433 (1984).
6. P.A. CHOLLET, F. KAJZAR, J. MESSIER, J.M. NUNZI and D. GREC
   Revue Phys. Appl. 22, 1221 (1987).
7. J.M. NUNZI and D. RICARD,
   Appl. Phys. B35, 209 (1984).
8. D. BLOOR and R.R. CHANCE (Eds),
   in Polydiacetylenes, Nijhoff, Boston (1985).
9. J.M. NUNZI and F. CHARRA, in the proceedings of the Organic Materials
   for Nonlinear Optics conference, held in Oxford (June 1988).

368

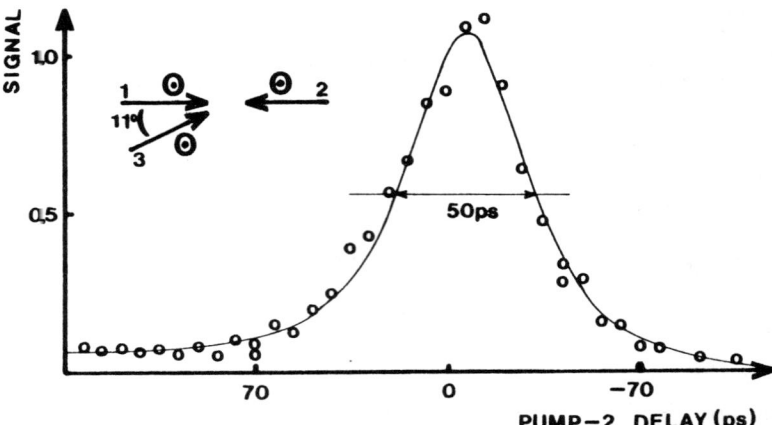

FIGURE 1. Signal evolution on delaying backward pump 2 in the parallel polarizations configuration (I). Solid line is a least squares fit of experimental points with a test function consisting of a Heaviside function superimposed on a sech function with 50 ps FWHM. Polarizations and angles of incoming beams are depicted at the upper left of the figure.

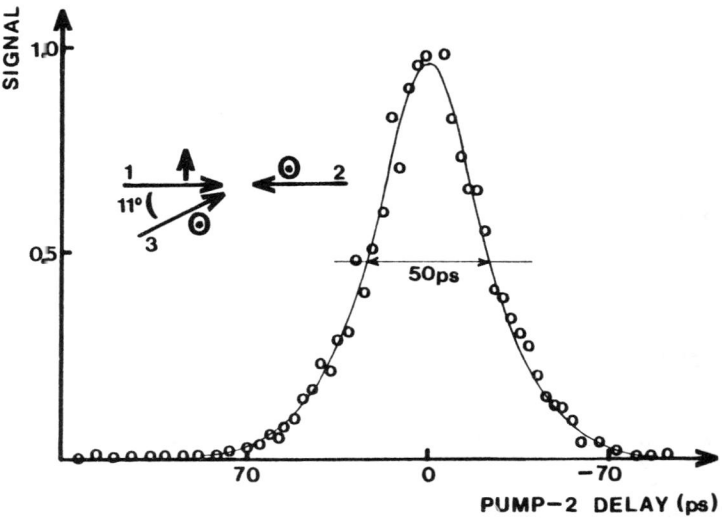

FIGURE 2. Signal evolution delaying backward pump 2 in the configuration (III) : erasure of the small intensity grating contribution. Solid line is a fit with a 50 ps FWHM sech function. Signal scale is the same as in figure 1.

PICOSECOND STUDIES OF OPTICAL STARK EFFECT IN POLYDIACETYLENES

F. CHARRA and J.M. NUNZI
CEA – CEN SACLAY – Laboratoire de Physique Electronique des Matériaux
IRDI/D.LETI/DEIN/LPEM – 91191 GIF-SUR-YVETTE – FRANCE

Irradiation of a medium with non-resonant light leads to a shift of the energy levels of the system known as optical Stark effect. This effect has been widely studied in atomic gases[1], in semiconductors[2], and in quantum wells[3]. Conjugated polymers which have been predicted to behave like one dimensional quantum confined systems[4] and which have a strong isolated excitonic dipolar transition[5] are excellent candidates in that point of view. In this paper, we present the first observed optical Stark effect in conjugated semiconducting polymers: polydiacetylenes.

The polydiacetylenes studied here are:
. An amorphous 4 BCMU 0.2% red gel prepared in chlorobenzene[6]. Its maximum optical density is 3 at 545 nm in a 1 mm thick quartz sample cell.
. A polycristalline 4 BCMU blue evaporated layer[7]. Its maximum optical density is 2 at 620 nm for a 3200 Å layer on fused silica.
The chains in the red gel are isotropized in 3 dimensions,[6] while in the blue thin layer, they lie in the substrate plane.[7]

The experiment is done using a pump-probe technique with 33 ps pulse duration.[8] The pump is the 1064 nm fundamental wavelength of a passive-active mode-locked Quantel Nd-YAG laser. The probe is obtained after frequency doubling and Raman shifting part of the laser beam in selected solvents.[9] In order to avoid efficient bleaching of the transition, the probe intensity is attenuated till ≈ 4 $MW.cm^{-2}$. Transmission is measured alternatively with and without the pump pulse and studied as a function of pump-probe delay, pump-probe polarization, and pump intensity in the two polymers. All the experiment which is referenced shot to shot at 1 Hz repetition rate is computer controlled. A typical time evolution of absorbance variations (δOD) is reported on figure 1. It has the same width as the laser intensity autocorrelation function. Thus, up to our experimental time scale, the response is faster than about 10 ps. A typical pump to probe polarization angle (Θ) dependence of absorption variations δOD at 545 nm for the red polymer at 1 $GW.cm^{-2}$ pump power is reported on figure 2. It fits δOD $\propto -(2 + \cos2\Theta)$ as it should for a one dimensional effect, linear in pump intensity. Figure 3 depicts the 1064 nm pump intensity dependence of the red polymer absorption changes at each probed wavelenght. There is no detectable effect at 630 nm probe wavelenght. Figure 4 depicts the same intensity dependence for the blue polymer.

To interpret those measurements in terms of physical quantities (shift, bleaching, broadening), we must fit the absorption peak with its position, height and width. The best fit of the studied absorption region is provided by a Gaussian. But as long as it concerns differences, fit of absorption changes using a Gaussian or a Lorentzian gives the same results.

369

J. Messier et al. (eds.), Nonlinear Optical Effects in Organic Polymers, 369–374.
© 1989 by Kluwer Academic Publishers.

For the red polymer, there is very little bleaching ($< 0.03\%$). The absorption band is shifted towards blue and broadened (fig. 5). Above 1.5 GW.cm^{-2}, the shift increases nonlinearily with pump intensity and reaches 0.5 nm at 2.2 GW.cm^{-2}. For the blue polymer, with our two probe wavelengths, we mainly observe a blue shift of the absorption. This shift is almost linear in the studied intensity region (fig. 4), and reaches 1.2 nm at 0.8 GW.cm^{-2}.

Theoretically, isotropization produces a 5 fold decrease of the linear effect in the red 3 dimensional amorphous gel, and a 8/3 fold decrease in the blue 2 dimensional polycristalline layer. Thus, for an electric field parallel to the polymer, the linear Stark shift may be evaluated to 0.75 rm/GW.cm^{-2} (3 meV) for the red polymer and to 4 nm/GW.cm^{-2} (13 meV) for the blue polymer.

Stark effect provides an intrinsic measurement of dipolar transition moments. The intensity dependent absorption-frequency shift for a two level system with a transition frequency ω_{01} is given by $\Delta\omega_{01} = \omega - \omega_{01} + \delta$, where

$$\delta = [(\omega_{01} - \omega)^2 + \Omega^2]^{1/2} \qquad (1)$$

$\Omega = 2\,\mu_{01}.E(\omega)/\hbar$ is the Rabi frequency and $1/2\,E(\omega).e^{i\omega t} + cc$ is the pump electric field.[1] From our experimental shifts, equation (1) gives an effective transition dipole moment for the exciton:

$$\mu_{01} \approx 2.8\ \text{Å}.e = 4.5\ 10^{-29}\ \text{C.m} = 13.5\ \text{D}$$

for the red form, and

$$\mu_{01} \approx 3.7\ \text{Å}.e = 5.9\ 10^{-29}\ \text{C.m} = 18\ \text{D}$$

for the blue form. Though these values are larger than those reported for Rhodamines,[10] they are half what is expected for polydiacetylenes.[5] Moreover, equ. 1 does not account for the nonlinear Stark shift positive curvature (fig. 5).
However, polydiacetylenes have proved to be fairly described by 3 level systems (fig. 6).[5,11] In that picture, the linear Stark shift expresses as:[1]

$$\frac{\hbar\Delta\omega}{E^2_{(\omega)}} = 2\mu^2_{01}.\left(\frac{1}{\omega_{01} - \omega} + \frac{1}{\omega_{01} + \omega} \right) + \mu^2_{12}.\left(\frac{1}{\omega_{01} - \omega_{02} - \omega} + \frac{1}{\omega_{01} - \omega_{02} + \omega} \right) \qquad (2)$$

Taking reasonable parameters for those polymers;[12]

$$\omega_{01} = 2.28\ \text{eV}, \quad \omega_{02} = 2.19\ \text{eV}, \quad \mu_{01} = 6\ \text{Å}.e$$

for the red form, and

$$\omega_{01} = 2.01\ \text{eV}, \quad \omega_{02} = 1.95\ \text{eV}, \quad \mu_{01} = 6\ \text{Å}.e$$

for the blue form, we can obtain an estimate of the ratio μ_{21}/μ_{01} from equ. 2, for both polymers. We obtain 3.8 for the red form and 3.9 for the blue form. Those ratios are in good agreement with frequency dependence studies of harmonic generation.[12]

Examining our results in view of equ. 2, we see that since μ_{21} is 4 times larger than μ_{01}, the Stark shift is extremely sensitive to resonance between the excited levels $1_u \rightarrow 2_g$ (fig. 6). So, Stark effect provides a measurement of the characteristics of excited one photon forbidden levels. In our experiments with polydiacetylenes, at 1064 nm pump wavelength, both terms of equ. 2 have the same magnitude and opposite signs. However, the level 1_u is closer to one photon $0_g \rightarrow 1_u$ resonance in the blue than in the red form ; this explains the larger shift and also the better accuracy of a two level system description (equ. 1) in the blue case.

Moreover, the large intensity broadening of the absorption in the red form (fig. 5) can be explained by an inhomogeneity of the 2_g level comparable to the absorption line width, making the two terms of equ. 2 compete differently and give different shifts for the different classes of systems. This effect is less sensitive in the blue form whose absorption line is closer to the pump frequency, making the first term of equ. 2 dominate the shift and leading to a weaker intensity broadening of the absorption.

A possible explaination of the non linear Stark shift positive curvature in the red form (fig. 5) is that the systems having the highest 2-photon 2_g levels are closer to 1064 nm two photon resonance ($0_g \rightarrow 2_g$) and can reach an anticrossing at 2 GW.cm^{-2}.[8] This anticrossing corresponds to the appearance of two photon absorption which is more effective at 1064 nm in the red than in the blue form.[6]

In conclusion, the demonstration of dynamic Stark shift in 1-dimensional semiconducting polymers, on a picosecond time scale, raises the interesting possibility of the switching of optical transitions on or off resonance. From this point of view, the magnitude of the optical Stark effect in polydiacetylenes compares well to quantum wells.[3] In addition, we recently developped a theory showing how all degenerate nonlinear electronic responses, related to index modulations, could simply be derived from optical Stark shifts.[8]

REFERENCES

1. A.M. Bonch-Bruevich and V.A. Khodovoi : Sov. Phys. Usp 10, 637 (1968).
2. D. Frohlich, A. Nothe and K. Reimann : Phys. Rev. Lett 55, 1335 (1985).
3. M. Joffre, D. Hulin, A. Migus, A. Mysyrowicz and A. Antonetti : Rev. Phys. Appl. 22, 1705 (1987).
4. B.I. Greene, J. Orenstein, R.R. Millard and L.R. Williams : Phys. Rev. Lett. 58, 2750 (1987).
5. R.R. Chance, M.L. Shand, C. Hogg and R. Silbey : Phys. Rev. B, 22, 3540 (1980).
6. J.M. Nunzi and D. Grec : J. Appl. Phys. 62, 2198 (1987).
7. P.A. Chollet, F. Kajzar and J. Messier : in Non Linear Optical and Electroactive Polymers, edited by P.N. PRASAD and D. ULRICH (Plenum Press, New York, 1988), pp. 121.
8. F. Charra and J.M. Nunzi : in the proceedings of the Organic Materials for Nonlinear Optics, held in Oxford (june 1988).
9. M.J. Colles : Opt. Com. 1 , 169 (1969).
10. P.C. Becker, R.L. Fork, C.H. Brito Cruz, J.P. Gordon and C.V. Shank : Phys. Rev. Lett. 60, 2462 (1988).
11. P.A. Chollet, F. Kajzar, J. Messier, J.M. Nunzi and D. Grec : Rev. Phys. Appl. 22, 1221, (1987).
12. J. Messier, NATO Advanced research workshop on Non Linear Optical Effects in Organic Polymers, Nice (June 1988).

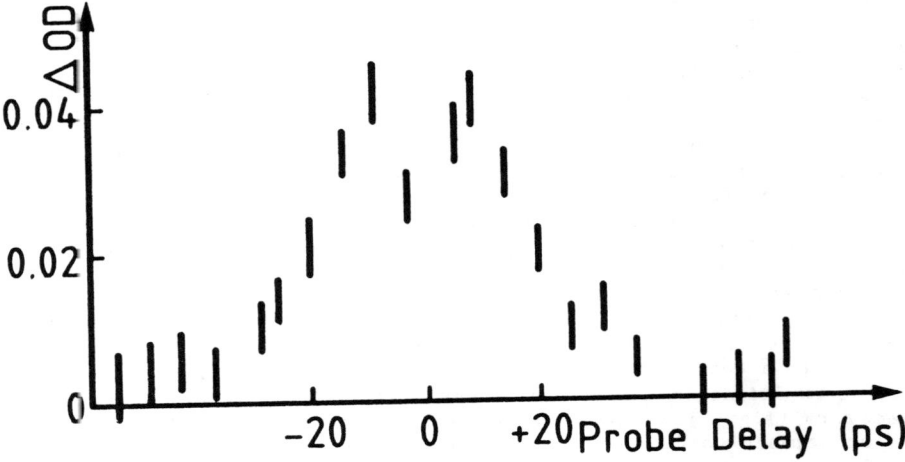

FIGURE 1. Time dependence of optical density variations of the red gel probed at 532 nm, with 1.7 GW.cm^{-2} pump intensity. The error bars represent 70 % confidence limits.

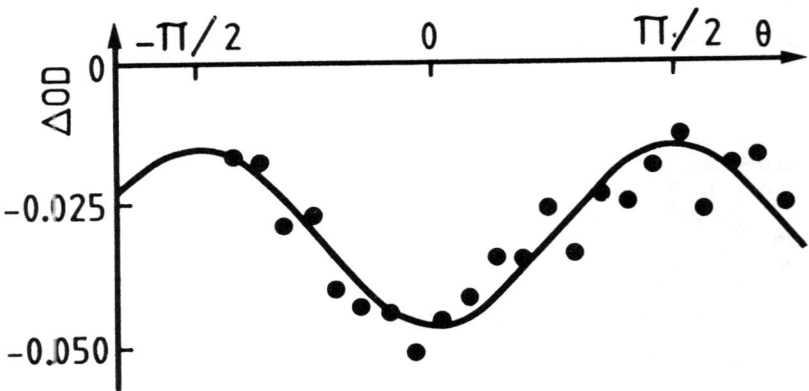

FIGURE 2. Pump to probe polarization angle dependence of optical density variations of the red gel probed at 545 nm, with 1 GW.cm^{-2} pump intensity. The solid line is a least squares fit of the parameter A with the test function : $-A.(2+COS\ 2\theta)$

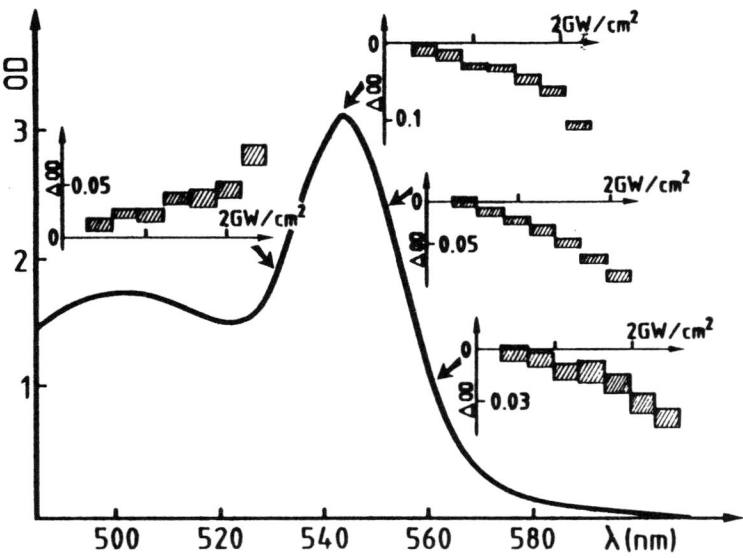

FIGURE 3. Red gel excitonic absorption spectrum. Insets depict pump intensity dependence of optical density variations at 532 nm, 545 nm, 551 nm and 562 nm probe wavelenghts. Rectangles represent 70 % confidence limits.

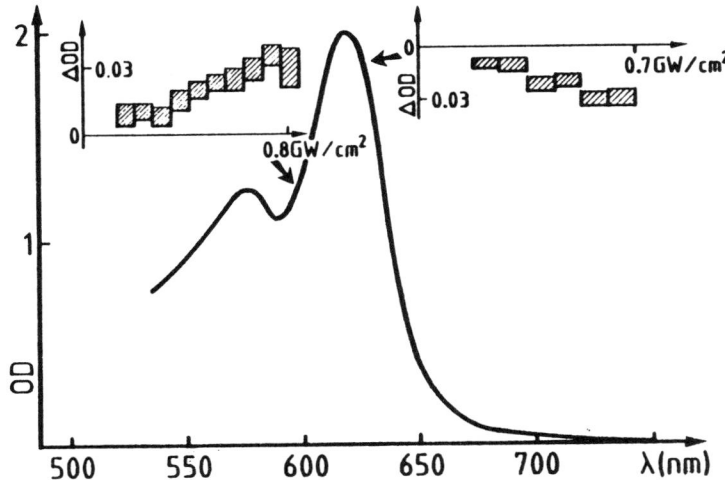

FIGURE 4. Blue thin layer excitonic absorption spectrum. Insets depict pump intensity dependence of optical density variations at 599 nm, and 627 nm probe wavelengths.

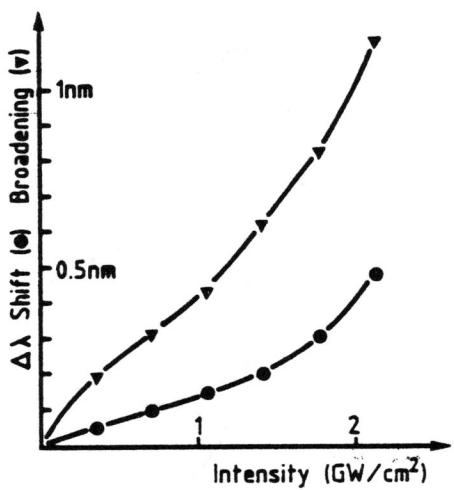

FIGURE 5. Pump intensity dependence of dynamic shifts and broadenings of the red gel absorption peak. The absorption in the probed region has been fitted with a Gaussian.

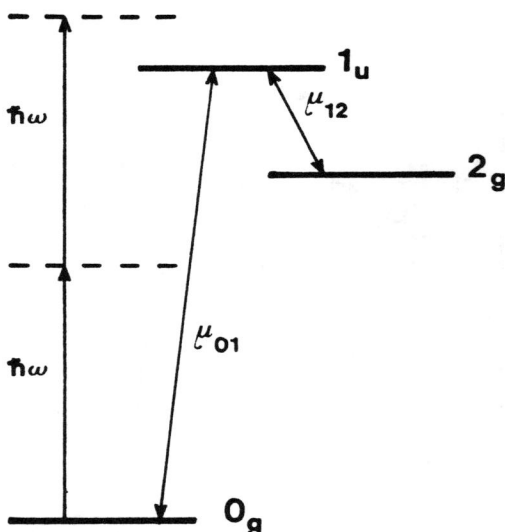

FIGURE 6. λ-type 3-level energy diagram for red and blue polydiacetylene forms. $\hbar\omega$ is the 1.17 eV pump energy.

ULTRAFAST PHENOMENA IN CONJUGATED POLYMERS

P.N. PRASAD

POINTS :

I.   HOW FAST IS ULTRAFAST
II. OBJECTIVES

     a. Ultrafast processes
     b. Mechanisms for ultrafast processes
     c. Application of ultrafast processes
     d. Materials requirements

III. CURRENT STATUS

IV. RECOMMENDATION

I. <u>HOWFAST IS ULTRAFAST</u>

   Rise time $\ll 10^{-12}$ s

   Decay time $< 10^{-11}$ s

              $\ll$ (less than electronics)

II. <u>OBJECTIVES</u>

   a. <u>Ultrafast processes</u>

      a.1. Nonresonant or preresonant NLO processes all optical

      a.2. Resonant processes in the presence of ultrafast nonradiative
           decay

           Examples : conjugated structures which permit nonlinear confor-
           mational defects

      a.3. Two-photon resonances

      a.4. Fast photoconductivity and charge carrier dynamics

      a.5. Photobleaching

b. Mechanisms for ultrafast processes

b.1. Any NLO processes involving virtual states are inherently ultra-fast, but the magnitude of the nonlinearity is what would determine whether a nonresonant NLO process is of significance.

This is where π-conjugated electronic structures come in play because they have a relatively large non-resonant nonlinearity. In addition, existence of one-dimensional delocalization in linear conjugated polymers enhances the non-resonant nonlinearity.

These virtual processes could involve interband transitions, excitons, structural deformations, charge transfers, etc...

One of the principal goal of future research in this area should be to identify which mechanisms are important for a specific polymer system.

b.2. Resonant processes require, specific conditions to be met in order to be ultrafast. Some of these are :

- Since radiative processes have fundamental limitation of speed, ultra-fast non-radiative relaxations (e.g. structural deformations, internal conversion, etc...) are required.

- Low dimensionality of electronic structure can enhance certain non-radiative relaxation pathways. For example the diffusion limited processes are enhanced in 1 D, conformational deformations, etc...

b.3. Two-photon resonances : resonance enhancement through coherence term (especially in $\chi^{(3)}$) without significant absorption.

b.4. Since hybrid (NLO + electronics) will likely be first use of NLO fast photo conductivity can play important role.

- Mechanisms : photo generation → fast
  Fast recombination
  diffusion
  High trap density, etc...

b.5. Photobleaching based on saturation of absorption or frustrated saturable absorption. Some of the conditions are :

. strong transitions
. fast nonradiative decay for a fast recovery.

c. Applications

1. All-optical switching and bistable devices (signal processing)
2. Sensor protection e.g. limiter action
3. Control of properties of excited species
4. Mode-locking and pulse shaping
5. Optical computing (very long term)
6. All optical gate
7. Image processing
8. Hybrid electrical/optical devices

d. Materials requirement

d.1. Non-resonant $\chi^{(3)}$ superior to 10^{-8} esu would be sufficient for many guided wave devices. For guided wave devices (for serial processing) the speed is limited by the nonlinear response which for non-resonant case is always ultrafast. The losses are mainly due to structural inhomogeneity which would have to be minimized for wave guides. Also, high optical damage threshold.

d.2. For resonant case, one has to define a figure of merit

$$\frac{\chi^{(3)}}{\alpha\tau}$$

which has to be maximized

large $\chi^{(3)} \approx 10^{-4}$ esu
low $\alpha < 10^{3}$ cm^{-1}
small $\tau < 1$ ps

Thermal stability
photostability
high thermal conductivity for high repetition rate operation.

d.3. A general requirement is processibility of materials into device structures with optical clarity, reduced defects and impurities. Control of defect centers by processing is an important parameter to acheive ultrafast speeds.

III. CURRENT STATUS

a. Non-resonant NLO processes have been demonstrated to have ultrafast speeds. Here to the best of our knowledge $\chi^{(3)} < 10^{-9}$ esu. Therefore, $\chi^{(3)}$ is not yet where one would like it to be at. However, theoretical modelling as well as studies of systematically engineered structure reveal that $\chi^{(3)} > 10^{-9}$ esu may be a realistic goal.

b. Ultrafast resonant processes have been demonstrated in many conjugated structures where many different mechanisms are invoked. e.g. excitonic processes in polydiacetylenes, phthalocyanines.

Photo generated conformational defects in conjugated polymers with bond alternation.

Photobleaching in many dyes ultrafast charge carrier dynamics in some organic photoconductors (polyacetylenes) have also been reported.

However, all reported resonantly enhanced $\chi^{(3)}$ with ultrafast speeds are $< 10^{-8}$ esu. In order for π-conjugated materials to compete with inorganic resonant $\chi^{(3)}$ materials, the currently reported values are not large enough.

However by controlling the excited state dynamics it should be possible to enhance $\chi^{(3)}$ appreciably and also acheive ultrafast speed. Therefore, by structural control, processing and control of excited state dynamics, one can maximize $\chi^{(3)}/\alpha\tau$.

IV. RECOMMENDATIONS

a. An increased data base on nonlinearities both in the nonresonant and resonant case is required. Dispersion effects of $\chi^{(3)}$ and careful temporal characterization are necessary.

b. Improvement of the theoretical understanding and better predictive capability for both non-resonant and resonant nonlinearities. In the latter case one needs to develop a better understanding of excited state dynamics.

c. Input from systematically derivatized and sequentially built structures are needed, e.g. synthesis, theory, measurement, processing control, etc...

d. One needs to develop or improve processing for optical control of materials (surface control, control of domain structure, trapping sites defects and impurities, dimensionality).

The ultrafast speed is the main attractive feature for optical processing. Conjugated polymers are inherently suitable for this task. An increased multidisciplinary effort in this area is strongly recommended !

DEVICE APPLICATIONS AND MATERIALS REQUIREMENTS

R. Lytel and G.I. Stegeman

Organic and polymeric materials have emerged in recent years as promising candidates for advanced device and system applications. This interest has arisen from the promise of extraordinary optical, structural, and mechanical properties of certain organic materials, and from the fundamental success of molecular design performed to create new kinds of materials. From an optical standpoint, organics offer temporal responses ranging over fifteen orders of magnitude, including nonresonant electronic nonlinearities (fsec-psec), thermal and motional nonlinearities (nsec-msec), configurational and orientational nonlinearities (msec-sec), and photochemical nonlinearities (psec-sec). Additionally, organic and polymeric materials can exhibit high optical damage thresholds, broad transparency ranges, and can be polished or formed to high-optical quality surfaces. Structurally, materials can be made as thin or thick films, bulk crystals, or liquid and solid solutions, and can be formed into layered film structures, with molecular engineering providing different optical properties from layer to layer. Mechanically, the materials can be strong and resistant to radiation, shock, and heat. When coupled with low refractive indices and low D.C. and microwave dielectric constants, the collective properties of these extraordinary materials show great promise towards improving the performance of existing electro-optic and nonlinear optical devices, as well as allowing new kinds of device architectures to be envisioned.

The field of organic and polymeric materials development is still rather young, compared to the long periods of development of inorganic crystals, such as $LiNbO_3$, $LiIO_3$, $LiTaO_3$, KDP, and KTP; semiconductors, such as GaAs, and even compound systems, such as the GaAs-AlGaAs multiple quantum wells. Initial excitement was provided in the late 1970's and early 1980's by the identification of a large electro-optic effect in single crystal MNA and a large, nonresonant response in diacetylene polymer crystals, such as PTS. Since that time, little improvement in single crystal growth has emerged, and the difficulty of crystal growth, common to organic and inorganic materials, have turned the synthesis direction away from crystals toward the development of clear, highly active polymer films. Since films of order 1-10 μm thick are easily produced by spin or spray coating, initial application to devices has been focused on waveguides, where the long path lengths required to achieve usable electro-optic or nonlinear optical phase shifts, can be realized. As such, the most significant advances in organic and polymeric devices have been in the waveguide arena.

This report reviews some of the key applications and materials requirements for second- and third-order nonlinear optical polymers. It is meant solely to provide some guidance to materials synthesizers so that their developments can be focused toward achieving the most critical materials parameters required for devices. Applications of resonant polymer systems are mentioned in connection with certain unique applications, but the focus is on nonresonant polymeric nonlinear materials.

There are many potential device applications for nonlinear polymers exhibiting second-order ($\chi^{(2)}$) and third-order ($\chi^{(3)}$) nonlinearities. In some cases, devices based on other materials will be replaced by polymeric devices with superior characteristics, and in other cases, nonlinear

J. Messier et al. (eds.), Nonlinear Optical Effects in Organic Polymers, 379–381.
© *1989 by Kluwer Academic Publishers.*

polymeric materials will make possible new classes of devices. Device-driven materials requirements are reasonably well defined for second-order devices, and with further materials development and improvement, fast (>10 GHz) and inexpensive phase and amplitude modulators are expected in the near term (3-5 years). Similarly, frequency doubling of semiconductor diode lasers would enhance their applications to optical data storage and xerography, and supplant other approaches to generating blue light. Other key applications awaiting a new generation of $\chi^{(2)}$ materials include parametric amplifiers and their application to communications. For third-order materials, however, the device implications are still being formulated, implying time scales of 10+ years for useful devices. The most promising applications currently appear to be to all-optical serial information processing at terahertz and larger data rates. Other potential applications, such as parallel processing of optical information and dynamic holography require many orders of magnitude improvement in third-order optical nonlinearities.

Many of the materials requirements depend on the specific application. However, a number of research issues common to all devices have been identified. Stability in the linear and nonlinear optical properties versus temperature and other environmental parameters for periods of years and under conditions of continuous illumination must be established. Second-order devices typically require tolerances and stability in the refractive indices of order 0.001, and dimensional tolerances of order ±30 to ±100 Angstroms. Another common requirement is for low loss and excellent optical clarity, with target losses, including absorption and/or scattering of order 0.1 dB/cm for second-order materials and 0.1 cm^{-1} for third-order materials. Finally, the material must be processable into the device structure format.

To realize the exciting potential of polymeric electro-optic devices, research is needed in a number of key areas. Principal among these is the stability of the poling or alignment of the molecules. It is suggested that the following are target parameters for the best material, based upon current knowledge: 5-10 year stability, electro-optic coefficient r >30 pm/V, loss α < 0.1 dB/cm, good performance at 0.85 μm, 1.3 μm, and 1.55 μm; ±50 Angstrom dimensional stability and control, low dispersion in the optical and microwave regimes, nonresonant material response, low dielectric loss and electrical conductivity, and high thermal conductivity. It should be stressed that this current wisdom will most certainly evolve in time, but that the requirements on electro-optic activity and optical loss are rather rigid and most critical.

There are additional research issues for second-harmonic nonlinear polymeric devices. The large, projected values of $\chi^{(2)}$, flexible refractive index control, and possibility of multiple layer and molecular engineering make SHG devices especially promising for frequency conversion at low powers. The requirement of phase matching places very strict tolerances on film thickness (±30 Angstroms), index uniformity (< ±0.001), and optical clarity. The ideal material would have a low birefringence and dispersion, a $\chi^{(2)}$ of order 10^{-7} esu, no two-photon absorption at operating wavelengths, a high degree of stability in all dimensional and optical parameters for periods of years, and the best possible transparency.

The area of third-order nonlinear polymeric devices is still in its early stages, which reflects both the infancy of third-order nonlinear device concepts, as well as the developmental stage of nonlinear polymeric materials. Although current interest is centered on all-optical signal processing, the are other interesting devices possible, and many more will emerge as the materials parameters improve. Using all-optical signal processing as an initial target, a number of materials-related issues have been identified. Nonlinearity and absorption (more generally, loss) are both critical parameters and values of n_2 >> 10^{-16} m^2/W and α < 0.1 cm^{-1} are needed in waveguide processable materials. Thermal nonlinearities, relative to the electronic, are currently too large, and materials (perhaps composites) with smaller dn/dT and no hydrogen (to eliminate vibrational overtones) should be investigated. Environmental stability is again a key, as it was for the $\chi^{(2)}$ systems. Other criteria which an "ideal material" should have are: $n_2 > 10^{-14}$ m^2/W and $\alpha < 0.1$ cm^{-1}, with optimum performance preferred for 1.3 μm and 1.55 μm. *These are critical numbers!* In addition, the materials must be processable so that ±100 Angstrom dimensional stability and control are possible. High thermal conductivity will help with thermal control and damage reduction.

We envision many other potential applications for third-order nonlinear optical materials. Each will have its own materials requirements. For example, for parallel signal processing, nonlinearities of order $n_2 = 10^{-7}$ m^2/W with relaxation times shorter than 1 μsec are needed, quite different from the serial processing case. Another example is a nonlinear filter for intense short-pulse lasers, which requires $n_2 = 10^{-14}$ to 10^{-16} m^2/W, $\alpha < 1.0$ cm^{-1}, and relaxation times under a picosecond. Again it is clear that intense materials research is needed.

In summary, the second-order nonlinear optical polymers represent a rapidly maturing technology, and significant application to devices is expected within the next five years. The development of third-order nonlinear optical polymers is still very much in its infancy, however, and very large improvements in $\chi^{(3)}$, optical clarity, and processability will be required before practical devices may be realized. In both cases, it is critical to achieve high electro-optic or nonlinear optical activity *simultaneously* with low loss. These are primary, critical requirements. Secondary critical requirements, such as thermal and environmental stability, must be achieved along with the primary requirements. We emphasize this point because many synthetic efforts are currently producing materials with larger nonlinearities, but very high optical losses, in systems that are not even stable in air. The efforts underway focusing on optical clarity, processability, and stability are expected to lead the way toward the development of materials suitable for practical and important device applications within the next decade.

POLYMER SYNTHESIS FOR NONLINEAR OPTICS AND CHARACTERIZATION

L.R. DALTON

Development of $\chi^{(3)}$ materials is at an immature stage although some impressive accomplishments have been realized in the past year in the investigation of "delocalized electron (π conjugat§ion), rigid red polymers" and in the study of phtalocyanines.

Symmetry perturbation and intramolecular charge transfer, as well as electron delocalization, have been shown to be important concepts for $\chi^{(3)}$ materials and the need for enlarging the consideration of absorption has been demonstrated. Doping induced transparence has been demonstrated as a potentially exploitable phenomenon.

The most important continuing need for $\chi^{(3)}$ materials is an improved characterization of existing materials, e.g., delocalized electron rigid rod polymers, phtalocyanines, etc... and a survey synthesis of new materials including polymers containing variable redox and mixed valence metals.

Materials do not currently exist to satisfy the diverse and severe requirements for fabrication of most anticipated devices.

Design and processing of $\chi^{(2)}$ materials is much more advanced than $\chi^{(3)}$ materials. Actual devices may be realized in the near future for $\chi^{(2)}$ polymers while $\chi^{(3)}$ materials are likely to require at least five years of basic research on fundamental phenomena before material characteristics appropriate for device fabrication can be realized.

Improvement in $\chi^{(2)}$ materials will likely focus upon improvement in nonlinear susceptibility and improvement in processing of ordered lattices. In particular, an effort appears to be required to improve device longevity by greater (longer) retention of order. In addition to developing materials with higher Tg for poling experiments, it is appropriate to explore other means of achieving order including the investigation of self ordering systems. Also, it is important to reduce optical losses.

In the preparation of $\chi^{(3)}$ materials attention must be given to factors such as optical transparency, resistance to optical damage, and resistance to environmental degradation. In turn, device physicists and engineers must focus more attention to "narrow" band or wavelength specific applications.

New synthetic routes affording improved processing options should be investigated e.g., the preparation of polymers via soluble processors, by sequential synthesis and systematic derivatization.

The investigation of new ultrastructure forms e.g. : polymer blends and alloys, composites, copolymers, self assembling systems, Langmuir Blodgett films should be pursued.

J. Messier et al. (eds.), Nonlinear Optical Effects in Organic Polymers, 383.
© 1989 by Kluwer Academic Publishers.

CHARACTERIZATION TECHNIQUES

Characterization of nonlinear optical media is a complex topic which could not be fully treated by the panel. However, an overview was undertaken which left the perception that, understanably in such a new and interdisciplinary field, better perspective and attention to detail is warranted. Clearly, the specific properties of interest and the degree of rigor which are required depend on one's motivation for conducting characterization experimentation. This motivation might arise from an interest in some scientific concept, from a need to quickly sample the magnitude and quality of nonlinear behavior in a materials scouting effort, from a need to determine the magnitude of various device related properties, etc. Given the ease with which lasers can now be acquired and the fact that various nonlinear optical phenomena are easily observed, the panel thought it to be essential that the importance of careful experimental design and interpretation of results be stressed. Both novice and expert have been guilty of reporting either incompletely interpretable or even fundamentally inappropriate experimental activities. Therefore, a list of questions was generated as guidelines.

. Of the numerous possible techniques, is one best for this purpose ?

. Can the technique give unambiguous results ? Is uncertainty related to imprecision in analysis, e.g. fundamental limitation of current theory (as in the problem of local fields or the approximation of a molecular polarizability in condensed media) or lack of rigor (as in the use of the naive two-level model as the basis for solvatochormic determination of second-order hyperpolarizability, as in neglect of dispersion, or as in neglect of important aspects of optical propagation) ? Is cuncertainty related to experimental limitations (such as the limitations of laser spectral, coherence, pulsewidth, etc... properties ; such as the sample's quality and characteristics ; or such as techniques not being fully developed, not being appropriate to the characterization goal or not being sufficiently facile) ?

. Are equipment and functioning experimental arrangements available ?

. Will results and interpretations be reproducible and verifiable by others ?

. Is the material of interest stable or in equilibrium, that is, are concentrations, pressure and temperature sufficient to identify the state of the material ? If the results depend on preparation and format of the materials, is anything known about the dependence ?

385

J. Messier et al. (eds.), Nonlinear Optical Effects in Organic Polymers, 385–387.
© 1989 by Kluwer Academic Publishers.

. Are techniques besides nonlinear optical processes needed for interpretation or to understand the physics of the material ? These probably include linear optical and dielectric properties, molecular conformations and arrangements (as determined by spectroscopies and diffraction), larger scale morphology (as determined by electron and optical microscopies, thermal analysis), macroscopic properties (such as elastic, diffusion, thermal parameters), and the presence and role of impurities and defects.

. Can limits of accuracy be placed on any quantities reported ?

There are currently several classes of materials being investigated. These include molecules, crystals, molecular thin films, second—order nonlinear ("$\chi^{(2)}$") polymers, and materials for all-optical (imprecisely termed "$\chi^{(2)}$") processes.

Molecules : Molecules are generally studied in liquids by the nonlinear techniques EFISH (dc-electric-field-induced second-harmonic generation) and THG (third-harmonic generation) and other techniques to determine information on molecular μ, α, β and γ tensors. They are fairly well understood techniques for which newer, simpler and more precise implementations are bieng invented. Unfortunately, there is a tendency to report values without regard for the complexities of determining molecular properties in condensed phases, a well-known problem in the older area of molecular dipoles. Difficulties remain in the problems of absolute calibrations, models of the liquid dielectric in molecular terms, dispersion and the limited amount of molecular information which can be obtained directly.

Crystals : Discounting polymeric crystals, the interest in molecular crystals is primarily for second-order purposes. This requires characterization of d, r and dielectric or indicatrix tensors. Scouting is often done by powder SHG. This being a very crude test, it is noteworthy that a single—crystallite method has recently been developed which allows phase—matching characteristics and d-tensor anisotropic behavior to be investigated without the need to grow large single crystals. The quality of crystals and growth methods is a key area of research, more so than nonlinear optical characterization.

Molecular thin films : Molecular thin films, many prepared by the Langmuir-Blodgett technique, are increasingly being studied by and prepared for nonlinear optics. Aspects of d, r and indicatrix tensors are mostly investigated by second-harmonic generation, ellipsometry and guided-wave techniques. Interest lies in molecular orientational distribution functions and magnitudes of molecular hyperpolarizabilities jointly determined, and in domain structure. Given the absence of an accepted molecular theory of dielectric behavior of such films, there is significant uncertainty in this field. Therefore, the distinction between data and interpretations is vital.

$\chi^{(2)}$ Polymers : Polymers prepared from materials containing large β-hyperpolarizable units and subjected to external orienting forces (e.g. electric fields) are widely pursed. Unfortunately, these processed materials are generally metastable. Aspects of d, r and indicatrix tensors, as well as their relaxation behavior are important. Techniques for characterization are well-known (SHG, clamped and unclamped electro—optic effects). Interest lies also in the orientational distribution functions, magnitudes of achieved and achievable nonlinearity, validity of simple microscopic pictures of the orientation, and in dispersion and mechanisms. The causes and factors affecting orientational stability are central.

$\chi^{(3)}$ Polymers : All optical processes are very diverse, especially when resonant behavior is encountered. THG and EFISH are good probes of "nonresonant" behavior. "n_2" methods such as power limiting, degenerate four-wave mixing, Kerr effects, nonlinear guiding, nonlinear transmission, etc are too numerous to list. They are also much more complex in their mechanistic origins, being highly sensitive to coupling of electronic properties to ionic and molecular deformations, and to generation of transient excitation. Careful studies, therefore, require care as to spectral and spatial dispersion and temporal behavior. It is judged that availability of spectrally vesatile, temporally short light sources and of improved, useful theoretical models for such experimentation are needed.

The summary recommendations of the panel are that greater effort is needed in considering, developing and using laser light sources, in the theory of many experiments, in the careful separation of data and analysis, and in the refinement of techniques. Particularly it is important to make techniques more accessible, reliable and understable for workers with an interest in materials. It is acknowledged that many properties, besides those cited above which allow to understand the nonlinear optical materials physics, are critical in evaluating newer polymeric materials for their real technological potential. These are such obvious, yet unknown, properties as processing potential, long term photostability of structures and compounds, play-off of color and nonlinearity enhancement, etc.

INDEX

Ab initio calculations
        coupled Hartree-Fock procedures   13
        of molecular hyperpolarizabilities   13-26
Acceptors   76,79,85-86,103,108
Acoustic response in DFWM   57-58
Acrylamid   24-25
Air contribution to THG   233
Alkanes   111-113
        branched -   112
Anisotropy
        - of cubic susceptibilities   32-33,308,358
Aromatic chromophores   197
Aromatic heterocyclic molecules   97-99
Aromatic tetraamines   137
        derivatized -   136
Auger processes   9
Average polarizability   25
Aza benzthiazole   114
Aza cyanine   114
Aza thiazole   114
1-aza 1,3,5,7-trans-octatetraene   20
4-aza 1,3,5,7-trans-octatetraene   20
Band state filling effect   4
Beam waist   231
Benzene   16
        - substituents   115
Beta carotene   114
Bipolaron   29,167,171
        - bands   165
Bond additivity   110-112
Bulk susceptibility   105
Butadienyl formamide   22
Butatriene   16-17
Bystander
        - effects   107-110
        - interference   109
Carbon disulphide   147
Charge transfer   84-87,101-103,108
        - bands   107
Chromophores
        aromatic -   197
        - functionalization   173-174
        - functionalized polystyrenes   177
        - in polymer chain   201-204
        pi-electron -   173
Cis-(CH)x   30-32
Cis-hexathiophene   56

Cis-octatraene	70
CMONS	119
CNDO	
C.I. calculations	62-76
finite field method	106
Coherence length	103,234,236
Composite materials	7,137-138
Conducting polymers	29,159
Configuration	166
excited states of the s -	41
Conformations	135
Conjugated chains	
longitudinal polarization of -	18
Conjugated polymers	
excitations in -	71
Conjugation length	103
Copolymers	137-139
flexible chains	139
Correlation effects	4-5,62-63,76
Coulomb correlations	17
Cyanines	114
Cyclic formamide dimer	25-26
Cyclic voltammetry	137
Damage threshold	327
optical -	134
DANS	107,118
1,5,9-decatriene-3,7-diyne	18-19
Degenerate four wave mixing	56-58,133-135,146
- in poly-4-BCMU	365
- in poly-p-phenylene vinylene	146
Diacetylenes	
chiral -	152
crystal structure of -	154-155
DSC spectrum of -	156
polymerization of	152
second harmonic generation	157
synthesis of -	150
Diazostylbene derivatives	
absorption of -	81
Langmuir-Blodgett films from -	81-84
linear refractive index of -	81
second order nonlinear optical prop.	81-84
Dielectric confinement effect	6-7
3,6-dimethylene-1,4-cyclohexadiene	18-19
Dipolar approximation	238
Dipole	
field dependend -	15
- moments	62-64,101-102,108,117-121
- polarization	105
Dithieno [3,4-b;3´,4´-d] thiophene (DTT´)	161-163
poly -	161-163
Disubstitued hydrocarbons	101-104
Donor	76,79,103,108

Drude model   7
Electric field induced second harmonic gen.   47,51,64-65,80,83,101.110,
                                                     116-118,202
Electric field poling   173
Electron affinity   106
Electron delocalization   101.104.123-124.130,161
          - effects   6,101-104,306.358.376
Electron distribution
          - calculation   72-73
          valence -   4-5
Electrooptic
          - linear coefficients   64
Electropolymerization   160
Evaporated thin films   49-51
Excitations
          energies   64
          in conjugated polymers   71
Excitons
          neutral -   38
Finite field method   14,93-95
Fletcher-Powell procedure   14
Fock operator   14
Formamide
          butadienyl -   22
          cyclic - dimer   25-26
          - monomer   25
Franz-Keldysh effect   3
Frozen gaz model   119
Genkin-Mednis approach   1-2
Glassy polymer systems   173
Hartree-Fock-Roothaan calculations   14
Hemicyanine dye
          absorption spectra   206
          Langmuir-Blodgett films from -   205,219-223
          reflectance spectra   221
          second harmonic generation   205-208,222-223
1,5-hexadiene-3-yne   18-19
1,3,5-hexatriene   114
Hückel approximation   6,17,23.106
Hydrocarbons   107-111
          disubstitued -   101-104
Hydrogen bonded systems   22-26
Hyperpolarizabilities   303
          ab initio calculations of molecular -   13-27
          - calculations of polythiophenes   53-55
          complex -   9,238-244
          cubic -   70-76,93-99,111-115,225,244
          finite field calculations of -   93-99
          frequency dependence   65-76
          perturbation calculations of -   51,65,70
          phase of cubic -   243
          quadratic -   65-69,79.84,101-104,106.117.
                                                     351,354

392

```
        topologically induced quadratic -        106
Kerr effect
        optical -                                 5-9
Kleinman symmetry                                 175,198
Ladder polymers                                   123-140
        FTIR of                                   130-131
        linear absorption of                     131-133
        synthesis                                 128-130
        TGA                                       126
        third order susceptibility               134
Langevin function                                 199-200
Langmuir-Blodgett films                           137,185-189,205,219-223
        polymerization of -                       185-186
        - of poly-4-BCMU                          360
        second harmonic generation in -          79-81,90
Linear formamide dimer                            25-26
Linear polarizabilities                           15-26,109-110
Liquid crystalline polymers                       153
Liquid crystals                                   152-153
Local field                                       6,64,67,198
        - tensor                                  120
Longitudinal polarizabilities                     18
Mach-Zehnder interferometter                      309
Macroscopic dipole polarization                   105
Maker fringes                                     233
Matrix contour diagrams                           72-73
Maxwellian field                                  120
Merocyanine                                       242,244
Metal particles                                   1,7-8
2,4-methylnitro-aniline (MNA)
        CNDO C.I. calculations                    62
        second harmonic generation in -          64-66
Methyl 2,3-trichlorosilyltricosanoate             214
Microscopic dipole polarization                   64,105
Millers rule                                      5
Modified wedged technique                         234
Molecular engineering                             225
Monolayers                                        209-215
MONS                                              119
Muconic acid                                      24
Multiple quantum wells                            268,299,301,310,312,316,321,325
Multiplexers                                      309,316,322
Naphtalenes                                       116-117
        - related compounds                      117
Neural networks                                   310,316,322,323
Nitroaniline                                      107-108
Nitrostilbenes                                    118
        4-substituted-4' -                        118-119
NMR                                               127,132
NPP                                               80
Nonlinear polarization                            32,175-176,231
Nonresonant susceptibilities                      30,61,75
Normalized harmonic intensity                     238
```

Nylon-6   23
Octadecyltrichlorosilane   209-213
Oligomers   17,75.125
One electron approach   5
One photon states   69
Optical absorption
        - of diazostylbene derivatives   81
        - of hemicyanine dyes   206
        - of ladder polymers   131-133
        - of 2,4-methylnitro-aniline (MNA)   66
Optical bistability   309,312,316,321
Optical damage threshold   134
Optical Kerr effect   5-9
        recovery time in -   8-9
        susceptibilities   8
Optical nonlinearity per valence electron   1
Optical Stark effect   47
Optical susceptibilities   1-11
        interband -   3
        intraband -   3
Optical switching   258-260,268-274,301.309.316,362
Orientational distribution function   198
Oxy-cyanocinnamate   201-202
Pariser-Parr-Pople
        - calculation of quadratic hyperpol.   107-108
        - hamiltonian   354
Perturbation calculations
        - of quadratic hyperpolarizabilities   65
        - of cubic hyperpolarizabilities   51,70
Phase conjugation   56-58,133-135
Phase matching   227
Phenylethylene   18-19
Phenylhydrazone   86-87
Photobleaching   375,377
Photodegradation   186
Photoinduced absorption   5
Photopolymerization   205
Phtalocyaninato-polysiloxane   187-191
Pi-electron
        - chromophores   173
        - delocalization   101,104,123-124,130,161
Planar conformations   24,135
Polar polymers   198
Polarizabilities
        average -   25
        linear -   15-26,109-110
        longitudinal -   18
        static electric -   13
Polarization
        dipole -   105
        nonlinear -   32,175-176,231
        longitudinal -   18
Polarized optical absorption   188

394

Polarons   29,351,359
Poled films   175,203
Poled polymers   278-288,291-296,301-303,311
Poling   279-288,328,383
        - fields   177,203
Poly-3-alkylthienylene   29
Poly-4-BCMU   361
        degenerate four wave mixing in -   365
        Langmuir-Blodgett films of -   360
        third harmonic generation from -   358
Poly-dithieno [3,4-b:3´,4´-d] thiophene (DTT´)   167-169
Poly-p-hydroxystyrene   173
Poly-p-phenylene vinylene   143-147
        degenerate four wave mixing in -   146
        third harmonic generation in -   144-145
        thin film preparation of -   143
Polyacetylene   29-38
        cis -   30-32
        trans -   30-42
        oriented -   34
Polycarbonate   140
Polycondensation   123-125
Polydiacetylenes   17,317
        - TCDU   17
Polyenes   73,96,114,119
        - derivatives   85
Polyglutamate   189
Polymerization
        electro-   160
        - of diacetylenes   152
        - of dithieno-thiophene (DTT)   162
        - of Langmuir-Blodgett films   185-186
        photo-   205
        topochemical -   23,149
Polythiophene   52-54,97,317,360
Polytoluenesulfonate (PTS)   34-38,254,379
Population redistribution   9
Prepolymers   125
Pyrrole   160
Quantum confinement   6-10
Quantum wells   268,299,301,310,312
Quantum zero-point fluctuations   37-42
Recovery time   258,308
        - of optical Kerr effect   8-9
Refractive index
        complex   40
        intensity dependant   47
Resonance valence bond ground state structure   165
Resonant susceptibilities   376
        cubic -   29,353,357-359,362
        quadratic -   102
Response time   301,311
        - in phase conjugation   57

Restricted Hartree-Fock-Roothaan method   14
Rigid rod polymers   93,383,317
Scalling laws for nonlinear susceptibilities   1-5
Second harmonic generation   278,302,329,337,380
        electric field induced -   47,51,64-65,80,83,101,110,
                                                 116-118,202
        - in Langmuir-Blodgett films   79-87,90
        - in 2,4-methylnitro-aniline (MNA)   64-66
        - in poled films   175-177,203-204
Second order nonlinear optical properties   301-302,327,380,386
        frequency dependence of -   64-68
        - in poled polymers   291-296
        temporal behavior   180-181
Self assembled monolayer   197,209-216
Self consistent field -CNDO   54,94
Self focusing   247
Semiconductors particles   1,7
Sigma-electron contribution   114
Solitons   29,36-42
        nonlinear -   30
        virtual -   37,41
Spin coating   143
Stark effect
        optical -   47,369-371
Stark shift   3
STO-3G basis   15-16
Structural anharmonicity factor   113
Structural relaxation   29
Structurally induced anharmonicity   112
Su-schrieffer mechanism   40
Sum over states method   93-94,352-354
Surface plasmons   304,360
Susceptibilities   303
        anisotropy in cubic -   32-33,308,358
        cubic -   93,238,306
        intraband/interband optical -   3
        Kerr -   5,8,357
        nonresonant -   30,61,75
        quadratic   201
        resonant   29,102,353
        scalling laws for nonlinear -   1-5
Symmetrical diacetylenes   150
Thiazole   114
Thiophene   160,355,358
Third harmonic generation   329
        air contribution to -   233
        - in poly-4-BCMU   358
        - in poly-p-phenylene vinylene   144-145
        - in polydiacetylenes   30-37
        - in Polytoluenesulfonate (PTS)   34-35
Third order nonlinear optical properties   347,351-352,362,305
        - applications   257,343
Three level model   51,58

Three photon resonance   42
Topochemical polymerisation   149
Trans-(CH)x   30-42
1,3,5,7,9-trans-decapentaene   18-19
1,3,5-trans-hexatriene   16,18-19,22
1,3,5,7-trans-octatetraene   20,70
Transition
        pi-pi* -   29-30,165
        virtual -   71-76
Transition dipole moments   2,64-65,71
Trichlorosilanes derivatives   209
Two photon absorption   57,247,250
Two photon resonances   42,52
Two photon states   63,69
Unsymmetrical diacetylenes   150-151
Van der Waals forces   87,305
        - radii   154
Vibronic side bands   52
Vinylacetylene   16-17
Vinylacrylamide   22
Vinylamine polymers   133
Vinylformamide   24-25
Virtual transitions   61,71-76
        solitons -   37,41
Wave guides   90,257-261,351,357,361,377,379
Wave guiding   263-272,279-288,291,301,309,
                                                312,316,328,332-335
Wedged liquid cell   231
Wedged Maker fringe method   64
Wedged technique   226
        modified -   234
Zero dimensionality   1,7
Zero emission window   228
Zero-point fluctuations   37-42